CEREBRAL BLOOD FLOW

Clinical and Experimental Results

Edited by

M. Brock · C. Fieschi · D. H. Ingvar

N. A. Lassen · K. Schürmann

With 113 Figures

Springer-Verlag Berlin · Heidelberg · New York 1969

ISBN 978-3-642-85862-8   ISBN 978-3-642-85860-4 (eBook)
DOI 10.1007/978-3-642-85860-4

Softcover reprint of the hardcover 1st edition 1969

Library of Congress

Catalog Card Number 71-95562.

Preface

This book is a survey of some aspects of current knowledge on regional Cerebral Blood Flow (rCBF), mainly as studied by the isotope clearance method. Although both theoretical and methodological problems are discussed, attention is mainly dedicated to data obtained from clinical studies.

The papers which make up this book were presented at the *International Symposium on the Clinical Applications of Isotope Clearance Measurement of Cerebral Blood Flow*, held in Mainz, Western Germany, on April 10–12, 1969.

The previous meetings on Cerebral Blood Flow, held in Lund in 1964* and in Lund and Copenhagen 1968**, had shown that the moment had come to concentrate on the possibilities of introducing rCBF measurements into clinical routine. This is why in the Mainz Symposium attention was initially focused on methodological aspects. This is also why theoretical problems of physiology of CBF were not emphasized. Finally, this explains why such topics as cerebrovascular disease, head trauma, coma, carotid surgery, brain tumors and intracranial pressure were given pride of place. However a survey of the clinical aspects of rCBF measurements would not be complete without an account of the application of such measurements to monitor cerebral circulatory changes during anesthesia and therapeutic procedures like, for instance, hyperventilation and hyperbaric treatment.

Furthermore, it is now possible to obtain data from correlative rCBF studies performed before, during and after surgical operations on the human brain. Such measurements, associated with the biochemical study of tissue fragments removed during surgery from focal and perifocal areas, will allow a better insight into the cellular metabolism. Important correlations between local blood flow, energy turnover and function will then provide most valuable information.

The chapters of this book correspond to the sections of the Mainz Symposium. The concluding remarks at the end of each chapter were made by the respective chairmen. At the close of the meeting Drs. KETY and LASSEN made final summaries underlining the main points presented and discussed. These commentaries have been reproduced in the present book because they will give the reader – as they did the participants in the Symposium – an illuminating survey of the material discussed.

---

* Regional Cerebral Blood Flow, Ed. D. H. INGVAR and N. A. LASSEN, Acta Neurol. Scand., Suppl. 14, 1965.

** Cerebral Blood Flow and Cerebro-Spinal Fluid, Ed. D. H. INGVAR, N. A. LASSEN, B. K. SIESJÖ and E. SKINHØJ, Scand. J. Clin. Lab. Invest., Suppl. 102, 1968.

Editing has been restricted to rearrangements and small corrections. No homogeneity of style was attempted. The editorial board considered that rapid publication should be their main objective and preferred to publish the results in this field of research while they were "still warm".

The next CBF-Symposium will be held in London, September 17-19, 1970. Informations are to be obtained from Dr. L. SYMON, National Hospital, Queen Square.

We are very grateful to all participants and collaborators. A special thank goes to Dr. Heinz Götze of Springer Verlag for his generous cooperation.

Mainz, April 1969   K. SCHÜRMANN

Contents

Chapter I : Methodology

Chapter II : Regulation of CBF

Chapter III : Cerebrovascular Disease

X Contents

Chapter IV : Carotid Surgery

Chapter V : Tumors, Intracranial Pressure

Chapter VI: Trauma, Coma, Alcoholism and Dementia

Chapter VII: Anesthesia and Therapy

Closing Remarks I

Closing Remarks II

List of Contributors

ACKER, H., Max-Planck-Institut für Arbeitsphysiologie, Dortmund

AGNOLI, A., Department of Neurology and
Psychiatry, University of Genoa

ALEXANDER, S.C., Department of Clinical Physiology, Bispebjerg Hospital, Copenhagen

ALLAN, R. N., Medical Physics Department,
Hammersmith Hospital, London

ARBUS, L., Department of Neurology, University Hospital, Toulouse

ARNOT, R. N., Medical Physics Department,
Hammersmith Hospital, London

BALDY-MOULINIER, M., Service de Eléctroéncephalographie et de Neurochirurgie,
Centre Hospitalier, University of Montpellier

BARTOLINI, A., Department of Neurology
and Psychiatry, University of Genoa

BATTISTINI, N., Department of Neurology
and Psychiatry, University of Genoa

BAUM, P., I. Medizinische Universitätsklinik,
Mainz

BAVA, G., Department of Neurology and
Psychiatry, University of Genoa

BEDUSCHI, A., Department of Neurosurgery,
Benfratelli Hospital, Palermo

BERGHOFF, W., Medizinische Universitätsklinik, Kiel

BÈS, A., Department of Neurology, University Hospital, Toulouse

BETZ, E., Institut für Angewandte Physiologie der Universität, Tübingen

DU BOULAY, G. H., National Hospital,
Queen Square, London

BOYSEN, G., Surgical Department, Rigshospitalet, Kopenhagen

BOZZA-MARRUBINI, M., Department of Anesthesiology, Ospedale Maggiore, Milano

BOZZAO, L., Department of Neurology and
Psychiatry, University of Genoa

BREGENTZ, S. E., Department of Surgery,
University of Göteborg

BROCK, M., Neurochirurgische Universitätsklinik, Mainz

BULL, J. W. D., National Hospital, Queen
Square, London

BUYNISKI, J. P., Department of Physiology,
Bowman Gray School of Medicine,
Winston-Salem, North Carolina

CASACCHIA, M., Department of Neurology
and Psychiatry, University of Genoa

CHIŞU, V. G., Neurosurgical Service, State
Hospital No 1., Timisoara

CHRIST, R., Neurochirurgische Universitätsklinik, Mainz

CHRISTENSEN LOU, H. O., Department of
Neuromedicine, Gentofte Hospital,
Hellerup

CIFFONE, D. L., Ames Research Center,
Moffet Field, California

CLARK, J. C., Medical Physics Department,
Hammersmith Hospital, London

CRANSTON, W. I., Department of Medicine,
St. Thomas's Hospital, London

CRONQVIST, S., Department of Neuroradiology, University Hospital, Lund

D'AMICO, P., Ospedale S. Gerardo dei Tintori, Monza

DAVIS, D. O., Edward Mallinckrodt Institute of Radiology, St. Louis, Missouri

DELPLA, M., Department of Neurology, University Hospital, Toulouse

DIECKHOFF, D., Physiologisches Institut der Universität, Göttingen

DIETZ, H., Neurochirurgische Universitätsklinik, Mainz

DITTMANN, J., Neurochirurgische Universitätsklinik, Homburg (Saar)

ECTORS, L. A., Neurological Clinic, University of Brussels

EICHLING, J. O., Edward Mallinckrodt Institute of Radiology, St. Louis, Missouri

EKSTRÖM-JODAL, B., University Lung Clinic, Renströmska Sjukhuset, Göteborg

ELLGER, M., Institut für Anaesthesiologie der Universität, Mainz

ENGELL, H. C., Surgical Department, Rigshospitalet, Kopenhagen

ERDMANN, W., Physiologisches Institut der Universität, Mainz

ESCANDE, M., Department of Neurology, University Hospital, Toulouse

FEINDEL, W., Cone Laboratory for Neurosurgical Research, McGill University, Montreal

FENSKE, F., Neurochirurgische Universitätsklinik, Mainz

FIESCHI, C., Department of Neurology and Psychiatry, University of Genoa

FISCHER, F., Institut für Anaesthesiologie der Universität, Mainz

FLOHR, H.-W., Physiologisches Institut der Universität, Bonn

FRÈREBEAU, PH., Service de Eléctroéncephalographie et de Neurochirurgie, Centre Hospitalier, Montpellier

GAMEL, J. W., Stanford University, Medical Center, Palo Alto, California

GLASS, H. I., Medical Physics Department, Hammersmith Hospital, London

GOLDBERG, H. I., Stroke Research Center, General Hospital, Philadelphia

GORDON, E., Department of Neuroanaesthesia, Karolinska Sjukhuset, Stockholm

GÖTT, U., Neurochirurgische Universitätsklinik, Bonn

GOTTSTEIN, U., Medizinische Universitätsklinik, Kiel

GROTE, J., Physiologisches Institut der Universität, Mainz

GUSTAVSSON, L., Department of Psychiatry I, University Hospital, Lund

HAAS, J. P., Röntgenologisches Institut der Universität, Mainz

HACKER, H., Neurochirurgische Universitätsklinik, Frankfurt a. M.

HADJIDIMOS, A. A., Neurochirurgische Universitätsklinik, Mainz

HADJIEV, D., Research Institute of Neurology and Psychiatry, Sofia

HÄGGENDAL, E., Klinisk fysiologisk Laboratorium, Sahlgrenska Sjukhuset, Göteborg

HARPER, A. M., University Hospital, Glasgow

HASE, U., Neurochirurgische Universitätsklinik, Mainz

HEIDENREICH, J., Physiologisches Institut der Universität, Mainz

HEIPERTZ, R., Neurochirurgische Universitätsklinik, Mainz

HEISKANEN, O., Neurosurgical Clinic, University Hospital, Helsinki

HEISS, W. D., Department of Psychiatry and Neurology, University Hospital, Vienna

HELD, K., Medizinische Universitätsklinik, Kiel

HERRMANN, H. D., Neurochirurgische Universitätsklinik, Homburg (Saar)

HODGE, C. P., Cone Laboratory for Neurosurgical Research, McGill University, Montreal

HOFFMAN, J. C., Department of Anesthesia University of Philadelphia, Pennsylvania

HOLBACH, K. H., Neurochirurgische Universitätsklinik, Bonn

HOYER, S., Pathologisches Institut der Universität, Heidelberg

HUTTEN, H., Physiologisches Institut der Universität, Mainz

INGVAR, D. H., Department of Clinical Neurophysiology, University Hospital, Lund

JAFFE, M. E., Stroke Research Center, General Hospital, Philadelphia

JANEWAY, R., Department of Neurology, Wake Forest University Winston-Salem, North Carolina

JENNETT, W. B., Department of Neurosurgery, Institute of Neurological Sciences, Killearn Hospital, Glasgow

KAASIK, A. E., Department of Neurology and Neurosurgery, State University, Tartu

KANZOW, E., Physiologisches Institut der Universität, Göttingen

KASSELL, N. F., Department of Neurosurgery, Hospital of the University, Philadelphia

KEMP, R., Department of Anesthesiology, Wake Forest University, Winston-Salem, North Carolina

KENNADY, J. C., UCLA School of Medicine, Harbor General Hospital, Los Angeles, California

KETY, S. S., Department of Psychiatry, Massachusetts General Hospital, Boston

KIENLE, G., Neurologische Klinik, Nordwest-Krankenhaus, Frankfurt a. M.

KLASSEN, A., University of Minnesota, Minneapolis

KNEBEL, U., Institut für Angewandte Physiologie der Universität, Tübingen

KONDO, A., Department of Neurology, Wayne State University, Detroit, Michigan

KOSMAOGLOU, B., Neurological Center, Polyclinic of Athens

KREUSCHER, H., Anaesthesiologisches Institut der Universität, Mainz

KVICALA, V., Department of Psychiatry and Neurology, University Hospital, Vienna

LADEGAARD-PEDERSEN, H. J., Surgical Department, Rigshospitalet, Kopenhagen

LANGFITT, T. W., Department of Neurosurgery, Hospital of the University, Philadelphia

LASSEN, N. A., Department of Clinical Physiology, Bispebjerg Hospital, Copenhagen

LAZORTHES, Y., Department of Neurology, University Hospital, Toulouse

LEDINGHAM, I. Mc. A., Department of Neurosurgery, Institute of Neurological Sciences, Killearn Hospital, Glasgow

LJUNGBERG, K., Department of Psychiatry, University Hospital, Lund

LÜBBERS, D. W., Max-Planck-Institut für Arbeitsphysiologie, Dortmund

MARC VERGNE, J. P., Department of Neurology, University Hospital, Toulouse

MARINOVA, Z., Research Institut of Neurology and Psychiatry, Sofia

MARSHALL, J., National Hospital, Queen Square, London

MARSHALL, W. H., Stanford University, Medical Center, Palo Alto, California

MARSHALL, W. J. S., Department of Neurology, Hospital of the University of Pennsylvania, Philadelphia

MAYNARD, C. D., Department of Radiology, Wake Forest University, Winston-Salem, North Carolina

MCHEDLISHVILI, G. I., Institute of Physiology, Georgian Academy of Sciences, Tbilisi

MCHENRY, L. C., JR., Stroke Research Center, General Hospital, Philadelphia

METZGER, H., Physiologisches Institut der Universität, Mainz

MEYER, J. S., Department of Neurology, Wayne State University, Detroit, Michigan

MEYER, M., University of Minnesota, Minneapolis

MILLER, J. D., Department of Neurosurgery, Institute of Neurological Sciences, Killearn Hospital, Glasgow

MINAZZI, M., P. Ospedale S. Gerardo dei Tintori, Monza

MIYAZAKI, M., Department of Internal Medicine, Kosaiin Hospital, Suita City, Osaka

MOHN, E., The EEG Research Institute, Gaustad Hospital, Oslo

MÜLLER-SCHAUENBURG, W., Institut für Angewandte Physiologie der Universität, Tübingen

NARDINI, M., Department of Neurology and Psychiatry, University of Genoa

NEIGH, J. L., Department of Anesthesia, University of Pennsylvania, Philadelphia

NEUMANN, L., Institut für Angewandte Physiologie der Universität, Tübingen

NGUYEN-DUONG, H., Institut für Angewandte Physiologie der Universität, Tübingen

NIEDERMAYER, W., Medizinische Universitätsklinik, Kiel

NILSSON, N. J., Klinisk fysiologisk Laboratorium, Sahlgrenska Sjukhuset, Göteborg

NOMURA, F., Department of Neurology, Wayne State University, Detroit, Michigan

NORBÄCK, B., Klinisk fysiologisk Laboratorium, Sahlgrenska Sjukhuset, Göteborg

O'BRIEN, M. D., Regional Neurological Centre, General Hospital, Newcastle

OECONOMOS, D., Neurological Center, Polyclinic of Athens

OLDENDORF, W. H., UCLA School of Medicine, Los Angeles, California

OLESEN, J., Department of Neurosurgery, Bispebjerg Hospital, Copenhagen

PALLESKE, H., Neurochirurgische Universitätsklinik, Homburg (Saar)

PALVÖLGYI, R., Radiological Clinic, Faculty of Medicine, University of Budapest

PASSERO, S., Department of Neurology and Psychiatry, University of Genoa

PAULSON, O. B., Department of Clinical Physiology, Bispebjerg Hospital, Copenhagen

PHILLIPS, K. M., Cone Laboratory for Neurosurgical Research, McGill University, Montreal

PÖLL, W., Neurochirurgische Universitätsklinik, Mainz

PRENCIPE, M., Department of Neurology and Psychiatry, University of Genoa

PRIORI, A. M., Department of Neurology and Psychiatry, University of Genoa

PROSENZ, P., Department of Psychiatry and Neurology, University Hospital, Vienna

PROSSALENTIS, A., Neurological Center, Polyclinic of Athens

RAPELA, C. E., Department of Physiology, Bowman Gray School of Medicine, Wake Forest University, Winston-Salem, North Carolina

REIVICH, M., Department of Neurology, University of Pennsylvania, Philadelphia

RESCH, J., University of Minnesota, Minneapolis

REULEN, H. J., Neurochirurgische Universitätsklinik, Mainz

RISBERG, J., Department of Clinical Neurophysiology, University Hospital, Lund

ROSENDORFF, C., Department of Medicine, St. Thomas's Hospital, London

ROSS RUSSEL, R. W., National Hospital, Queen Square, London

ROSSANDA, M., Department of Anesthesiology, Ospedale Maggiore, Milano

SAKAMOTO, K., Department of Neurology, Wayne State University, Detroit, Michigan

SAMII, M., Neurochirurgische Universitätsklinik, Mainz

SANGUINETTI, I., Ospedale S. Gerardo dei Tintori, Monza

SANO, N., Department of Neurology, University of Pennsylvania, Philadelphia

SAPIRSTEIN, L. A., Stanford University, Medical Center, Palo Alto, California

SCHULZ, V., Physiologisches Institut der Universität, Mainz

SCHÜRMANN, K., Neurochirurgische Universitätsklinik, Mainz

SCHWARZ, W., Physiologisches Institut der Universität, Mainz

SEM-JACOBSEN, C. W., The EEG Research Institute, Gaustad Hospital, Oslo

SHAPIRO, H. M., Department of Neurosurgery, Hospital of the University, Philadelphia

SKINHØJ, E., Department of Neurology, Bispebjerg Hospital, Copenhagen

SLATER, R., Department of Neurology, University of Pennsylvania, Philadelphia

SMELLIE, G. D., Department of Neurosurgery, Institute of Neurological Sciences, Killearn Hospital, Glasgow

SMITH, A. L., Department of Anesthesia, University of Pennsylvania, Philadelphia

STOSSECK, K., Max-Planck-Institut für Arbeitsphysiologie, Dortmund

STYRI, O. B., The EEG Research Institute, Gaustad Hospital, Oslo

SVEINSDOTTIR, E., Department of Datalogy, Institute of Mathematics, University of Copenhagen.

SYMON, L., National Hospital, Queen Square, London.

TAYLOR, A. R., Department of Neurological Surgery, Royal Victoria Hospital, Belfast

TER POGOSSIAN, M. M., Edward Mallinckrodt Institute of Radiology, St. Louis, Missouri

TERAURA, T., Department of Neurology, Wayne State University, Detroit, Michigan

TOOLE, J. F., Department of Neurology, Wake Forest University, Winston-Salem, North Carolina

TROUPP, H., Neurosurgical Clinic, University Hospital, Helsinki

TSCHABITSCHER, H., Department of Psychiatry and Neurology, University Hospital, Vienna

TSCHETTER, T., University of Minnesota, Minneapolis

VAPALAHTI, M., Neurosurgical Clinic, University Hospital, Helsinki

VEALL, N., Guy's Hospital Medical School, London

WEINSTEIN, J. D., Department of Neurosurgery, Hospital of the University, Philadelphia

WELCH, M. J., Edward Mallinckrodt Institute of Radiology, St. Louis, Missouri

WILKINSON, I. M. S., National Hospital, Queen Square, London

WINSÖ, I., Department of Anesthesiology, Sahlgrenska Sjukhuset, Göteborg

WITCOFSKI, R. L., Department of Radiology, Wake Forest University, Winston-Salem, North Carolina

WITSCHEFF, E., Research Institute of Neurology and Psychiatry, Sofia

WODICK, R., Max-Planck-Institut für Arbeitsphysiologie, Dortmund

WOLF, R., Röntgenologisches Institut der Universität, Mainz

WOLLMAN, H., Department of Anesthesia, University of Philadelphia, Pennsylvania

VON WOWERN, F., Department of Neuromedicine, Gentofte Hospital, Hellerup

WÜLLENWEBER, R., Neurochirurgische Universitätsklinik, Bonn

YAMAMOTO, Y. L., Cone Laboratory for Neurosurgical Research, McGill University, Montreal

ZÄTTERSTRÖM, U., Department of Psychiatry, University Hospital, Lund

ZUPPING, R., Department of Neurology and Neurosurgery, State University, Tartu

Chapter I: Methodology

Dependence of ^{85}Kr(β)-Clearance rCBF Determination on the Input Function

H. HUTTEN, W. SCHWARZ, and V. SCHULZ

Physiological Institute, University of Mainz

Two procedures have been proposed for the quantitative determination of rCBF by analysis of radioactive inert gas clearance curves, following intra-arterial injection:

1. the slope-method [1],
2. the height-over-area-method [4].

The derivation of the slope-method is based upon an essential assumption, namely that there are no concentration gradients during the clearance process. Therefore, INGVAR and LASSEN originally suggested that the tissue should be saturated by infusing the indicator until a steady state is reached. After infusion has been stopped, the above mentioned condition can be considered as fulfilled as regards the initial part of the clearance curve [2]. However, this method has also been subsequently used when the indicator was injected as rapidly as possible [3].

A more general procedure was derived by ZIERLER. By limiting his method to the case in which the input function is identical to the delta-function, he developed the simple height-over-area-method. The input function is defined as the function of time of the indicator's entry into the tissue as seen by a detector. However, if the input function differs from the delta-function, the application of ZIERLER's method depends upon the following:

1. That the time course of the real input function is known.
2. That the mathematical operation of deconvolution can be carried out.

At the same time it must be taken into consideration that the injection time function generally does not coincide with the input function, mainly due to:

a) The indicator transport in the blood and the velocity profile in the vessels.

b) The distribution of the indicator between the individual capillaries seen by the detector.

c) The vessels acting as an "windkessel", which can not be excluded, especially if a very fast injection is applied.

A clear indication that there is no impulse function, even in the case of an instantaneous injection, is shown by the fact that even for sufficiently small time constants of ratemeter and recorder, the activity increase up to its maximum, takes longer than the injection.

The authors are indebted to M. BROCK and W.-R. PALMER, Neurosurgical Dept. (Prof. Dr. K. SCHÜRMANN), Univ. of Mainz, for their aid in the animal experiments.

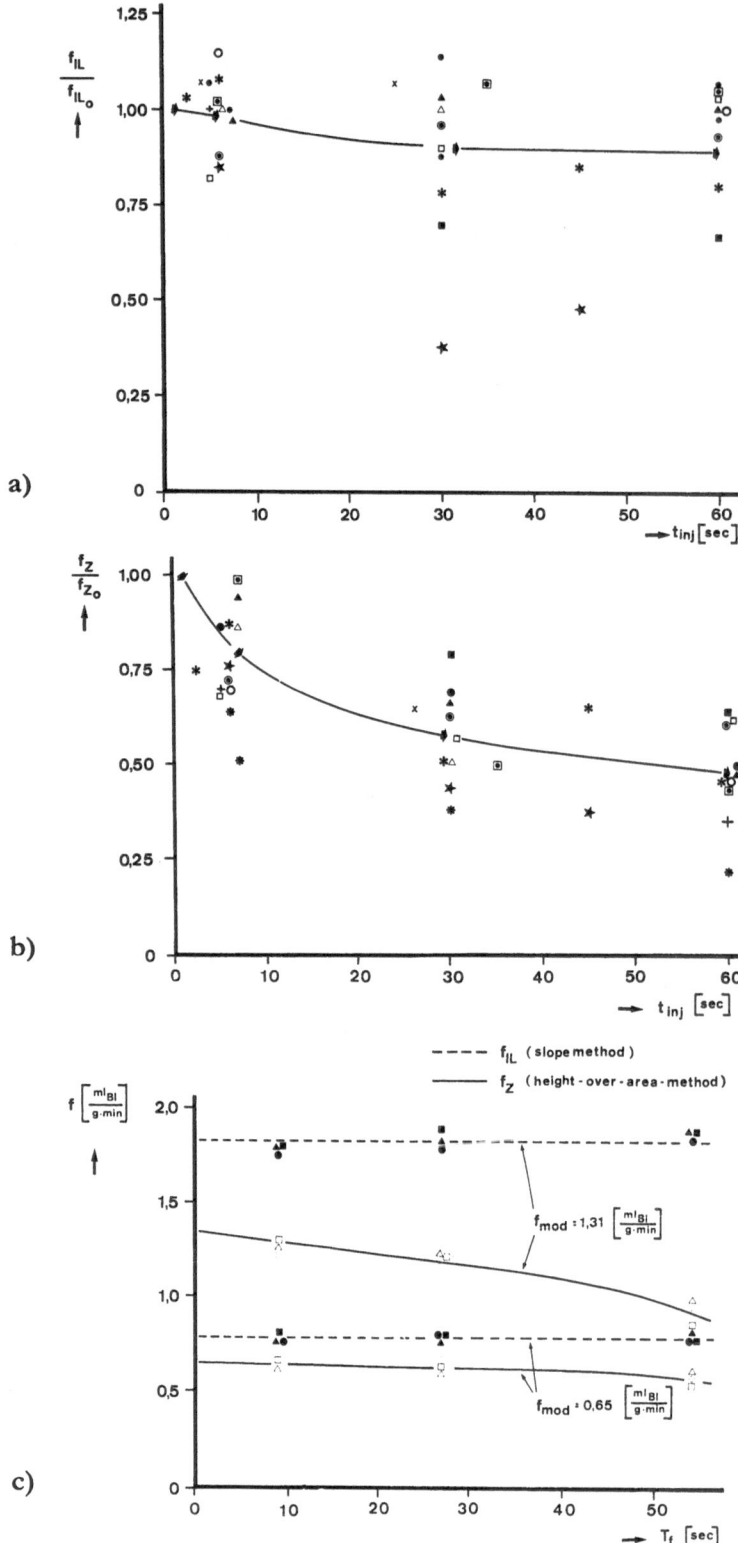

Fig. 1. a) Dependence of the rCBF values, determined by the slope-method and referred to the 1-second-injection-value, on the injection duration; b) Dependence of the rCBF values, determined by the height-over-area-method and referred to the 1-second-injection-value, on the injection duration; c) Dependence of the slope-method (dotted lines) and the height-over-area-method (solid lines) for two different blood flow values on the duration and the shape of the input function (□ = square-wave, △ = triangle, ○ = positive alternation of sinusoidal wave)

The following two questions now arise:

1. Does the shape of the clearance curves and, hence, the rCBF determined by the slope method, depend on the shape of the input function?

2. How large is the error when the height-over-area-method is applied to an input function that is different from the delta-function?

To answer these questions, a total of 50 $^{85}Kr(\beta)$-clearance curves were recorded from the brains of 11 cats, the injection duration being varied between 1 and 60 sec. The clearance curves were evaluated by means of both methods. The rCBF value, determined for the shortest injection time of about 1 sec, was used as a reference value for the same animal.

Fig. 1a shows the results determined by the slope-method, as a function of the injection duration. The values obtained from each animal are characterized by the same symbol. For the evaluation the biological half-time was always taken from that part of the clearance curve which best fitted a monoexponential decay. In spite of the variations in the individual values, the mean values do not depend on the injection time.

Fig. 1b shows the results determined from the same clearance curves, using the height-over-area-method, as a function of the injection duration. In spite of the variation of the individual values, the mean values distinctly decrease with increasing injection time. Even an injection time of 7 sec gives, as an average, a rCBF value 20% lower than the instantaneous injection.

The dependence of the blood flow values on the injection time has been used because in biological experiments it is not possible to measure the real input function. Therefore additional studies were performed with an electrical analogue computer, based on the analogy between the diffusion equation and the equation of current flow in an electrical network, in order to determine the effect of the input function on the shape of the clearance curves, as well as on the flow values obtained from them. The KROGH's cylinder for a single capillary surrounded by tissue was used as a model. The square-wave, the triangle, and the positive alternation of the sinusoidal wave were used as input functions. Fig. 1c shows the results for two different blood flow values, as a function of duration and shape of the input function. The values determined by the slope-method are indicated by dotted lines, and those determined by the height-over-area-method by solid lines. Although the slope-method leads to an overestimation of the real flow values, it is evident that this method does not depend on the shape and the duration of the input function. The values determined by the height-over-area-method, on the other hand, show a distinct dependence on the duration of the input function. The longer its duration the greater the underestimation of the blood flow. Moreover, this dependence is more pronounced for high blood flow values than for low ones, if the duration of the input function is the same. However, there is no sign of a clear dependence on the shape of the input function.

References

1. INGVAR, D. H., and N. A. LASSEN: Regional Blood Flow of the Cerebral Cortex Determined by Krypton 85. Acta physiol. scand. **54**, 325 (1962).
2. HUTTEN, H.: The influence of Diffusion of Inert Gases on the Determination of Blood Flow by the Clearance Method. III. International Symposium: Cerebral Blood Flow and Cerebro-Spinal Fluid. Lund-Copenhagen, Scand. J. Clin. Lab. Invest., Suppl. **102**, p. II-C, 1968.
3. LASSEN, N. A., K. HØEDT-RASMUSSEN, S. C. SØRENSEN, E. SKINHØJ, S. CRONQUIST, B. BODFORSS, and D. H. INGVAR: Regional Cerebral Blood Flow in Man Determined by Krypton 85. Neurology **13**, 719 (1963).
4. ZIERLER, K. L.: Equations for Measuring Blood Flow by External Monitoring of Radioisotopes. Circulat. Res. **16**, 309 (1965).

1*

Comments on the Inert Gas Elimination Method for the Determination of Cerebral Blood Flow

B. Ekström-Jodal, E. Häggendal, and N. J. Nilsson

Departments of Clinical Physiology, University of Göteborg

In the compartmental analysis of inert gas elimination curves from the brain the curves are customarily resolved into two components, regarded as representing flow in grey and white matter, respectively. However, very often a careful analysis of the initial part of curves obtained after short intra-arterial injections reveals another component, which may be relatively large in quantity, but which, because of its high disappearance rate, is evident only for a very short time. This fast initial component, described earlier by us [1], has attracted little attention, although, apart from its obvious theoretical interest, it is of definite practical importance for the calculation of mean cerebral blood flow, either from the initial slope, where it will tend to make the slope steeper, or by the height-over-area-method, where its presence may give a considerable uncertainty to the determination of the initial amplitude of the curve.

The initial component is seen occasionally in curves registered with a scintillation detector collimated over the entire brain. It appears almost regularly and with greater amplitude in β-recordings from the exposed cerebral cortex. As to its origin several possibilities may be considered.

1. Arterio-venous shunts, either as actual blood channels or as a possibility for gas diffusion through the vessel walls, may be present in the area from which activity is recorded.

2. Part of the tissue may have an extremely rapid perfusion.

3. The detector may record activity from large arteries supplying tissues farther away from the detector.

4. The indicator gas may be eliminated not only by the blood flow but also by rapid diffusion from the tissue into the air.

In this paper the results of some experiments on this phenomenon are presented.

Methods

The experiments were made on dogs in pentobarbital anesthesia, "curarized" and artificially ventilated; some measurements were made after the death of the animal. Cerebral blood flow was determined according to the inert gas elimination principle, using Kr^{85} in saline, given either in small rapid injections into the non-ligated vertebral artery or into the cortical tissue. For the recording of the activity, scintillation detectors, Geiger-Müller end-

Supported by the Swedish Medical Research Council under Grant K 66:881–23x–1044–01.

window and channelled tubes and in some cases a needle detector [3] were used, often two or three detectors simultaneously. Details of the methodology have been given in a previous publication [1].

Results and Comments

Ad I and II, Shunts or Rapidly Perfused Tissues

In elimination curves from the brain of patients with known arterio-venous shunts an initial "peak" can be observed [2]. It is clear that this peak must be present also in a corresponding curve from the venous blood in such a case. The same would be true if, in part of the tissue, the indicator gas would not reach complete diffusion equilibrium, or the flow rate would be very high. In our experiments, however, when recording the activity in blood drawn from the superior sagittal sinus through a channelled GM tube, we have never observed the fast initial component.

Ad III, Activity in Arteries

The half-time of the initial component, 0.1–0.2 min, is about the same as that of blood-borne indicators, e.g. iodinated serum albumin.

Since it is practically impossible to find areas as large as the window diameter of the ordinarily used end-window tubes, i.e. about 1 cm, where no larger blood-vessels are found, we have tested the influence of the large vessels by covering them with strips of lead, about 1 mm thick, cut to dimensions fitting the arteries in the individual detector fields. When this

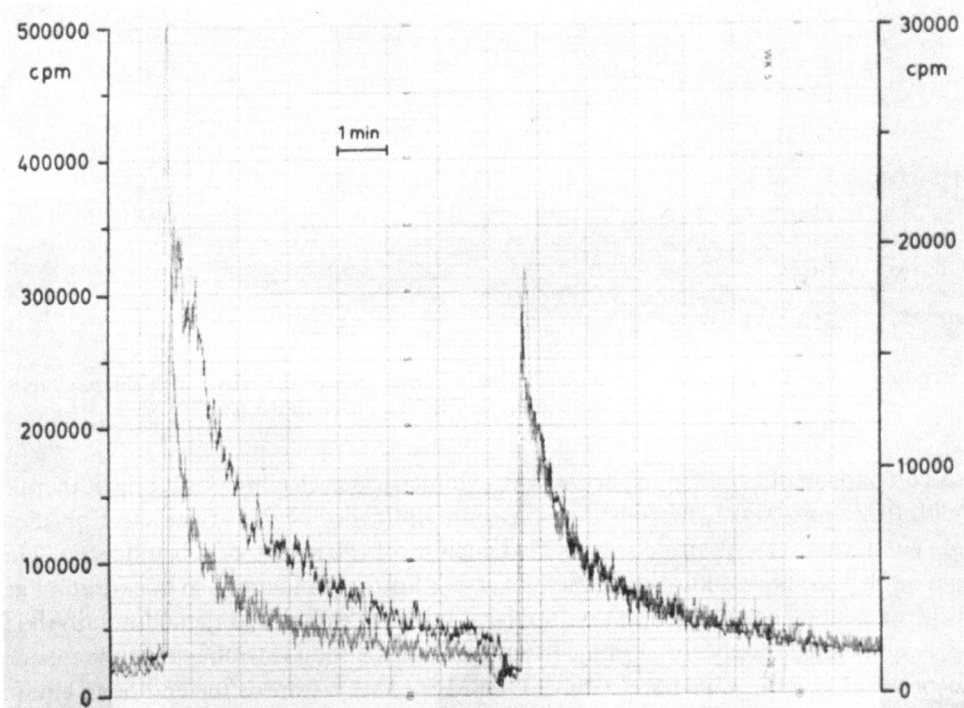

Fig. 1. Simultaneous GM tube recordings over symmetrical parts of the cortex with and without lead shielding of visible arteries. Scales to the left for the left part of the figure and to the right for the right part of the figure. Left part of the figure: lead shield over arteries on the left side corresponding to the curve with slower ascending phase; right part of the figure: lead shield over the right side, also corresponding to a slower ascending phase

covering was carefully performed, it was possible to obtain an elimination curve without any initial fast component (see Fig. 1). Fig. 2 shows how the indicator appears later in the area free from visible arteries than in the symmetrically located area on the other hemisphere, where the arteries are not covered. The curve from this latter area demonstrates a fast initial component; a considerable part of its total activity has been eliminated before maximal activity is reached in the area with the vessels covered.

The fact that this component is not always observed is probably in part due to different approaches to the curve analysis, but may also be connected with the technique used by many investigators of covering the brain surface with a sheet of Mylar. This is done to protect the cortex against desiccation. It has also been stressed that Mylar is gastight and would therefore prevent loss of indicator by diffusion from the brain surface.

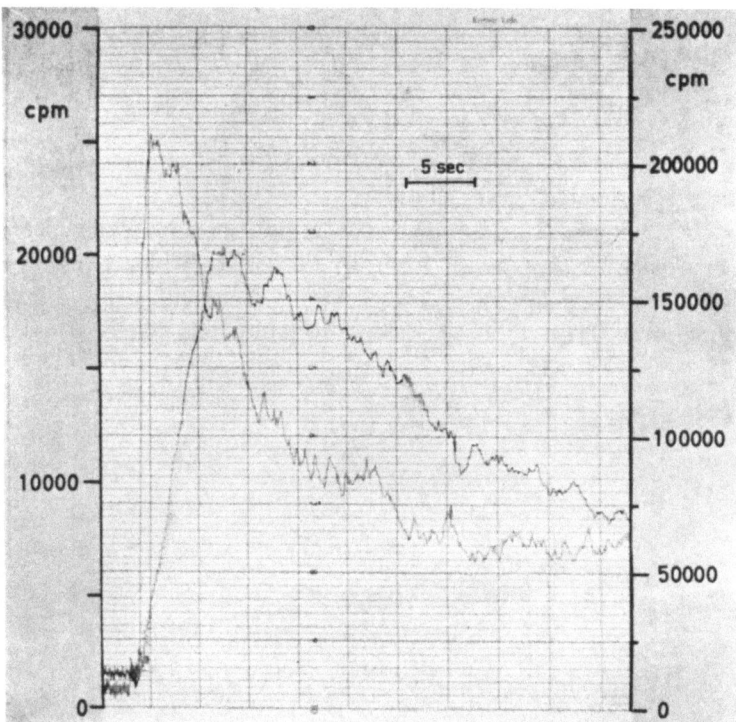

Fig. 2. Same as Fig. 1 with higher paper speed. Lead over arteries in the curve with slowest onset. Scale for slow curve to the right, for fast curve to the left

A comparison of elimination curves from symmetrical parietal regions, one of them covered with 0.023 mm Mylar, demonstrates a definite influence of the Mylar sheet on the curve form. We have not as yet made any detailed analysis of the nature of this influence. The fact, however, that krypton diffuses into the Mylar will lead to an increase in background activity and exert a general damping effect on the curve. It must be observed that, although the Mylar mass is quite small, its position immediately in front of the GM tube strongly enhances its influence. In a few cases we have removed the Mylar sheet towards the end of an elimination curve and observed a reduction of the activity by about 30%. The krypton elimination from isolated Mylar sheets in three cases had half-times of 36, 39 and 41 minutes. Application of a Mylar film in the course of an elimination begun without it caused a rise in activity persisting for several minutes; after its removal the activity returned to about its previous level and continued to fall. Renewed application again led to a rise.

Ad IV, Diffusion into the Air

To study the diffusion of the indicator gas from the tissue into the air we have made intra-arterial injections post mortem. The initial part of the β-curves in this case had a varying and often irregular curve form, which may be connected with the flushing with saline after the injection. The main part of the β-curves always showed an exponential decline with half-time values of 15–25 min. When studying these problems one must observe the effect of gravity, which will cause a dislocation of fluid within the vessels during some time after the injection, tending to diminish the activity in the upper and to increase it in the lower parts of the organ.

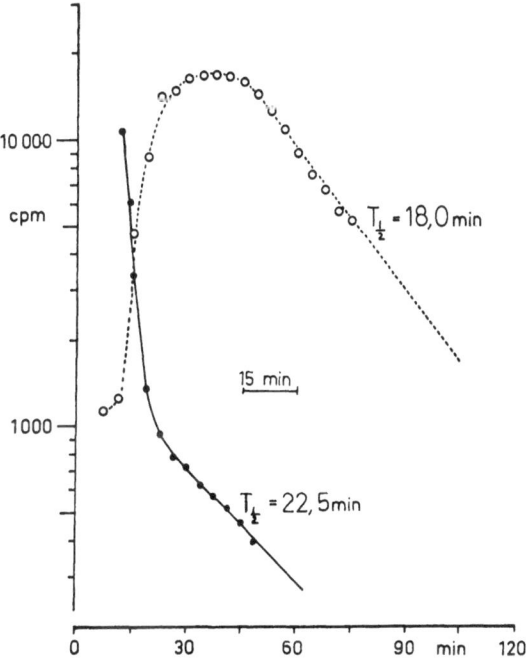

Fig. 3. Simultaneous post-mortem recordings from two GM tubes, one from above (drawn out line), and one from below (broken line) a dog's head lying on its side. Recordings from symmetrical parts of the cortex. At time zero, 25 min after an intra-arterial injection, the head was turned upside down

This can be demonstrated by recording the activity simultaneously in the upper and the lower parts of the brain with the dog's head lying on its side (Fig. 3). After about half an hour, however, both detectors show the slow fall mentioned above, which we interpret as due to diffusion into the air. It is not seen in γ-curves from the whole brain.

Our conclusion therefore is that the initial fast component probably emanates from activity in arteries on the brain surface.

References

1. HÄGGENDAL, E., N. J. NILSSON, and B. NORBÄCK: On the components of Kr[85] clearance curves from the brain of the dog. Acta physiol. scand. **66**, Suppl. **258**, 5 (1965).
2. —, D. H. INGVAR, N. A. LASSEN, N. J. NILSSON, G. NORLÉN, I. WICKBOM, and N. ZWETNOW: Pre- and postoperative measurements of regional cerebral blood flow in three cases of intracranial arterio-venous aneurysms. J. Neurosurg. **20**, 1 (1965).
3. LAUBER, A., and B. ROSENCRANTZ: A needle-type p-i-n junction semiconductor detector for in vivo measurement of beta tracer activity. Rep. AE 162 from AB Atomenergi. Stockholm 1964.

Further Studies on Exponential Models of Cerebral Clearance Curves

M. Reivich, R. Slater, and N. Sano

Spiller Neurological Unit and the Research Laboratories of the Department of Neurology of the University of Pennsylvania

In a previous communication [2] we have described a new mathematical model for the analysis of cerebral clearance curves consisting of a bimodal Gaussian distribution of exponentials. The purpose of the present communication is to further evaluate this bimodal Gaussian model and to compare it with the two compartment model using cerebral clearance curves synthetized from regional cerebral blood flow data obtained in cats.

The bimodal Gaussian model was developed since it is known from autoradiographic studies that the brain represents a multicompartment system in regard to its perfusion rates. An explicit function precisely describing a bimodal distribution of perfusion rates was derived as follows: let the desired function, $f(t)$, be represented by a sum of n exponentials

$$(1) \qquad f(t) = \sum_{i=1}^{n} W_i e^{-K_i t}$$

Since the rate constants of these exponentials are normally distributed, the relationship between the weights, W_i, and the rate constants (perfusion rates), K_i, is described by the expression for the

$$(2) \qquad W = \frac{1}{\sigma \sqrt{2\pi}} e^{-\frac{(\bar{K} - K)^2}{2\sigma^2}}$$

where K is the mean and σ is the standard deviation of the distribution. Substituting equation (2) into (1) and assuming that there is a continuous distribution of flows in the brain, produces

$$(3) \qquad f(t) = \frac{1}{\sigma \sqrt{2\pi}} \int_0^\infty e^{-\frac{(\bar{K} - K)^2}{2\sigma^2}} e^{-Kt} dK$$

integrating equation (3) produces

$$(4) \qquad f(t) = \frac{1}{2} e^{-\bar{K}t} e^{\frac{\sigma^2 t^2}{2}} erfc\left(\frac{\sigma^2 t - \bar{K}}{\sigma \sqrt{2}}\right)$$

which is then an explicit function describing the clearance of an indicator from a multicompartment system containing a normal distribution of perfusion rates. For a bimodal distribu-

Supported in part by a U.S.P.H.S. Research Grant (NB-06314-04) from the National Institute of Neurological Diseases and Blindness. Dr. Reivich is recipient of Career Research Development Award 5 K 03 HE 11896-04.

tion of flows, the sum of two such terms properly weighted would describe the clearance curve. A subroutine to the SAAM program of Berman et al., 1962 [1], was written to fit this equation to cerebral clearance curves. A previous theoretical analysis had shown that the errors in predicting the mean rate constants and weights of the fast and slow components of the cerebral clearance curves were smaller with the bimodal Gaussian model than with the two exponential model.

We have now compared these two models using regional cerebral blood flow data obtained in 10 cats by the C^{14}-antipyrine autoradiographic technique [3]. Flow data was obtained for 25 regions of the brain. The relative weights of these regions were determined by dissecting and weighing the regions from a perfusion fixed cat brain. With this information it was possible to calculate the composite cerebral clearance curve from the whole brain in response to a step function change in inert gas concentration in the arterial blood. This composite clearance curve was then analyzed by means of both the bimodal Gaussian and the two exponential models, and the mean gray and white matter flow rates and weights predicted by these two models were compared with the actual known values at these parameters.

In the 10 animals the mean white matter flow rate varied from 17.5 to 43.3 ml/100 gm/min and the mean gray matter flow rate varied from 83.3 to 155.8 ml/100 gm/min. The relative weight of the white matter was 25% and of the gray matter 75%. Fig. 1 illustrates the actual distribution of flows in the brain of one cat in the form of a histogram with the distribution of flows predicted by the bimodal Gaussian model superimposed. The values for the mean white (K_1) and gray (K_2) matter flows and their respective relative weights (W_1 and W_2) are compared with those predicted by both models.

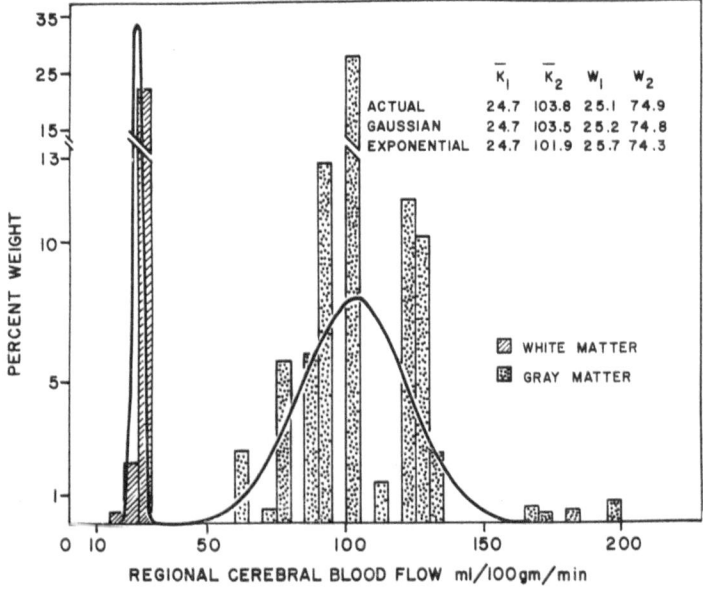

Fig. 1. The histogram represents the actual distribution of flows in the brain of an awake cat while the curve represents the distribution predicted by the bimodal Gaussian model. The values for the mean white ($z\mathcal{y}_1$) and gray ($z\mathcal{y}_2$) matter flows and their respective weights (W_1 and W_2) are shown along with the values predicted by the Gaussian and exponential models

The mean errors in predicting the mean flow rates and weights of the white and gray matter structures are shown in Table 1. It will be noted that these mean errors are all relatively small. The largest individual error was 8% and occurred in the estimation of the weight of white matter by the exponential model. The Gaussian model predicted the mean gray matter flow rate with a significantly smaller error than the exponential model. There was no significant difference between the two models in predicting the mean white matter flow rate or the relative weights of gray and white matter. In addition to these data the Gaussian model provided information concerning the standard deviation of the distribution of flows in the white and gray matter. The mean $\pm$ S. E. of these values were 0.6 ± 0.3 and 27.4 ± 3.2 ml/100 g/min respectively.

Table 1. *Percent error in predicted values*[1]

	Exponential Model	Gaussian Model
K_1	-0.5 ± 0.6	-1.3 ± 0.4
K_2	-3.7 ± 0.5	-1.3 ± 0.2[2]
W_1	0.7 ± 1.5	-3.4 ± 0.7
W_2	-0.2 ± 0.5	-1.1 ± 0.2

[1] means $\pm$ S.E.

[2] significantly different (p < 0.01 paired t-test)

The goodness of fit of these two models to the cerebral clearance curves was evaluated by comparing the root mean square deviation of each model about the curve. In three cases there was no significant difference between the fits while in the other seven the Gaussian model produced a significantly better fit to the data.

In conclusion, on the basis of this data it appears that the two exponential model is a very good description of cerebral clearance curves with a maximal error of about 8%. However, data obtained under conditions of low cerebral blood flow have not been analyzed and theoretical considerations suggest that the errors may be somewhat larger under these conditions. In spite of the good description provided by the exponential model, the bimodal Gaussian model predicts the mean gray matter flow rate with significantly less error. In addition, it produces a significantly better fit to the data and provides information regarding the standard deviation of the distribution of gray and white matter flows.

References

1. Berman, M., E. Shahn, and M. F. Weiss: The routine fitting of kinetic data to models: A mathematical formalism for digital computers. Biophys. J. **2**, 275 (1962).
2. Reivich, M.: Observations on experimental models of cerebral clearance curves. In: Research on the Cerebral Circulation. J. S. Meyer (Ed.). Springfield: Charles C. Thomas 1969. III.
3. —, J. Jehle, L. Sokoloff, and S. S. Kety: Measurement of regional cerebral blood flow with C¹⁴-antipyrine in awake cats. J. appl. Physiol., 1969, in press.

Evaluation of the rCBF Method of Lassen and Ingvar

L. C. McHenry, Jr., M. E. Jaffe, and H. I. Goldberg

Stroke Research Center, Philadelphia General Hospital

The development of a method by Lassen and Ingvar [4] to measure regional brain blood flow has added a new dimension to the studies of the cerebral circulation. Lassen, Ingvar and their associates [5], and Glass and Harper [1] first used the isotope clearance method for measurements over the intact skull in man.

In our laboratory, regional cerebral blood flow (rCBF) measurements are carried out in a manner similar to that of Lassen and Ingvar and in conjunction with cerebral angiography. Under fluoroscopic guidance a small catheter is placed into the internal carotid artery. Xenon-133 in saline, 2 millicuries, is then injected into the catheter. The uptake and clearance of the isotope is monitored by 8 scintillation detectors with $1/_2 \times 1/_2$ inch NaI crystals. The probes are contained in a lead collimator block at the side of the subject's head. Each probe is set $1^1/_2$ inches from the face of the block and has an opening $1/_2$ inch in diameter. With this collimation each probe "sees" an area of brain approximately $1^1/_2$ inches in diameter. Data from all probes is recorded on magnetic tape. Each channel is replayed through a scaler and the Xenon clearance curves are plotted on a linear chart recorder. Points on clearance curves are transferred to punch cards for analysis on the IBM 7040 computer. The computer program is a Fortran translation of Sveinsdottir's ALGOL program [10]. Data is given as corrected and uncorrected (for pCO_2) stochastic (10-min and infinity) values and as 2-compartmental values.

To obtain data on normal rCBF values with our equipment, rCBF measurements were performed in 5 neurologically normal individuals. Two were normal volunteers, 2 controlled epileptics and the other an alcoholic 1-week post withdrawal seizures. The mean uncorrected values were: stochastic $rCBF_{10}$ 54.4 cc/100 gm/minute with a S.D. 2.7; stochastic $rCBF\infty$ 48.7 cc/100 gm/min with a S.D. of 4.0; compartmental rCBF was 51.6 cc/100 gm/min with a S.D. of 3.2. Mean gray and white matter flows were 77.0 and 21.9 cc/100 gm/min, respectively, with a gray matter weight of 54%. Although this is a small group, these mean values are similar to those reported by others. Høedt-Rasmussen's [2] stochastic $rCBF\infty$ for his group was 46.7 cc/100 gm/min. When individual rCBF values from the 8 probes were compared in these normal cases, inter-regional differences in rCBF were found to be less than 10% (coefficient of variation).

In order to test the reproducibility of the rCBF method, serial measurements were performed in the same individual. The mean stochastic $rCBF_{10}$ values for repeat measurements on 10 neurological patients were 33.8, and 33.4 cc/100 gm/min, respectively, with a correlation coefficient of 0.97 (p < 0.001). This confirms a similar serial study by Høedt-Rasmussen [2] and demonstrates that the method of Lassen and Ingvar is highly reproducible. In a compara-

Supported by grants from the U.S. Public Health Service, NB-06520, and from Marion Laboratories.

tive study rCBF measurements were made on 11 patients that also had CBF determined by the Krypton desaturation modification [6] of the Kety-Schmidt technique. The mean stochastic $rCBF_{10}$ was 32.6 and the CBF was 31.2 cc/100 gm/min with a correlation coefficient of 0.90 (p < 0.001). The serial and comparative studies were carried out on patients with stable neurological deficits. In both studies the pCO_2 did not vary over 4 mmHg between individual measurements. A linear plot of the 2 studies is given in Fig. 1 and details are reported elsewhere [8].

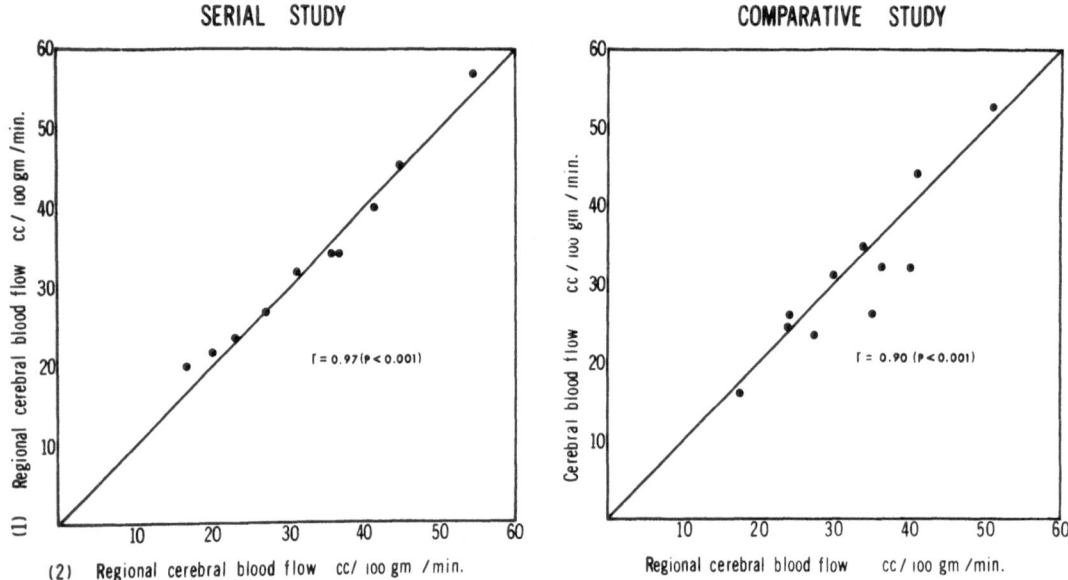

Fig. 1. Individual values from serial rCBF measurements are plotted on the left. Shown on the right is a linear plot of rCBF values and the results of CBF measurements by the Krypton desaturation modification of the Kety-Schmidt technique in the same patient

In the comparative group, these 11 cases were used because of the nature of the neurological disorder, that is, most patients had diffuse neurological disease. Also, the CBF data was compared to uncorrected stochastic $rCBF_{10}$ values rather than to stochastic $rCBF\infty$ or compartmental values. The comparison would not have been as close if the latter values were used. Under the circumstance chosen, therefore, it does appear that values from the rCBF method are comparable to Kety-Schmidt technique data. On the other hand, the 2 methods may not give comparable quantitative data in the presence of focal disease of the brain. The Kety-Schmidt method measures an average portion of total brain blood flow, whereas the rCBF method can detect very high or low areas of blood flow. Depending on the collimation used and location of the probes, the mean rCBF values might not be the same as the "average portion" value.

The applicability of the rCBF has been demonstrated in studies of a variety of neurological diseases [3, 7] and of the effects of pharmacological agents on the cerebral circulation. Six patients with vascular occlusion and cerebral infarction demonstrated angiographically had rCBF measurements before and 10 min after the intravenous injection of 100 mg of papaverine. In the 6 cases there were 20 probes over ischemic or infarcted areas (dotted lines in Fig. 2) and 22 probes over normal angiographic regions (solid lines). A significant increase in the *mean* $rCBF_{10}$ occurred after papaverine in all cases, but the changes in regional flow values varied considerably with individual probes and were not consistent from case to case. In all 6 cases 8 individual probes had a significant increase in rCBF in the regions of cerebral infarction; in no areas were there significant decreases or "stealing" of regional blood flow. There was a marked increase in blood flow in one probe (HH) over an area of hyper-perfusion. In probes over normal areas blood flow increased significantly in 8 of these regions following papaverine, but in 2 instances (GW, EP) there were significant decreases in flow. These studies confirm previous reports [9] of an increase in CBF following parenteral papaverine. Regional CBF measurements, however, show that in cerebrovascular disease there are not only different flow rates in the brain but involved and uninvolved regions may respond differently to vasodilating agents. The factors that result in these differential flow responses have yet to be clearly defined.

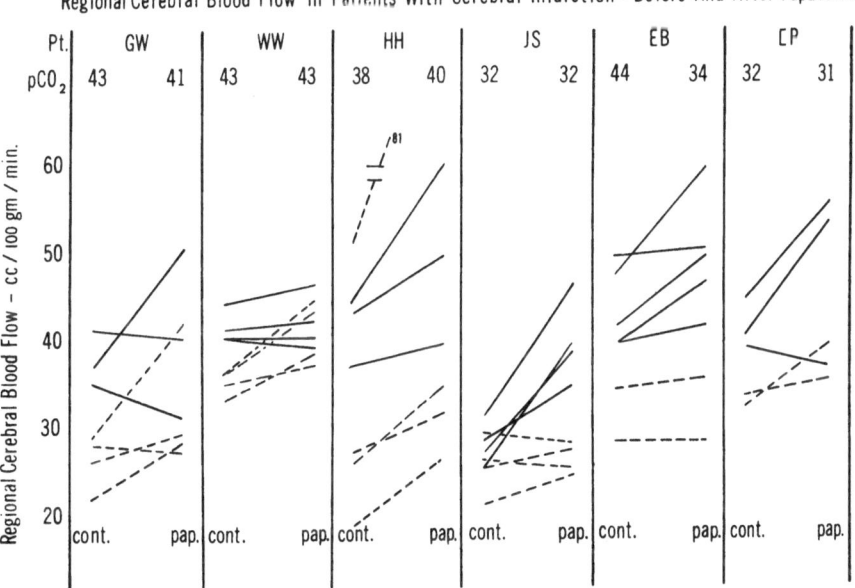

Fig. 2. The results of rCBF measurements before and after the intravenous administration of 100 mg of papaverine in 6 patients with cerebral infarction are shown. Each line represents the change in the rCBF value. The broken lines are probes over areas of cerebral ischemia or infarction; solid lines represent probes over regions considered to be angiographically normal

References

1. Glass, H. I., and A. M. Harper: Measurement of regional blood flow in cerebral cortex of man through intact skull. Brit. med. J. 1, 593 (1963).
2. Høedt-Rasmussen, K.: Regional cerebral blood flow, the intra-arterial injection method. Acta neurol. scand., Suppl. 27, Vol. 43 (1967).
3. Jaffe, M. E., L. C. McHenry, Jr., and H. I. Goldberg: Regional cerebral blood flow measurements with small probes, II. Clinical application of the method (submitted for publication); abstract in Neurol. 18, 280 (1968).
4. Lassen, N. A., and D. H. Ingvar: The blood flow of the cerebral cortex determined by radioactive Krypton[85]. Experientia 17, 42 (1961).
5. —, K. Høedt-Rasmussen, S. C. Sorensen, E. Skinhøj, S. Cronquist, B. Bodforss, and D. H. Ingvar: Regional cerebral blood flow in man, determined by Krypton[85]. Neurol. 13, 719 (1963).
6. McHenry, L. C., Jr.: Quantitative cerebral blood flow determination, application of Krypton[85] desaturation technique in man. Neurol. 14, 785 (1964).

7. —, M. F. Jaffe, and H. I. Goldberg: Regional cerebral blood flow measurements with small probes. I. Evaluation of the method. Neurol., submitted for publication.

8. — In: Cerebral Vascular Diseases. J. F. Toole, R. G. Siekerat, and J. P. Whisnant (Eds.). New York: Grune and Stratton 1968. p. 204.

9. Meyer, J. S., F. Gotoh, J. Gilroy, and N. Nara: Improvement in brain oxygenation and clinical improvement in patients with strokes treated with papaverine hydrochloride. JAMA **194**, 957 (1965).

10. Sveinsdottir, E.: Clearance curves of Kr^{85} and Xe^{133} considered as a sum of monoexponential wash-out functions. Acta neurol. scand., Suppl. **14**, 69 (1965).

Problems in Control Studies for the Evaluation of Drugs

H. HACKER

Neurosurgical Department, University of Frankfurt a. M.

In investigating the effects of drugs on the CBF, repeated examinations of blood flow before and after medication under otherwise constant conditions are mandatory. This may be easily done in an anesthesized animal, but difficulties arise when testing is performed in the awake man.

In our experiments we have studied the effect of Theophyllin-Ephedrin, which is being used for the treatment of cerebral vascular diseases but has been proven not to alter CBF according to GOTTSTEIN, using the Kety-Schmidt method. We monitored the decrease of radiation for 15 min after the injection of 133 Xenon into the internal carotid artery using four one in. crystals and in a few cases a 2.5 in. crystal. A specially adapted "Nucleopan" by Siemens was used for detecting and recording the radiation.

After a control determination of CBF had been completed, we started an infusion of 340 mg Theophyllin-Ephedrin and 672 mg Oxyethyltheophyllin in 500 ml Ringers' solution. The drip was regulated so as to increase systolic blood pressure by 25 mmHg and to maintain this level. The pulse rate increased only slightly in 4 patients and end-tidal CO_2, determined by infrared absorption, decreased by a small percentage in 4 patients. A steady state was reached after 10–15 min. Thirty min from the beginning of the infusion the second Xe study was performed. In 9 cases examined we noted a decrease of 13% (SD $\pm$ 2.5%) in CBF. This result was unexpected and possible reasons inherent in the experimental set-up had to be looked for.

Two series of studies were undertaken under the following conditions:

1. The infusion consisted of Ringers' solution only, given at the same rate.

2. No medication at all was given in the 30 min interval and the patient was kept very quiet, avoiding any disturbance.

After administration of Ringers' solution in 5 cases CBF increased by a mean of 7% (SD $\pm$ 2.6%). End-tidal CO_2, arterial BP and pulse rate were stable.

In the second group of 4 patients we avoided any disturbance of the patient, and the second study after 30 min was undertaken in a most gentle way. In average end tidal CO_2 decreased by 19% ($\pm$ 3.3%), three patients had a fall of systemic BP of 5–10 mmHg and CBF was lowered by 17% ($\pm$ 1.7%).

This decrease of CBF is not the general rule, as proven by the following investigation: In 3 patients with increased intracranial pressure an infusion of 100 g Sorbitol (Tutofusin S 40, PFRIMMER) was given. MABP did not change, end-tidal CO_2 was slightly lowered but CBF increased by 28% ($\pm$ 2.8%). This increase could also be demonstrated by angiography. We assume that release of intracranial pressure and hemodilution enhanced the CBF.

The differences in CBF in these four groups cannot be fully explained by changes in BP, pulse rate and end-tidal CO_2. A decrease of CO_2 was noticed concomitant with an increase of CBF in the patients receiving Sorbitol. The fact is that in all such studies the first Xenon-test is performed shortly after carotid puncture with the patient in a state of anxiety. 30 min later the patient has usually calmed down and the Xe-injection is no longer troubling him. Respiration is more regular and deeper, more CO_2 is exhaled and CBF is decreased – both facts may be in part independently related to the emotional state. In this context we have to consider the influence of mental activity on CBF as shown by Ingvar and Risberg. We have made similar observations of high blood flow in hysteria and acute psychosis. Therefore we interpret our findings in groups 2 and 3 as representing the difference between a CBF examination during emotional activity and an examination in relaxation.

As a conclusion we may say that great care should be taken in the interpretation of control studies with differences in CBF of less than 20% decrease. Positive reactions seem to be less critical.

Table 1. *Changes in mean rCBF, end-tidal* CO_2, *systemic arterial blood pressure and pulse rate in the various groups of patients studied*

Drug	N	Mean rCBF Diff. % S. D.	CO_2 End-tidal	ABP syst.	Pulse rate
Theophyll.-Ephedrin	9	-13 ± 2.5	$(-)$	$+ 25$ Torr	$(+)$
Ringers' sol	5	$+ 7 \pm 2.6$	$=$	$=$	$=$
No medic.	4	-17 ± 1.7	$-$	$- 10$ Torr	$=$
Sorbitol	3	$+28 \pm 2.8$	$(-)$	$=$	$=$

$(-)$ slight decrease

$-$ decrease

$=$ unchanged

References

1. Eichhorn, O.: Probleme in der Therapie der Hirndurchblutungsschäden durch sogenannte vasoaktive Substanzen. Therpiewoche **10**, 618 (1960).
2. —, K.-H. Auell u. L. Dörfler: Statistische Untersuchungen über Ergebnisse der Schlaganfallbehandlung. Med. Klin. **60**, 2100 (1965).
3. Gottstein, U.: Der Hirnkreislauf unter dem Einfluß vasoaktiver Substanzen. Heidelberg: Dr. Alfred Hüthig, 1962.
4. Hacker, H., u. A. Alonso: Karotisangiographie nach Entwässerung – bessere Darstellung von Hirntumoren. Neurochirurgie, 1969, in press.
5. Ingvar, D. H., and J. Risberg: Increase of regional cerebral blood flow during mental effort in normals and in patients with focal brain disorders. Exp. Brain Res. **3**, 195 (1967).

The Heterogeneity of Blood Flow throughout the Normal Cerebral Hemisphere

I. M. S. Wilkinson, J. W. D. Bull, G. H. du Boulay, J. Marshall, R. W. Ross Russell, and L. Symon

National Hospital, Queen Square, London

A study was designed to examine the homogeneity of blood flow throughout the normal cerebral hemisphere. 10 subjects, in whom carotid angiography was indicated on clinical grounds, were investigated. In all cases there were no abnormal physical signs referable to the hemisphere studied, and the accompanying carotid angiogram was normal. 5–8 mC of ^{133}Xenon was injected as a single slug into the upper cervical part of the internal carotid artery. The clearance of ^{133}Xenon from the hemisphere was monitored by 16 regional detectors situated on the lateral aspect of the head. Each detector was made as regional as possible by collimation and lower limit discrimination at 77 KeV.

Two-compartmental analysis of the 16 clearance curves for each subject was performed to yield regional values for:

Fg = rate of perfusion of grey matter
Fw = rate of perfusion of white matter
Wg= relative weight of grey matter.

To allow comparison of the 10 cerebral hemispheres, it was necessary to express all the regional results as percentages of the mean for the hemisphere.

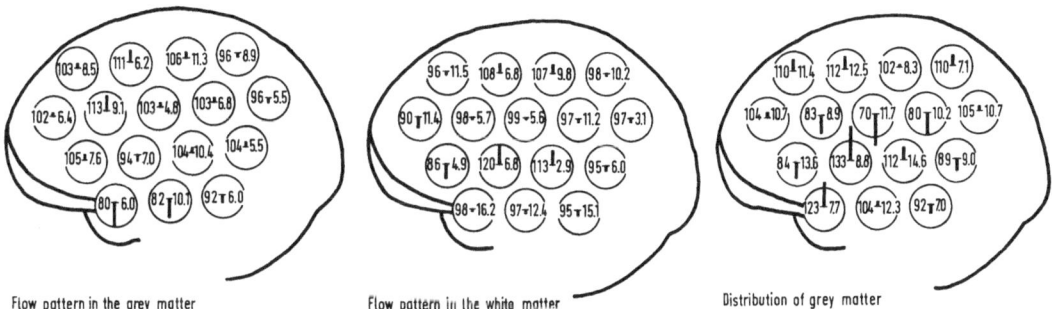

Flow pattern in the grey matter   Flow pattern in the white matter   Distribution of grey matter

Fig. 1. The accumulated results in 10 normal cerebral hemispheres. Within each circle the mean regional value is written on the left (expressed as a percentage of the mean for the hemisphere) and the S.D. is written on the right

The accumulated results for the 10 subjects are represented in Fig. 1. Perfusion of the grey matter was found to be significantly lower in the temporal region and higher in the precentral region than throughout the rest of the hemisphere. The white matter was more highly perfused in the region of the internal capsule than elsewhere in the hemisphere. (Low values for white matter perfusion were found in the frontal region, but these may be artefactual due to slow clearance of [133]Xenon from poorly perfused extracerebral tissues supplied by the supra-orbital, supratrochlear and meningeal branches of the ophthalmic artery.)

The percentage of grey matter in different regions of the hemisphere was not uniform. In the region of the insula and basal ganglia, and over the convexity of the hemisphere, high values were obtained (corresponding to 50–60% grey matter: 40–50% white matter). In the intervening regions over the corona radiata and corpus callosum, the relative weight of grey matter was much less (values corresponding to 30–40% grey matter: 60–70% white matter). Post-mortem dissection of a normal cerebral hemisphere into its component grey and white matter was performed on a regional basis, giving values which correlated well (numerically and in terms of pattern) with the regional distribution of grey matter derived from [133]Xenon clearance.

Conclusions

1. There is a definite regional pattern throughout the normal cerebral hemisphere for the perfusion of grey matter, perfusion of white matter and distribution of grey matter.

2. This lack of homogeneity can be demonstrated by the intracarotid [133]Xenon injection method using highly regionalized detectors.

3. This normal pattern must be established for the correct interpretation of regional blood flow studies in the abnormal cerebral hemisphere.

4. The mean flow value for any region of the hemisphere depends on 3 independent variables, F_g, F_w and W_g in that region. As CBF detectors become more regional, mean flow values become less informative, and the greater is the need to identify F_g, F_w and W_g separately.

The Two-Minutes-Flow-Index (TMFI)

H. Hutten and M. Brock

Physiological Institute and Neurosurgical Department, University of Mainz

While performing clinical rCBF functional tests it is difficult to maintain the patient in a stationary condition over the entire clearance period. For this reason it has become a common practice to restrict the recording of the clearance curves to the first 2 min [1, 2, 3]. This gives rise to one question:

1. Is it possible to evaluate the information contained in these 2 min of the clearance curve so as to obtain a quantifiable measure for rCBF?

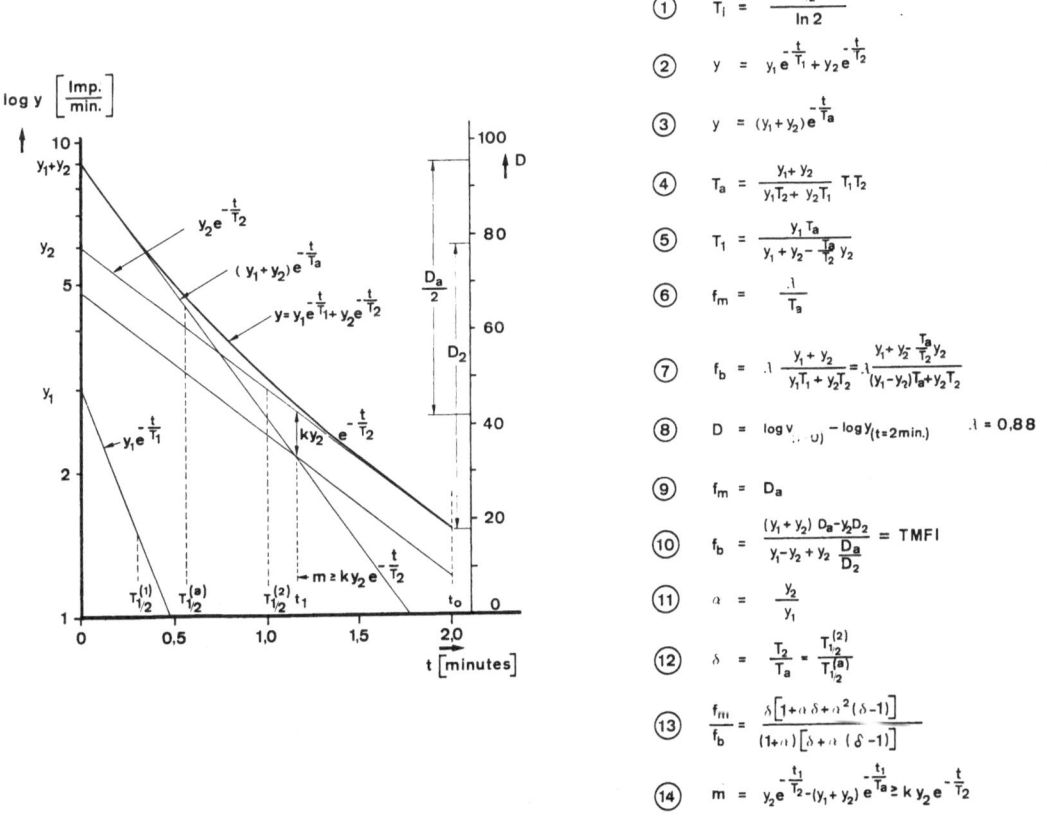

Fig. 1. Representation of a biexponential clearance curve. Abscissa: time t; Ordinate: activity (written logarithmically) $\log y$, respectively D (% of one decade). y_1 and y_2: intercepts; T_1, T_2 and T_a: time constants; $T_{1/2}^{(1)}$, $T_{1/2}^{(2)}$ and $T_{1/2}^{(a)}$: biological time constants; $t_0 = 2$ min: measuring time for the TMFI; D_2 and D_a: CBF values directly obtainable for $\lambda = 0.88$ and $t_0 = 2$ min; t_1: measuring time necessary for recognizing a bi-exponential shape; k: confidence factor admitted to recognize a biexponential shape; m: deviation of the monoexponential from the biexponential shape

2*

We have departed from the height-over-area-method of Zierler. Thus, for the determination of the comprised area, an extrapolation becomes necessary which presupposes the approximation of the clearance curve by certain mathematical functions. This raises two further questions:

2. What functions are to be adequately employed for the approximation?
3. Is it possible to determine the error introduced by arbitrariness into this approximation?

The second question is easiest to answer, since the shape of the clearance curves practically implies approximation by means of exponential functions. Most of these clearance curves can be described by one, occasionally by two, exponential functions with sufficient exactness during the first 2 min. The application of exponential functions also makes sense from the mathematical point of view, since the time integral of these functions always converges for $t \to \infty$. The approximation by exponential functions, at the same time, avoids the disadvantage of the height-over-area-method, namely the dependence on the input function.

The answer to the 3rd question, concerning error determination, is more difficult, since sometimes and without mathematical verification the decision is not possible whether one or two exponential functions are needed for approximation.

In order to determine the error thus originating, we suppose (Fig. 1) a biexponential curve described by:

$$(1) \qquad y = y_1 e^{-\frac{t}{T_1}} + y_2 e^{-\frac{t}{T_2}}$$

The time constants T_i can be calculated in a known way by Eq. (2):

$$(2) \qquad T_i = \frac{T_{1/2}^{(i)}}{\ln 2}$$

y_1 and y_2 are the corresponding intercepts. For small t the curve may also be approximated by the monoexponential course given by Eq. (3):

$$(3) \qquad y = (y_1 + y_2) e^{-\frac{t}{T_a}}$$

In this, T_a represents a common time constant which, according to:

$$(4) \qquad T_a = \frac{y_1 + y_2}{y_1 T_2 + y_2 T_1} T_1 T_2$$

depends on the values T_1, T_2, y_1 and y_2 of Eq. (1). Inversely, this expression, after being transformed into:

$$(5) \qquad T_1 = \frac{y_1 T_a}{y_1 + y_2 - \frac{T_a}{T_2} y_2}$$

can also be employed to determine the fast time constant T_1 directly, without peeling, from the values T_2, T_a, y_1 and y_2. If the curve given by Eq. (1) is erroneously considered to be monoexponential, rCBF is obtained from Eq. (6):

$$(6) \qquad f_m = \frac{\lambda}{T_a}$$

If, on the other hand, a biexponential curve is supposed, rCBF is obtained from Eq. (7):

$$(7) \qquad f_b = \lambda \cdot \frac{y_1 + y_2}{y_1 T_1 + y_2 T_2} = \lambda \frac{y_1 + y_2 - \frac{T_a}{T_2} y_2}{(y_1 - y_2) T_a + y_2 T_2}$$

As practiced by others [1, 2, 3], considering the blood brain partition coefficient λ equal to 0.88, a factor D can be introduced, which can be directly calculated from the difference between $t = 0$ and $t = 2$ min by means of Eq. (8):

(8)
$$D = \log y_{(t = 0)} - \log y_{(t = 2\,\text{min})}$$
$$\lambda = 0.88$$

when the clearance curve is logarithmically recorded on linear paper. Instead of the Eqs. (6) and (7) the following Eqs. (9) and (10) result:

(9)
$$f_m = D_a$$

(10)
$$f_b = \frac{(y_1 + y_2)\, D_a - y_2\, D_2}{y_1 - y_2 + \dfrac{D_a}{D_2} y_2} = TMFI$$

Eq. (10) being the TMFI proposed by ourselves for biexponential clearance curves. This turns into Eq. (9) for monoexponential curves ($y_2 = 0$). With the abbreviations:

(11)
$$\alpha = \frac{y_2}{y_1}$$

and

(12)
$$\delta = \frac{T_2}{T_a} = \frac{T_{1/2}^{(2)}}{T_{1/2}^{(a)}}$$

it is possible to construct a diagram with the help of which the relation between f_m/f_b, and thus the possible error in the determination can be estimated (Fig. 2a). On the abscissa there is α, on the ordinate δ, the curve parameter being f_m/f_b. It can be seen that when α is a constant the relation f_m/f_b increases with increasing δ and that when δ is a constant it increases with increasing α. This is easily understandable since in both cases the result is an increasing deviation from the monoexponential course.

Analysis of 20 typical ^{133}Xe (γ)-clearance curves, recorded from different areas in 5 patients has shown that most of the α values are within the range from 1.6 to 6.0, and that most of the δ values are in the range from 1.4 to 2.2. For 17 clearance curves the parameter values f_m/f_b were within 1.2 to 1.7.

This diagram is of use only if, at the same time, it is possible to state the degree of reliability with which a biexponential decay can be distinguished from a monoexponential one. For this purpose it has been supposed, that a differentiation is possible when

(13)
$$m = y_2 e^{-\frac{t_1}{T_2}} - (y_1 + y_2) e^{-\frac{t_1}{T_a}} \geq k y_2 e^{-\frac{t_1}{T_2}}$$

If $k = 0.1$ is considered a reasonable numerical value for clearance curves from the human brain, the diagram of Fig. 2b is obtained. Subdivision of abscissa and ordinate corresponds to the one of Fig. 1a. The curve parameter is $T_{1/2}^{(a)}/t_1$, where t_1 is the measurement time. From this diagram it can be deduced what measurement time is necessary for a given α, δ and $T_{1/2}^{(a)}$ in order to fulfil Eq. (13), and that deviation of a biexponential curve from a monoexponential one can be unequivocally verified. The relation $T_{1/2}^{(a)}/t_1$ increases with increasing α when δ remains constant and with increasing δ when α remains constant, since in both cases the result is an increase in deviation from the monoexponential shape. This means, furthermore, that when the statement of Fig. 2b is transferred to Fig. 2a, that usually with increasing f_m/f_b, that is with increasing over-estimation of rCBF, as is the case when a monoexponential curve is erroneously supposed, a shorter measurement time will be sufficient to recognize the biexponential course.

The analysis of the 20 typical clearance curves has shown that only in one single case the distinction between a monoexponential and a biexponential course was not possible with the necessary reliability. In this case the measurement time of 2 min was not sufficient. Although the relation f_m/f_b was 1.4, this clearance curve must be regarded as monoexponential during its first 2 min.

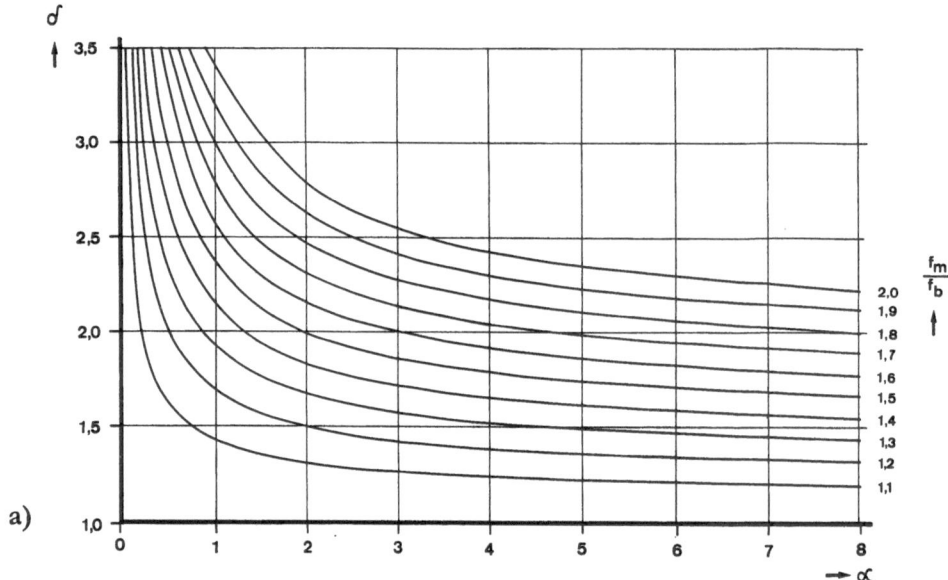

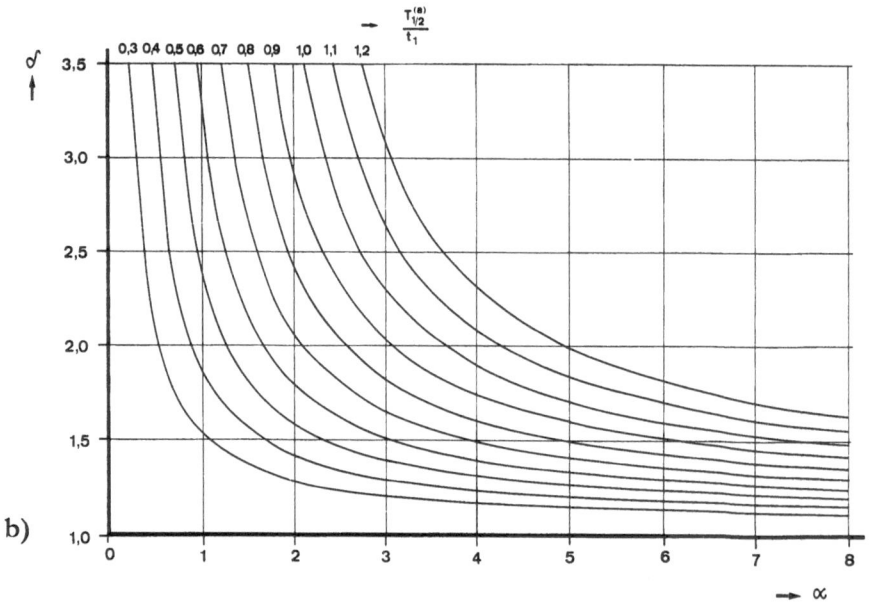

Fig. 2. a) Increasing deviation from the monoexponential configuration (α and δ increase) causes increasing overestimation of flow values when the clearance curve is erroneously considered monoexponential. b) With increasing deviation from the monoexponential configuration (α and δ increase) the time t_1 necessary for reliable recognition of a biexponential configuration decreases

Thus, it is now possible to answer the first question: Despite a certain possibility of arbitrariness in the curve approximation, the TMFI represents a reliable and quantifiable measure for rCBF, if the restriction imposed by the short recording time is taken into account.

References

1. HØEDT-RASMUSSEN, K., E. SKINHØJ, O. PAULSON, J. EWALD, J. K. BJERRUM, A. FAHRENKRUG, and N. A. LASSEN: Regional cerebral blood flow in acute apoplexy. Arch. Neurol. **17**, 271 (1967).
2. PAULSON, O. B., S. CRONQUIST, J. RISBERG, and F. I. JEPPESEN: Regional cerebral blood flow. A comparison of 8-detector and 16-detector equipments. J. nucl. Med., in press.
3. PAULSON, O.: Regional cerebral blood flow at rest and during functional tests in occlusive and non-occlusive cerebrovascular disease. Presented at the International CBF Symposium, Mainz, April 1969 (this volume, p. 111).

Stochastic Analysis and Slope Determination of Linear and Semilogarithmic Clearance Curves Respectively. Practical Considerations

D. Oeconomos

Neurological Center, Polyclinic of Athens

With the use of the stochastic analysis on the one hand and that of the initial slope index on the other, the CBF calculation has become an easy procedure in routine clinical studies.

Although extensive description of both the theoretical and the practical aspects of these methods has already appeared in the literature, there are practical problems which seem to be worth reconsideration.

In studying our material we tried to check the correspondence of the data of one method to those of the other and to see whether a meaningful correlation could be established between them.

The stochastic and the slope values were independently calculated by two different observers and their respective findings subsequently compared.

1. Difficulties in the Stochastic Analysis

Two main problems often interfere with the accuracy of the final results:

a) The first concerns the maximum initial height of the clearance curve. Either because of the short time constant (0.5 sec) used during the recording of the first part of the curve, or, mainly because of the existence of an initial peak activity, the starting point of the wash out phase cannot be accurately recognized.

Extrapolation to zero time, used by others as a standard procedure to fix the initial maximum height, would have resulted, in the presence of a high initial flow, in a marked over-estimation of the flow.

In such cases we found that, by looking at the semilogarithmic curves, one can more easily recognize on the curve a bending just after the initial peak; this bending apparently represents the starting point of the wash out phase. The presence of this bending point can also be checked on the curves obtained following the functional tests. By transposing this point to the linear curves according to its temporal momentum from the zero time line, one can finally locate the maximum initial height, used in the H/A formula.

The stochastic analysis also meets with another delicate problem, namely the amount of total area activity in relation to that of the background, which includes recirculation and/or the so called remaining activity.

As could be expected for statistical reasons, we have noticed that the results of the stochastic analysis better fitted those of the initial slope index, when the total area to background activity ratio was high.

This is illustrated in Table 1 where the $\frac{\text{area}^{10}\ \text{activity}}{\text{background activity}}$ was plotted against the mean differences between the rCBF (stochastic) and the initial slope index values.

When $\frac{A}{B}$ is of the order of five to one, the mean variation is 4.4 while for values of 10 to 1 and higher it drops down to 1.4.

As the corrections used by others for the background and the remaining activity seem to be encumbered with a good deal of arbitrariness, we finally adopted the solution of rising the isotope dose to 4 and 5 mCi per injection in order to obtain a high counting rate. Repeated injections on the other hand should not be made before the background activity drops down near to the pre-injection levels. This takes usually from 20 to 40 min in normally respirating subjects.

Table 1. $\dfrac{Area^{10}\ Activity}{Background\ Activity}$ plotted against mean difference between $rCBF_{10}$ and slope index

$\dfrac{Area^{10}\ Activity}{Background\ Activity}$	$\leq \dfrac{5}{1}$	$\dfrac{6}{1} - \dfrac{10}{1}$	$\dfrac{11}{1} - \dfrac{20}{1}$	$\geq \dfrac{20}{1}$
	n = 25	n = 25	n = 25	n = 25
$rCBF^{10}$ — slope ind.				
Mean	4.4	2.3	1.4	1.45
S. D.	6.2	1.95	1.38	0.96

2. The Initial Slope Index

With this fast and simple way for calculating the flow rates on humans, introduced by LASSEN and coworkers, it has became possible to shorten the recording time of the clearance curve down to 2 min. The advantage of such a procedure is evident in routine clinical work, where the functional tests require a long and tiring immobilization of the patient.

Nevertheless this method has its drawbacks and difficulties as it presupposes good shaped semilogarithmic curves. Here, again, the ratio between the maximum activity and the background rate is of prime importance for the good shape of the curve.

Schematically most of the curves are monoexponential (76.9%). 63.6% of them agree well with the linear values within 2 units, 13.3% giving values with more than 2 units deviation from those of the linear curves.

20.5% of the curves are clearly biexponential; 16.6% of them are concordant and 3.9% non-concordant, these latter terms being understood as mentioned in the previous paragraph.

1.4% of the curves are triexponential while about 1% have apparently more than three components.

Taking the data of the linear as reference we arrived to the following practical conclusions concerning the calculation of the slope indices.

For the non-concordant but evidently monoexponential curves no real solution can be offered for a better fitting with the stochastic results. On the other hand for the calculation of the slope index in bi- or triexponential curves, the value of the final slope index has to be deducted from the respective slopes of the different components. Since each separate component is imprinting its own slope index on the final value, its contribution has to be calculated accordingly during the 2 min period. This duration is consequently expressed as an approximative fraction of the 2 min interval, fraction by which this particular slope index is multipied. The sum of the partial slope indices, corrected for their time parameter, gives the final initial slope index in such cases of clearly bi- and triexponental curves.

On the other hand, in some of the triexponential as well as in the multiexponential curves, the best solution is the use of the cord of arc built by the different components provided this arc is a shallow one.

The above mentioned empirical solutions for the calculation of the initial slope indices cover most problems which may arise in this field. There are, nevertheless, some semilogarithmic curves with a very steep slope (high flows) which remain difficult to calculate with a fair degree of accuracy.

Regional Cerebral Blood Flow Measured by Multiple Probes: An Oscilloscope and a Digital Computer System for Rapid Data Processing

E. Sveinsdottir, N. A. Lassen, J. Risberg, and D. H. Ingvar

Department of Datalogy, Institute of Mathematics, University of Copenhagen, Department of Clinical Physiology, Bispebjerg Hospital, Copenhagen, and Department of Clinical Neurophysiology, University Hospital, Lund

Experiences from our laboratories as well as from others in recent years have shown that the intra-carotid Xenon-133 injection technique for measurements of regional cerebral blood flow may be used with advantage to study local circulatory events in the brain under normal and pathological conditions. So far the method has been used with 4, 8 or 16 individual detectors, and it has been established that a high number of probes enables an analysis in detail of e.g. the events which accompany a focal cerebrovascular lesion [2]. In this paper two new systems are described briefly by which rCBF are measured with 35, respectively 32, individual detectors, using two principally different solutions for the data processing. Before the systems are described, the reasons for not choosing other techniques available will be summarized.

An obvious alternative to the use of multiple individual scintillation detectors for rCBF measurements is the so called *gamma camera* with digital processing of the counting rate as a function of time in the various regions studied. This type of instruments consists essentially of one large crystal and the maximal counting rate, which at present lies in the order of 300,000 c.p.m., is limited by this fact. This rate may even be lowered by the characteristics of the data storage system used together with the camera. The systems available so far do not allow a rapid calculation of rCBF values, i.e. within the first hour following the isotope injection.

We have found it necessary to record a maximim counting rate per region of some 50,000 c.p.m. in order not to loose essential information during the initial 30 sec of the clearance period. If one wants to measure from 32–45 regions simultaneously, it is therefore not possible to use the gamma camera, which, in addition, has a realatively coarse localization power for low energy gamma radiation, apart from being quite expensive when equipped with a proper digital data system.

The *image intensifier system* admits practically unlimited counting rates. However, present models are not sufficiently linear to yield precise digital information on the counting rate in different regions as a function of time.

The *autofluoroscope* represent a condensed multidetector system containing 35 photomultipliers. The total maximal admissible counting rate appears to be in the order of 2,000,000 c.p.m., which would suffice for the purpose outlined above. Kvist [1] has recently reported successful rCBF measurements with this instrument. However, this instrument which is also expensive, is at present equipped with a tape recorder for data storage. Extensive personal experience with multichannel tape recording have firmly convinced us of the unpracticality of this approach when measuring rCBF from more than 16 regions simultaneously.

D. H. I. and J. R. were supported by the Swedish Medical Research Council (project B 70–21X–84–06 B) and the Wallenberg Foundation, Stockholm.

A 35 detector system for recording the "initial slope" of Xenon-133 clearance curves using an oscilloscope display*

This system contains 35 individual small cylindrical scintillation detectors mounted parallel in a stack, each equipped with a 12 mm in diameter and 40 mm long lead collimator tube. In each detector is a cylindrical NaI crystal 12 mm in diameter and 10 mm in length. All 35 detectors are equipped with their own discrimination circuits and logarithmic ratemeters with a time constant of 1 sec. The output from each probe is fed to a multiplexer unit and then displayed onto an oscilloscope screen. On the screen 35 traces are recorded and the location of the traces corresponds to the position of the probes within the field of measurement. The oscilloscope screen is photographed by a Polaroid camera and the sweep of the scope is set to picture the first 2 min of the clearance curve.

The advantages of this relatively simple and inexpensive multi-probe system are:

1. The information obtained is confined to the initial part of the clearance curve during which abnormalities of clinical significance are best discerned. If, however, later parts of the curve are of interest, a second exposure of the oscilloscope screen readily records this.

2. The rCBF data are obtained and can be interpreted immediately in their final form and data storage offers no problems.

3. With the results of each study so rapidly available, it is easy to plan a following one based directly upon the results of the first isotope injection.

A 32 detector system for on line recording and rapid processing of Xenon-133 clearance curves by means of a small computer

In this system 32 small scintillation detectors (14 mm in diameter, 15 mm in length) are used, each equipped with its own pulse height analyzer. The output from the probes are directly fed to the "multiplexer" input of a small digital computer (Varian 620/i) with a core memory with a capacity of 12 K and a disc memory of 128 K.

The computer program is designed to register and store the 32 clearance curves. Since the maximum counting rate does not exceed about 1000 c.p.s. there is ample time for simultaneous data processing such as finding maximum, accumulating counts for area calculation, finding initial slope, etc. Hence, it is possible after 2 min to put out rCBF values calculated with the initial slope method from 32 channels on the teletypewriter. Similarly after 10 min, while still registering counts, the computer puts out 32 flow values calculated after the 10 min height over area formula. And after 15 min when the study is finished the rCBF values using extrapolation to infinity are available. In addition a biexponential resolution of each curve will be obtained with estimates of the flows in the grey and white matter as well as the relative weights in each region studied.

The advantages of this more complex and costly system are

1. A rapid and very detailed evaluation of the various quantitative aspects of the 32 Xenon-133 clearance curves.

2. The original data can be easily stored for later analysis.

3. The instrument is readily applicable to other types of regional isotope measurements of the brain (or other organs).

References

1. PAULSON, O. B., S. CRONQVIST, J. RISBERG, and F. I. JEPPESEN: Regional cerebral blood flow. A comparison of 8-detector and 16-detector equipments. J. nucl. Med., **10**, 164 (1969).
2. KVIST, G.: Personal communication, 1969.

* Developed according to our specifications by Meditronic Ltd., Lyngby, Denmark.

Gamma-Camera and Multichannel Analyser for Multilocular rCBF Measurement

W.-D. Heiss, V. Kvicala, P. Prosenz, and H. Tschabitscher

Department of Psychiatry and Neurology, University Hospital, Vienna

The equipment we use for regional CBF measurement consists of a Scintillation camera (Pho-Gamma/III, Nuclear Chicago), a Dual ADC (Ridl, model 22–3), a 1600 Word Memory (Ridl, model 24–3) and a computer compatible digital Magnetic Tape System (model 24 202 Ampex TM 7291). The mode of operation of this set up has been already described [2, 3].

The resolution of the Pho-Gamma/III has been determined by many investigators and lays in the range of 10–15 mm [1, 3, 4, 5, 6]. Due to the parallel hole collimator the depth resolution is superior to that of cone shaped single probes [3, 5]. Thus, within the range of resolution, the overlapping of the areas observed is slight.

On ground of calculations based on the deadtime of the set up (Pho-Gamma/III 4 μs incl. pile up, ADC-1600 channel store 7 μs) and on ground of own experimental investigations, the highest count rate within linear registration range (less than 2% counting loss) amounts to 300000 cpm for the whole crystal and to 2400 cpm for 1 channel, provided one channel samples from a area of 12×12 mm lateral length.

That means that one cannot increase the Xe-activity over this limit of 300000 cpm in order to improve the counting statistics of the single regional curves. The maximal count-rate of 2400 cpm over these areas of 12 mm laterial length results in a standard deviation of 5–7% for a sampling period of 2.4 sec, as we use it for the first 2 min of registration. Thus the determination of the peak is sometimes difficult.

For reasons of resolution and counting statistics it is obviously not possible to register from areas smaller than of 12 mm lateral length. About 80–100 of such areas can be distributed over a human brain.

After the development of a computer program the tape recorded data could be fed directly to the computer. The regional blood flow values were automatically evaluated and printed out corresponding to their localisation on a cerebral blood flow map.

We could observe well marked areas of diminished blood flow, corresponding to angiographically demonstrated vascular occlusions. Regional reduced blood flow values were also detectable in patients with neurological disturbances, but practically normal angiograms.

In some cases remarkably increased blood flow values could be registered in the surroundings of such infarctions. In single cases these areas with increased blood flow showed angiographically demonstrable collateral path ways and accelerated passage.

Tumours showed distinct regional variations in the perfusion rate. The haemodynamic disturbances raised by a. v. malformations could be registered as well as typical shunt peaks over the malformation.

Within the above mentioned limitation, arising mainly from the restriction of counting rate, the Scintillation camera-Multichannel store is able to provide cerebral blood flow maps (Fig. 1). The thereby found alterations in regional blood flow correspond completely with clinical and angiographical findings.

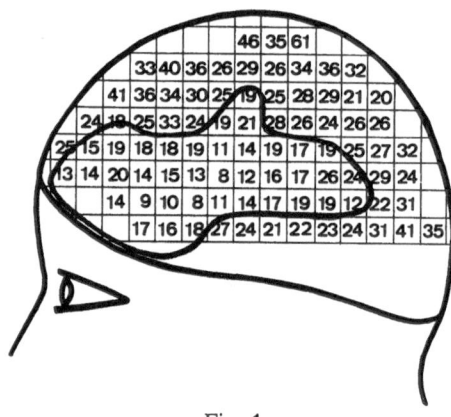

Fig. 1

References

1. Anger, H. O.: Scintillation camera. Rev. Sci. Instr. **29**, 27 (1958).
2. Heiss, W.-D., P. Prosenz, A. Roszouczky, and H. Tschabitscher: A quantitative gamma-camera technique. Scand. J. Lab. clin. Invest., Suppl. **102**, p. XI: L (1968).
3. — — — u. H. Tschabitscher: Die Verwendung von Gamma-Kamera und Vielkanalspeicher zur Messung der gesamten und regionalen Hirndurchblutung. Nucl. Med. **4**, 297 (1968).
4. — — u. H. Tschabitscher: Die Auswertung von Gamma-Kamera-Szintigrammen durch Vielkanalspeicher, Computer und Farbfernsehsystem. Fortschr. Röntgenstr. **1**, 108 (1969).
5. Lorenz, W. J., u. W. E. Adam: Digitale und analoge Auswertung von Aufnahmen mit der Szintillationskamera. Nucl. Med. **7**, 367 (1967).
6. Westermann, B. R., and H. J. Glass: Physical specification of a gamma-camera. J. nucl. Med. **9**, 24 (1968).

Measurements of rCBF by Intravenous Injection of ^{133}Xe. A Comparative Study with the Intra-Arterial Injection Method

A. Agnoli, M. Prencipe, A. M. Priori, L. Bozzao, and C. Fieschi

Department of Neurology and Psychiatry and Center of Numeric Calculus, University of Genoa

Clearance methods with radioactive inert gases injected into the internal carotid artery can be simplified. Actually, these methods seem essentialy limited to studies of clinical physiology.

It is possible, however, to develop methods which do not involve direct puncture of the carotid artery. These methods are atraumatic and can be repeated over several days. Such methods offer a unique opportunity to learn more about problems such as the time pattern of the derangement of cerebral circulation after an ischemic episode.

We focused our attention on the inhalation of radioactive inert gases, along the line proposed by Veall and Mallet [2], and re-examined by Obrist el al. [1].

Two main problems are apparent when dealing with these techniques:

1. The input is not a delta-function and thus the head curves must be corrected for recirculation, and

2. the brain clearance curve is affected by the superimposed extracerebral tissue clearance curve.

The initial criticism on the precision of the results was eliminated by Obrist et al., but only by use of a long and sophisticated computer program. To avoid this complication, Veall has recently suggested the use of an analogue computer.

Since in the usual inhalation technique, brain saturation is never reached, our initial trial was a forced single breath (about 1000 ml of ^{133}Xe dissolved in air, activity 6–8 millicuries). By this means the duration of arterial recirculation could be shortened and thus also the weight of the very slow component of the extracerebral tissues (about 5–10 percent according to Veall), can be reduced.

In the experiment, we injected a bolus of ^{133}Xe solution i.v. (1–3 ml of ^{133}Xe dissolved in saline, activity 6–8 millicuries).

We preferred intravenous route of administration both because of ease of administration (patient collaboration is not needed) and because any initial damage to the lung parenchyma, though it may decrease ventilation, leaves perfusion unchanged.

Because of the brief duration of the arterial recirculat on, we monitored arterial blood continously; we discarded end-tidal air activity assessment because in patients with lung disease, such measurements do not reflect arterial values.

Supported by C.N.R. contract, no. 115/2422.

The determination of the CBF was then effected as described in Fig. 1, and the arterial curve is shown in Fig. 2.

A correction for the deformation for the sampling system is necessary. The curve is described by a two exponential system, and the half-times of the two components are illustrated.

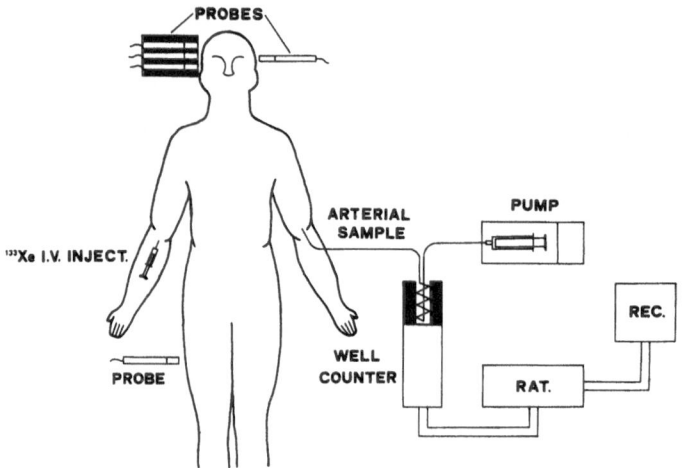

Fig. 1

	Mean	Min	Max
T ½(1) sec	12,0	6,7	21,5
T ½(2) sec	45,5	25,0	76,5

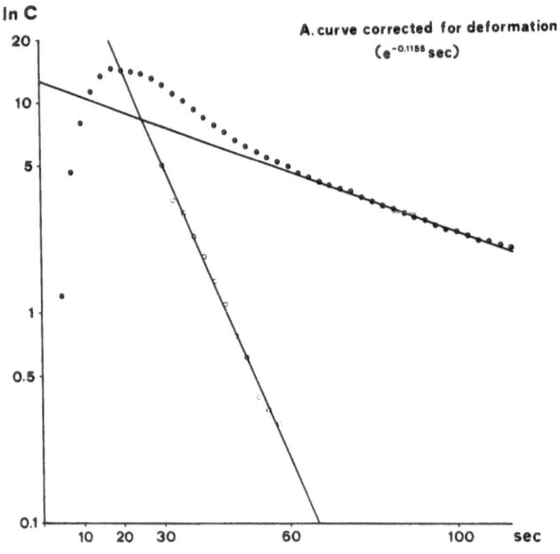

A. curve corrected for deformation
$(e^{-0.1155} sec)$

Fig. 2

The short duration of such recirculation is evident. Despite this kind of recirculation one should observe that the head curves are notably deformed. Fig. 3 shows deformation of the same head curve using the shorter and the longer recirculation curves observed.

Therefore, correction for recirculation was determined point by point, finding a numerical solution of the integral equation by the Newton rule. A C.I.I. 10070 computer was used.

The residual curve, corresponding to a delta function input, must be analyzed taking into account also the clearance of extracerebral tissues.

Because a biexponential analysis of the curve gave underweighted results, the head curve can be accurately fitted by a three component exponential function. It is necessary, however, to assume for the extracerebral tissue a mono-exponential decay, at least in the first 30 min. The

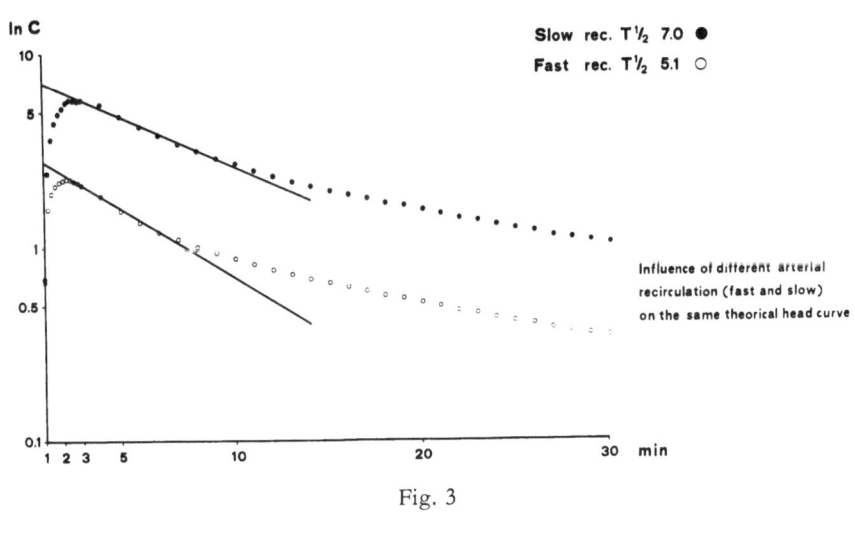

Fig. 3

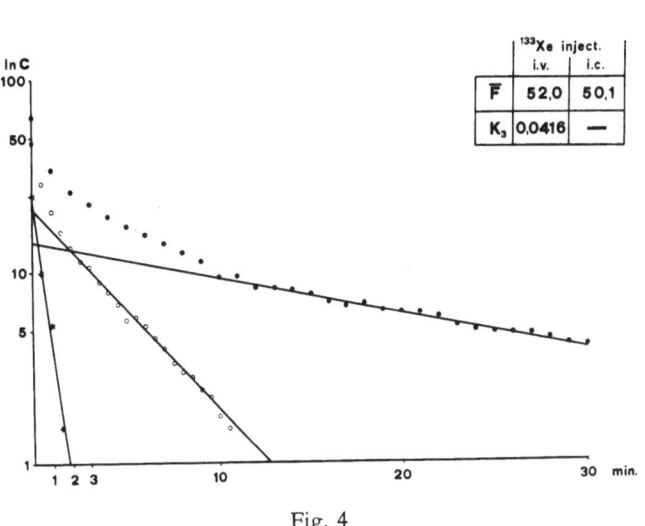

Fig. 4

results coming from such an analysis are compared with those obtained from the intracarotid injection in the same subject (Fig. 4). These results are a first step to a judgment of the accuracy of the method. Certainly these techniques must be pursued. The direction in which we plan to move is the following:

1. Use of high energy isotopes such as ^{135}Xe or ^{85m}Kr;
2. Focusing collimation in order to avoid the radioactivity coming from extracerebral tissues;

3 Cerebral Blood Flow

3. Using a calculation procedure which will allow shortening of the recording time. In this respect it is possible to introduce in the computer program a fixed weight and flow for the third component, or measuring the flow of a tissue with a monoexponential decay similar to the K_3.

References

1. OBRIST, W. D., H. K. THOMSON, C. H. KING, and H. S. WANG: Determination of Regional Cerebral Blood Flow by Inhalation of 133-Xenon. Circulat. Res. **20**, 124 (1967).
2. VEALL, N., and B. L. MALLET: Regional Cerebral Blood Flow determination by ^{133}Xe inhalation and external recording: the effect of arterial recirculation. Clin. Sci. **30**, 353 (1966).

Comment on the Paper by Agnoli et al.

N. Veall

Guy's Hospital Medical School, London

The possibility of using a very short ^{133}Xe inhalation time in order to simulate a bolus injection is a tempting one. We tried this modification and abandoned the idea because it introduces a major source of error. Provided the arterial input function is measured and the deconvolution carried out on the head curve the values obtained for the fast component clearance rate are independent of the breathing period if it is varied between 1 and 15 min. For shorter periods the values obtained for the fast component are higher. This is to the fact that in order to achieve adequate count rates it is necessary to use a much higher concentration of ^{133}Xe in the breathing mixture. Under these conditions the contribution to the observed count rate due to scattered radiation from ^{133}Xe in the nasopharynx, which is usually negligible, is considerably accentuated. This contribution can be measured quite simply by inhaling 50 to 100 ml of ^{133}Xe at the end of a normal inspiration so that only the respiratory dead space is filled with the tracer.

There is a good deal of evidence to suggest that the curve describing extracerebral tissue activity is not a simple exponential. Any method used to overcome this difficulty must be based on unverifiable assumptions. From this point of view, the use of the normalised high count is as good a method as any. We have a rather simpler procedure which appears to work well. We merely assume that the extracerebral component can be described by two exponentials of equal amplitude and rate constants k and k'. k is measured from the curve ($T^1/_2 = 20$–30 min) and k' is assumed to be $^1/_{15}$ of k. This merely requires a slight modification to the analogue computer which we use for analysing the data, and does not necessitate the cost of an additional counting channel.

Application of the [133]Xenon Clearance Method to the Measurement of Local Blood Flow in the Conscious Animal

C. ROSENDORFF and W. I. CRANSTON

Department of Medicine, St. Thomas's Hospital, Medical School, London

There are some situations where a compartmental analysis of regional cerebral blood flow is inadequate to provide sufficient information about local blood flow conditions in animal experiments. A counting system which uses multiple external collimated crystals, each looking at a "core" of brain tissue provides an estimate of regional cerebral blood flow; and increasing the number and improving the geometry of the counters will improve the resolution of the system. However, localization is still relatively crude, and it is impossible to distinguish easily between, say, flow in the cortex and in non-cortical grey matter in the same counting field.

During the course of studies on the hypothalamus in the rabbit it became necessary to adapt existing techniques to the measurement of local flow in the conscious animal. This has lead to a re-appraisal of local [133]Xenon injection techniques.

Methods

All experiments were performed on adult chinchilla rabbits, weighing 2.5–3.0 kg. The headplate used, an adaptation of that described by MONNIER and GANGLOFF [7], was screwed to the skull in a constant topographic relationship to the cranial sutures. Experiments were performed at least 2 weeks later. Using a modification of the MONNIER and GANGLOFF stereotaxic technique, a fine cannula was introduced into the area of brain studied, usually the hypothalamus, of the conscious, lightly restrained rabbit [9]. Volumes of 2.5–10 μl of a saline solution of ^{133}Xe containing 5 mCi/ml were injected, and the γ emission was measured by an external collimated scintillation probe, and a conventional amplifier, pulse height analyser, and ratemeter assembly. The output of the amplifier was recorded on an ultraviolet recorder and on a digital data logging system. Injections were repeated at intervals of 15 to 30 min.

Local flow was calculated using the formula

$$F \text{ (ml/g tissue/min)} = \frac{\lambda \log_e 2}{T \frac{1}{2}}$$

where $T \frac{1}{2}$ is the half-decay time derived from the initial slope of the semilogarathmic plot of the ^{133}Xe clearance, and λ is the brain tissue : blood partition co-efficient for [133]Xe. This was derived for rabbit hypothalamus as follows. The simultaneous rates of clearance from the hypothalamus of injected [133]Xenon and [125]iodo-antipyrine was measured. Since both tracers

We are grateful to the National Fund for Research in Crippling Diseases, Roche Products Ltd., and Imperial Chemical Industries Ltd., for their financial support.

are cleared by the same flow, there is a simple linear relationship between the ratio of their rates of clearance and the ratio of their λ values. Since ^{125}I-antipyrine does not come out of solution when exposed to air, in the same way as [133]Xenon, it is then relatively simple to determine the value of λ for ^{125}I-antipyrine by an in vivo technique, and hence to derive λ for [133]Xenon.

In seven experiments autoregulation in the hypothalamus was assessed by raising and lowering mean arterial blood pressure by infusing angiotensin and by controlled bleeding respectively.

The local effects of drugs could be examined by a variety of methods. Intraventricular injections were made into the lateral cerebral ventricle via a cannula inserted in the ventricle by the MONNIER-GANGLOFF method. The effect of local hypothalamic injections of drugs was assessed by the use of two hypothalamic cannulae, both filled with [133]Xenon-in-saline, and used for consecutive measurements of flow on the two sides of the midline. One side was used as a control and on the other side nor-adrenaline (NA) or 5-hydroxy tryptamine (5-HT) was added to the [133]Xenon-saline injectate.

Results

Fig. 1 shows the result in an experiment where flow in the cerebral cortex was measured using the local injection technique; in the same experiment total cerebral flow was measured by the injection of [133]Xenon via the internal carotid artery. Compartmental analysis of the "intracarotid" clearance curve shows a fast component which is of the same order of magnitude as the clearance of the locally injected [133]Xenon.

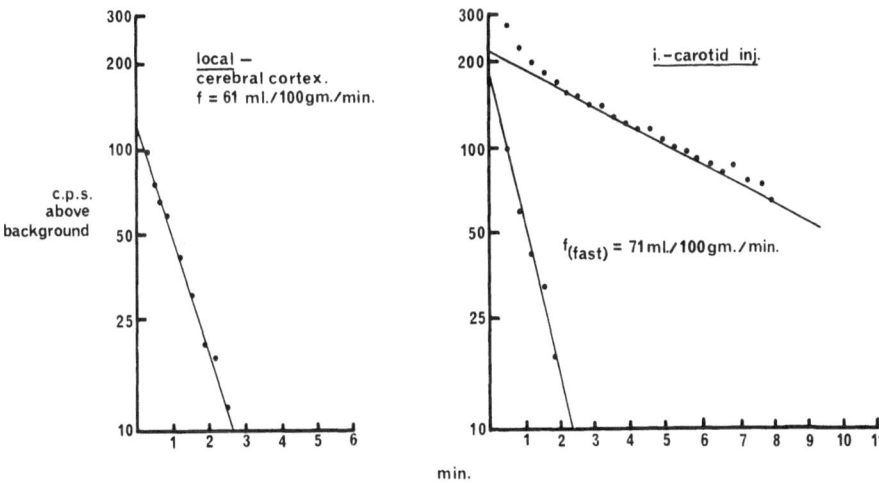

Fig. 1. A comparison of the rate of clearance of locally injected [133]Xenon from the cerebral cortex of the rabbit (left), and the fast component of a biexponential clearance of [133]Xenon injected into the internal carotid artery in the same animal (right). Ordinate: Counts per second above background. Abscissa: time in minutes

Hypothalamic clearance of locally injected ^{133}Xe was monoexponential giving values for local flow of 25–56 ml/100 g/min. In spite of the presence of the cannula (0.4 min O.D.) and repeated injections, both autoregulation within the mean arterial B.P. range of 60 to 140 mmHg and the response to changes in arterial pCO_2 were maintained.

Vasoactive drugs varied in their action, depending upon the route of administration and the dose. Intraventricular 5-HT (200 μg) had a small vasoconstrictor effect, whereas local injections of 20 to 80 μg of 5-HT increased local blood flow by 24–69%. Intraventricular NA (20 μg) and local injections of 1 μg NA caused a significant increase in local flow, while 10–200 μg injections of NA into the hypothalamus were vasoconstrictor.

Discussion

Local injection methods have been described [4, 6, 8] but only for cortical and subcortical white matter flow measurements. The present study has shown that the local injection of ^{133}Xenon can be used for the measurement of local blood flow in the forebrain, in the conscious animal. Further, in spite of the presence of a cannula, and repeated injections of small volumes (2.5–10 μl) of the Xenon in saline solution, autoregulation and the vasodilator response to inhaled CO_2 are maintained.

One advantage of this technique over others designed to measure local flow, such as heat clearance [2] or H_2 polarography [1] is that the "micropharmacology" of the relevant area may be investigated by adding drugs to the ^{133}Xe-saline injectate. While the pharmacology of the cerebral circulation has been intensively studied [10] most studies have been on the effects of drugs administered systemically. The local micro-injection of substances which are normally found in high concentrations in the hypothalamus may provide a clearer indication of the possible role of these compounds in the physiological control of local tissue perfusion. The opposite effects on blood flow of local injections of NA and 5-HT are of great interest, particularly in view of the opposite effects of local hypothalamic injections of these two amines on body temperature [3, 5].

References

1. Aukland, K.: Measurement of local blood flow with hydrogen gas. In Blood flow through organs and tissues. Ed.: W. H. Bain, and A. M. Harper. Edinburgh: E. & S. Livingstone Ltd. 1968.
2. Betz, E.: Local heat clearance from the brain as a measure of blood flow in acute and chronic experiments. Acta neurol. scand. 41, Suppl. 14, 29 (1965).
3. Cooper, K. E., W. I. Cranston, and A. J. Honour: Effects of intra-ventricular and intrahypothalamic injection of noradrenaline and 5-HT on body temperature in conscious rabbits. J. Physiol. 181, 852 (1965).
4. Espagno, J., and Y. Lazorthes: Measurement of regional cerebral blood flow in man by local injections of Xenon^{133}. Acta neurol. scand. 41, Suppl. 14, 58 (1965).
5. Feldberg, W., and R. D. Myers: Effects on temperature of amines injected into the cerebral ventricles. J. Physiol. 73, 226 (1964).
6. Häggendal, E., N. J. Nilsson, and B. Norbäck: On the components of the Kr^{85} clearance curves from the brain of the dog. Acta physiol. scand. 66, Suppl. 258, 5 (1965).
7. Monnier, M., and H. Gangloff: Atlas for stereotaxic brain research on the conscious rabbit. Rabbit Brain Research, Vol. 1. Amsterdam: Elsevier Publ. Co. 1961.
8. Nilsson, N. J.: Observations on the clearance rate of β radiation from Krypton^{85} dissolved in saline and injected in microlitre amounts into the grey and white matter of the brain. Acta neurol. scand. 41, Suppl. 14, 53 (1965).
9. Rosendorff, C., and W. I. Cranston: Measurement of hypothalamic blood flow in the conscious rabbit by a radioactive inert gas clearance technique. Proc. 5th Europ. Conf. Microcirculation, Gothenburg. Basel: Karger, in press.
10. Sokoloff, L.: The action of drugs on the cerebral circulation. Pharm. Rev. 2, 1 (1959).

Problems Concerning the H₂ Inhalation Technique to Determine the Cerebral Blood Flow by Means of Palladinized Pt-Electrodes

D. W. LÜBBERS, R. WODICK, K. STOSSECK, and H. ACKER

Max-Planck-Institut für Arbeitsphysiologie, Dortmund

Experimental work in the last years has shown, that H_2-polarography can be used to measure local and total cerebral blood flow [1–10]. The high diffusion coefficient of H_2 and the possibility to miniaturize the H_2-sensing device is sometimes an important advantage. VEALL [12] has studied in which way the inhalation of Xenon can replace the injection technique and developed a formula which permits calculating the arterial concentration from the Xenon concentration in the end expiratory air. In order to investigate in which way the inhalation of hydrogen influences the clearance curves we have monitored simultaneously the pH_2 of the arterial blood and the tissue (or venous) pH_2 clearance curves of animal brain (cat, rat, guinea pig).

A first series of experiments has been carried out together with BETZ by inserting two electrodes (100 μ of diameter) permanently in the white matter of a cat brain close to the hypothalamus. A ring-shaped electrode on the wall of the common carotid artery monitored the arterial pH_2. The H_2-gas mixture was given to the cat sitting in a cage (ca. 0.20 m³). The electrodes were stable over a period of 78 days, then the experiment was ended for other reasons. The average cerebral blood flow of this part of the white matter was between 20 and 30 ml/100 g/min. The pH_2 "clearance curve" on the wall of the carotis would correspond to a blood flow range of 80–200 ml. Therefore it was possible to measure only the flow range of 0–40 ml by this procedure. But in higher flows the effect of the arterial pH_2 cannot be omitted. In this case the arterial "clearance curve" and the curve of the tissue (or the vein) overlap considerably. In order to evaluate such overlapping curves we assume – in the same way as KETY [11] did – that an exact balance exists between the influx of H_2, the H_2 solved in the tissue and the outflow of H_2.

$$(1) \qquad p(t) + \frac{M}{W} \int_0^t p(u)\, du - \frac{M}{W} \int_0^t q(u)\, du = 0.$$

M is the flow which enters the volume W, $p(t)$ is the hydrogen pressure in the tissue (or the vein) and $q(t)$ the pH_2 of the influx. Multiplying the equation by W and αH_2 (solubility coefficient) the first term corresponds to the amount of hydrogen *in* the volume W, the second term to the amount of hydrogen which *leaves* the volume W in the time interval between 0 and t and the third term to the amount of hydrogen which is *transported into* the volume W by the

Supported by a grant of the Deutsche Forschungsgemeinschaft.

40   D. W. Lübbers, R. Wodick, K. Stosseck, and H. Acker:

perfusion M in the time interval between 0 and t. The equation has the following solution

$$(2) \qquad p(t) = \exp\left(-\frac{M}{W}t\right)\left[A + \int_0^t \frac{M}{W}q(u)\exp\left(\frac{M}{W}u\right)du\right]$$

but M/W cannot be expressed by known functions. Therefore we calculated numerically the curves for all expected values of cerebral blood flow. By comparing the measured curve with the calculated one, it is possible to find the best fitting curve which gives the cerebral blood flow. This procedure can easily be carried out by a computer using the GAUSS method calculating the squares of the differences between the curves and looking for its minimum.

$$(3) \qquad U(a) = \sum_{r=1}^{m}\left[p(t_r) - \frac{\sum_{i=1}^{u}p(t_i)Q(a,t_i)}{\sum_{i=1}^{m}Q^2(a,t_i)}Q(a,t_r)\right]^2$$

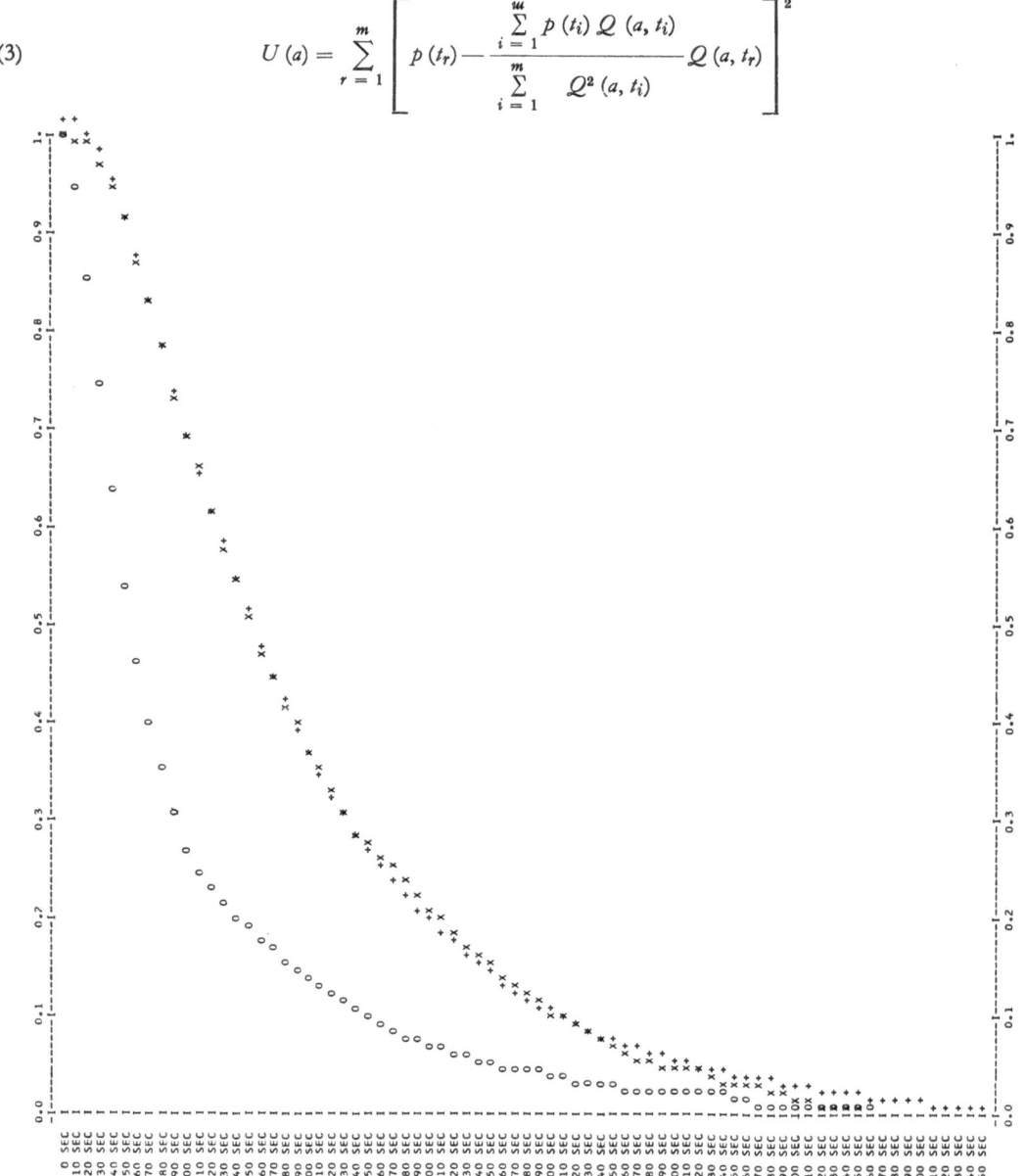

Fig. 1. pH$_2$-Clearance Curve (cat brain). Ordinate: pH$_2$ (full saturation $= 1.0$); Abscissa: time (every 10 sec a value). Symbols: $\bigcirc$ = arterial pH$_2$, $\times$ = measured values (pial vein, $\varnothing$ 100 μ), $+$ = calculated values (vein). The calculated curve ($+$) fits very well the measured one ($\times$)

$$(4) \qquad Q(a, t) = \exp(-at) \left[1 + \frac{1}{p(0)} \int_0^t aq(u) \exp(au)\, du \right].$$

This method allows also to take into account an electrode drift B, which can be measured by the zero values before and after the clearance.

$$(5) \qquad U(a, A, M, B) = \sum_{i=1}^{m} \left[p(t_i) - AQ(a, t_i) - Mt_i - B \right]^2.$$

This calculation does not involve any special assumption about the form of the arterial pH$_2$ curve, but needs the exact knowledge of its values. As mentioned first we tried to measure the arterial pH$_2$ by monitoring the pH$_2$ on the arterial wall of the carotis. The wall is like an additional membrane and the effect of the thickness can be calculated from the calibration curves equilibrating the total animal by different H$_2$-gas mixtures. But unfortunately the form of the response of the electrode is also influenced by the H$_2$ solubility of the surrounding tissue and therefore is difficult to measure *in situ*.

The measurements in small pial arteries showed that the arterial pH$_2$ of the carotid is additionally changed by the further passage of the blood through the arteries and by the contact with the cerebellar fluid. A brain with an exposed surface loses a considerable amount of H$_2$ to the surroundings. Therefore in order the measure the local blood flow it is necessary to monitor the arterial input function as close as possible to the entrance of the arteries into the brain tissue and the venous pH$_2$ close to the exit. Using such values the calculation by the above explained method can provide a good fitting between the measured and the calculated curve, as shown in Fig. 1.

References

1. AUKLAND, K.: Hydrogen polarography in measurement of local blood flow; theoretical and empirical basis. Acta neurol. scand., Suppl. **14**, 42 (1965).
2. — Measurement of Local Blood Flow with Hydrogen Gas. In: W. H. BAIN, A. M. HAPRER, W. A. MACKEY: Blood Flow through Organs and Tissues, p. 157. Edinburgh and London: E. & S. Livingstone Ltd. 1968.
3. —, B. F. BOWER, and W. BERLINER: Measurement of local blood flow with hydrogen gas. Circulat. Res. **14**, 164 (1964).
4. CLARK, L. C., JR., and L. M. BARGERON: Detection and direct recording of right to left shunts with a hydrogen electrode catheter. Surg. **46**, 797 (1959).
5. FIESCHI, C., L. BOZZAO, and A. AGNOLI: Regional clearance of hydrogen as a measure of cerebral blood flow. Acta neurol. scand. Suppl. **14**, 46 (1965).
6. — — —, M. NARDINI, and A. BARTOLINI: The hydrogen gas method to measure local blood flow in subcortical structures of the brain with a comparative study with the ^{14}C-antipyrine method. To be published.
7. GOTOH, F., J. S. MEYER, and M. TOMITA: Hydrogen method for determining cerebral blood flow in man. Arch. Neurol. **15**, 549 (1966).
8. HYMAN, E. S.: Linear system for quantitating hydrogen at a platinum electrode. Circulat. Res. **9**, 1093 (1961).
9. LÜBBERS, D. W., M. KESSLER, KR. KNAUST, D. G. McDOWALL u. R. WODICK: Die Verwendung von Wasserstoff und Sauerstoff zur Messung der lokalen Gewebedurchblutung in situ mit der Platinelektrode. Pflügers Arch. ges. Physiol. **289**, R 99 (1966).
10. — Regional cerebral blood flow and microcirculation. In: W. H. BAIN, A. M. HARPER, W. A. MACKEY, Blood Flow through Organs and Tissues. pp. 162. Edinburgh and London: E. & S. Livingstone Ltd. 1968.
11. KETY, S. S., and C. F. SCHMIDT: The determination of cerebral blood flow in man by use of nitrous oxide in low concentrations. Amer. J. Physiol. **143**, 53 (1945).
12. VEALL, N., and J. C. W. CRAWLEY: The ^{133}Xe inhalation technique. Scand. J. clin. Lab. Invest. Suppl. **102**, XI:F (1968).

Local H$_2$-Clearance and PO$_2$-Measurements in Microareas of the Brain Tissue in Rats

H. Metzger, W. Erdmann, and J. Heidenreich

Physiological Institute, University of Mainz

Lübbers [5], as well as others, measured the oxygen partial pressure of the exposed cerebral cortex by means of small platinum microelectrodes. They registered low pO$_2$ values between 1 and 3 mmHg in certain areas of the cortex, despite the fact that the EEG was of normal pattern. In order to decide whether these low pO$_2$ values are a result of low rCBF, or high local oxygen consumption of brain tissue, we developed a new method. Using the same platinum needle, we measured both local pO$_2$ values and H$_2$-clearance curves in microareas of the brain tissue. Cater and Silver [2], Lübbers [4, 5], and others used glass-insulated platinum microelectrodes to measure local pO$_2$. Further, Aukland, Bower and Berliner [1], Lübbers [4, 5], and others employed such electrodes for recording H$_2$-clearance curves. Combining these 2 methods, we are now able to measure local pO$_2$ and local blood flow, with the same microelectrode.

The platinum microelectrodes were covered with methylsilicon and had tip diameters of between 5–10 μ. The registered H$_2$- and O$_2$-polarogramms showed definite limiting currents. The plateau of hydrogen lays between $+300$ and -100 mV, and that of O$_2$ between -600 and -800 mV. As a reference we used either an Ag/AgCl or a calomel electrode.

To obtain the pO$_2$ profiles of the brain, the platinum microelectrodes were moved forwards, perpendicular to the brain surface, 10 μ at a time. At significant pO$_2$ maxima or minima the polarisation of the platinum electrodes changed from -800 mV to about $+200$ mV. Thereafter, the H$_2$-clearance was registered at the same point.

For this purpose a gas mixture of 20.85 vol% O$_2$, 60 vol% H$_2$ and 19.15 vol% N$_2$ was administered by inhalation. After saturation of the brain tissue the inspired gas mixture was abruptly changed to normal air. The half-time of the fast component was calculated from the recorded H$_2$-clearance curve.

The following preliminary results were obtained:

1. the experiments showed reproducible pO$_2$ values after changing from O$_2$ to H$_2$;
2. the pO$_2$ values of the frontal and occipital brain of the rats amounted to 0–25 mmHg;
3. $T^1/_2$ values amounted to 2.15–4.45 min for the fast component.

The above experiments were performed under Nembutal anesthesia, arterial pO$_2$ and pCO$_2$ values, as well as arterial blood pressure were registered in a series of control experiments and remained within the normal range.

The low pO$_2$ values described by Lübbers were repeatedly measured, but did not agree with the values calculated using a Krogh's cylinder. Moreover, continuous registration of tissue pO$_2$ values revealed large fluctuations at the same point within the tissue in some cases, especially at high pO$_2$ values, whereas in other cases only small changes were observed. Such fluctuations might be caused by changes in respiration and/or blood flow.

The brain tissue pO$_2$ values change significantly in experimental hypoxia and hyperoxia. In one case tissue pO$_2$ changed from 11 mmHg to 20 mmHg, after the inspired gas mixture had been changed from air to 100 vol% O$_2$, and from 20–8 mmHg, after a change from 100 vol% O$_2$ to 5.4 vol% O$_2$.

Alternate measurements H$_2$ and O$_2$ using the same electrode showed that it is possible to measure both local pO$_2$ values and H$_2$ clearance curves at the same point within the tissue.

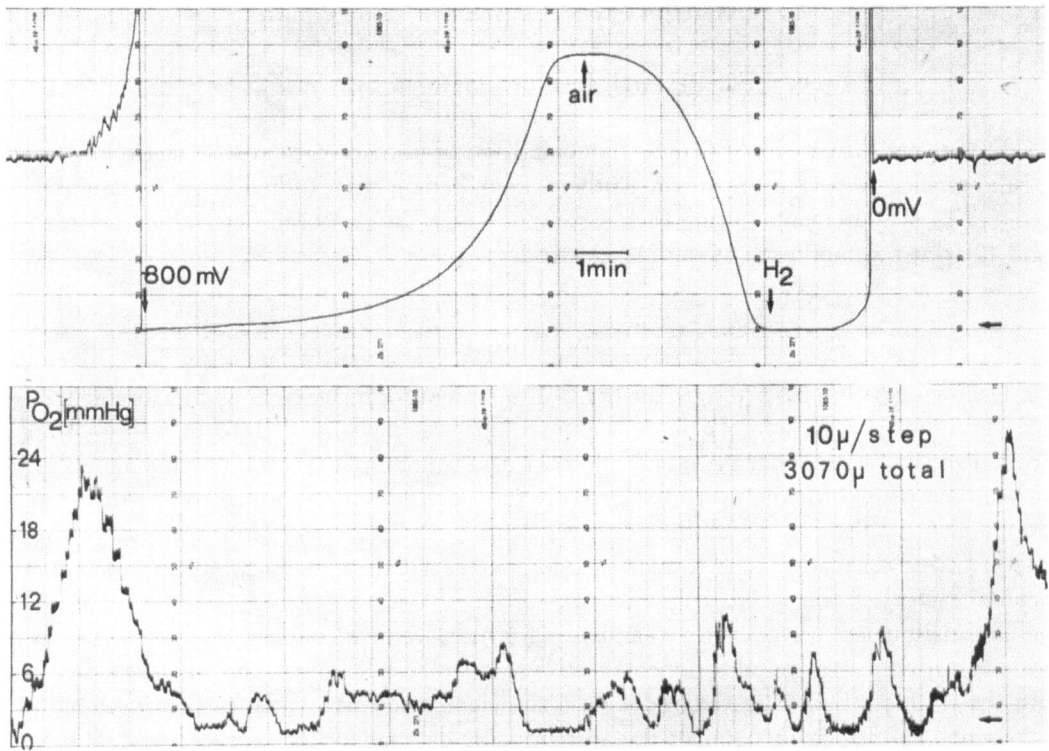

Fig. 1. The upper part of the figure shows a change from pO$_2$ (−800 mV) to a H$_2$-clearance (0 mV) measurement. The arrows indicate changes in polarisation voltage as well as changes from H$_2$ containing gas mixtures to normal air. The lower part shows an example of pO$_2$ measurement in the brain tissue of the rat. The micromanipulator was moved forwards, 10 μ at a time

References

1. AUKLAND, K., B. F. BOWER, and R. W. BERLINER: Measurement of Local Blood Flow with Hydrogen Gas. Circulat. Res. **14**, 164 (1964).
2. CATER, D. B., and I. A. SILVER: Microelectrodes and Electrodes used in Biology. In: Reference Electrodes. Eds.: J. G. IVES and G. JANZ. p. 512. New York: Academic Press 1961.
3. LÜBBERS, D. W.: The Oxygen Pressure Field of the Brain and its Significance for the Normal and Critical Oxygen Supply of the Brain. In: Oxygen Transport in Blood and Tissue. Eds.: D. W. LÜBBERS, U. C. LUFT, G. THEWS, and E. WITZLEB. p. 124. Stuttgart: G. Thieme Verlag 1968.
4. — Regional Cerebral Blood Flow and Microcirculation. In: Blood Flow through Organs and Tissues. Eds.: W. H. BAIN and A. M. HARPER. p. 162. Edinburgh and London: E. & S. Livingstone, Ltd., 1968.
5. — Capillary Pattern and Oxygen Tension of the Cerebral Cortex. Acta neurol. scand. Suppl. **14**, 92 (1965).

Simultaneous Focal Intracerebral Blood Flow Measurements in Man Around 18 Chronically Implanted Electrodes

C. W. SEM-JACOBSEN, O. B. STYRI, and E. MOHN

The EEG Research Institute, Gaustad Sykehus, Oslo

During the past year rCBF measurements have been made around 18 electrodes in each of 12 patients, 10 parkinsonian and 2 others. Each patient was studied in 9–22 sessions with from 2–15 measurements during each session. Thus, rCBF has been determined between 60 and 120 times around each of the 210 electrodes.

Technique

Following preliminary studies, the DY 2010 B system for instantaneous measurement of H_2 gas saturation and clearance data on a punch tape was developed. Information is sampled from up to 25 electrodes at the rate of up to 11 data sources per second. On line digital printing of data is used for monitoring. Currently the electrodes are sampled every 20th second. Supplementary data about blood pressure, heart rate, and respiration are also recorded. Mathematical calculation is performed by a computer facilitating rapid print-out of blood flow around each electrode.

A technique has been developed and used for measurements of changes of short duration, such as during rest, physical and mental activity, as well as CO_2 breathing and sleep. By keeping the platinum hydrogen electrodes on a sufficiently negative level, disturbances induced by oxygen are practically eliminated, and repeatable data have been recorded several times a week for several months.

After an initial stabilization period of 30–45 min, the patient breathes air with 3 per cent H_2 for 35–45 min to a saturation plateau. This is followed by periods of 4 or 5 min of clearance. During these short periods responses to physical activity, mental activity, or other physiological changes are measured.

Results

The figure illustrates the complexity of data obtained from repeated recordings. For many electrodes the 2 calculations gave identical flow values, the second set of measurements constantly gives lower flow values. The response pattern varies from electrode to electrode. CO_2 in the breathing air elicits major blood flow changes around certain electrodes, while hardly any change is found around others.

Motor and mental activity gave rise to different changes in certain areas, and these blood flow changes can be extremely focal.

Supported by the United States Government through the Department of the Army and the Atomic Energy Commission.

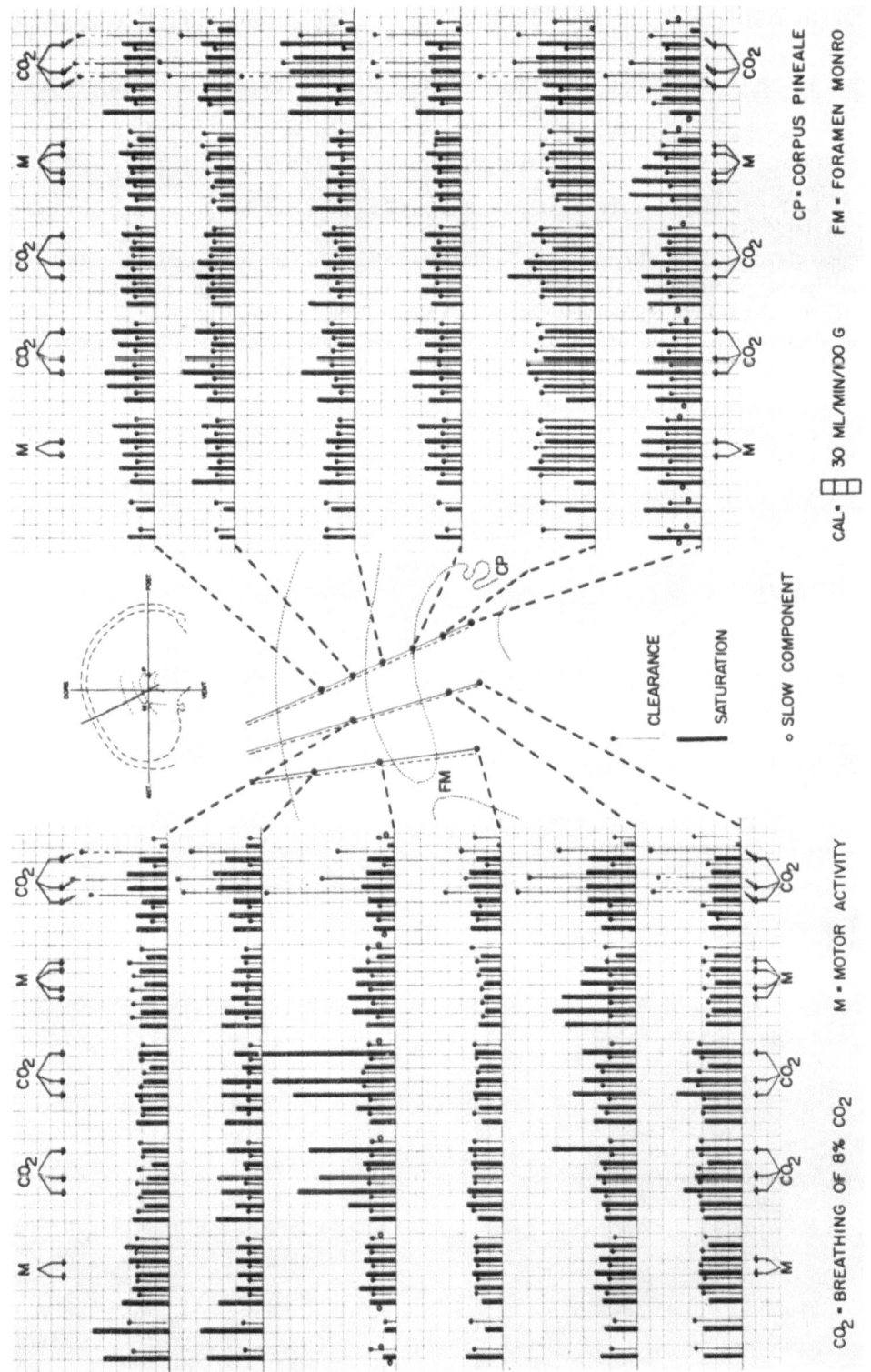

Fig. 1. Fluctuation in focal cerebral blood flow in man

For the past year, small solid state radiation sensors, 0.8 mm in diameter and 1.5 mm long, have been adapted for implantation into the brain together with the platinum electrodes. The solid state radiation sensors will be used to monitor focal blood flow using Krypton-85 and Xenon-133.

The first 4 sensors have been implanted in the brain after preliminary studies in animals.

References

1. Symposium on Intracerebral Electrography: Proceedings of the Staff Meetings of the Mayo Clinic. Vol. **28**, 145 (1953).
2. Sem-Jacobsen, C. W.: Depth-electrographic stimulation of the human brain and behavior from 14 years of studies and treatment of Parkinson's disease and mental disorders with implanted electrodes. Ed. W. Horsley Gantt. Springfield: Charles C. Thomas Publishing Co, III., 1968.
3. —, K. Aukland, and S. Akre: Attempts to localize gray and white matter around depth electrodes in human with hydrogen polarography. Electroenceph. clin. Neurophysiol. **19**, 617 (1965).

Gas and Heat Clearance Comparison and Use of Heat Transport for Quantitative Local Blood Flow Measurements

W. Müller-Schauenburg and E. Betz

Institute of Applied Physiology, University of Tübingen

An inert gas spreads in any medium according to the same law as heat. It is possible to find for each term of diffusion a corresponding term to heat conduction (Table 1).

Table 1

Heat conduction:	Diffusion of gas:
temperature ϑ	pressure p
heat content per volume	concentration
specific heat per volume ϱc	solubility α
heat conductivity λ	KROGH's constant K
$\varrho c \dfrac{\partial \vartheta}{\partial t} = \mathrm{div}\,(\lambda \;\mathrm{grad}\;\vartheta)$	$\alpha \dfrac{\partial p}{\partial t} = \mathrm{div}\,(K \;\mathrm{grad}\;p)$
$\dfrac{\partial \vartheta}{\partial t} = D\,\varDelta \vartheta$	$\dfrac{\partial p}{\partial t} = D\,\varDelta p$
$D = \dfrac{\lambda}{\varrho c} \approx 5 \cdot 10^{-3}\ \mathrm{cm^2/sec}$	$D = \dfrac{K}{a} \approx 2 \cdot 10^{-5}\ \mathrm{cm^2/sec}$
(heat in brain)	(H_2 in brain)

$$D_{\mathrm{heat}}/D_{\mathrm{H_2}} \approx 250$$

The point is that this correspondence includes the laws of spreading, i.e. the diffusion equation and the heat conduction equation which are presented above for inhomogeneous media and for the simpler case of a homogeneous medium. In homogenous media one single constant governs the spreading. This is the diffusion constant D of heat or gas. It is the order of magnitude of the constants which makes the difference between heat and gas. As examples we took a comparison of H_2 diffusion and heat diffusion in brain tissue, which differ by a factor of more than 200 (Table 1).

For blood flow measurements with heat or radioactive gas clearance additional differences e.g. in the measuring probes or the injection mode have to be taken into account. Temperature is recorded by thermocouples or thermistors which can be built with a smaller size than β or γ counters. The injection type yields the greatest difference between the CBF measurements by heat transport and by gas transport. Heat is applied locally by heating a nearly point shaped heater in the tissue. In contrast to that, the injection of gas is in no case as local as heat injection.

The usual way of measuring flow by heat clearance is a steady state technique and gives relative flow values. Gas clearance is applied intermittently and gives blood flow in ml blood

---

Supported by the Deutsche Forschungsgemeinschaft.

per ml tissue and minute, i.e. absolute blood flow has the dimensions of the inverse of time. Absolute blood flow measurements are measurements of time constants.

It is possible by a suitable, nonstationary injection mode, to find a local version of absolute blood flow measurements by taking time constants of heat clearance. Because of the local application and of the large diffusion constant of heat it is impossible to neglect heat conduction. Therefore we need a scheme to separate heat convection by blood flow from heat conduction. This discrimination is possible for homogeneous flow on the basis of the usual differential equation (1) (Fig. 1).

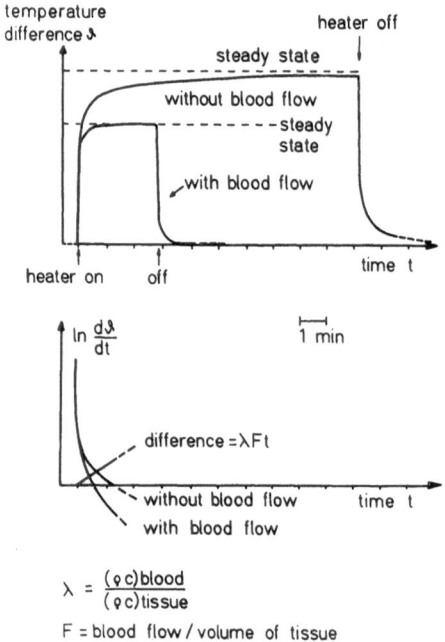

Fig. 1. Scheme for the separation of heat conduction and heat convection by blood flow for the determination of absolute blood flow values from heating and cooling curves

We use local and intermittent heating. The difference to the usual heat clearance method is that we switch on and off the implanted tiny and nearly point shaped heater. After switching the heating current on or off we record until the heater temperature is in a steady state.

We now separate heat conduction and heat convection by drawing the derivative of the heating or cooling curve in a semilogarithmic plot. In this plot the effect of homogeneous blood flow is additive to pure conduction. The difference of two curves of different but homogeneous blood flow at the same site will give a straight line with the slope $\lambda \cdot \Delta F$. ΔF is the difference of blood flow per tissue volume and λ is the partition coefficient of heat. If one takes the heating or cooling curve of the dead brain as reference curve, the difference will give the absolute blood flow.

Four comments have to be made on this scheme.

1. Heating and cooling curves are symmetrical and contain the same information.

2. The cooling curve $\vartheta(t)$ of a point shaped heater in a homogeneous medium is proportional to the inverse of the square root of time. Therefore its derivative $d\vartheta/dt$ is proportional to $t^{-3/2}$ and $t^{3/2} \cdot d\vartheta/dt$ is a constant. In a homogeneous part of a living brain $t^{3/2} \cdot d\vartheta/dt$ will be an exponentially decreasing function, i.e. a straight line in a semilogarithmic plot. Real curves differ from that behaviour for small t and for large t.

3. It is possible to understand why it is the *derivative* of the heating curve which separates heat conduction from convection by blood flow. The derived heating curve corresponds to local slug injection of heat. In case of local slug injection of heat it can be shown from the usual differential equation that blood flow enters only by a factor $e^{-\lambda Ft}$, therefore the time derivation reduces our local heat saturation curve to a local slug injection curve from which we can separate the blood flow factor $e^{-\lambda Ft}$ by a semilogarithmic plotting.

4. If blood flow is not homogeneous and the heated tip of the measuring probe is for instance a few millimeters from a larger vessel our method gives information about the order of magnitude of the distance of that inhomogeneity: locally applied heat needs a certain time for its transport in a measurable amount from the heated spot to a distant point. The slug injection interpretation of our method shows that the slope of the heating curve cannot be influenced by heterogeneities which are not yet reached by the heat. The information that there is an inhomogeneity needs a certain time to become manifest in the heating curve. From that time and the diffusion constant we calculate the distance.

If, therefore, in the immediate surrounding of the heated tip of the probe no heterogeneity of flow is recorded, this initial part of the heating curve gives an absolute value of flow in this surrounding.

One of the most important aspects of our method seems to be the following: It is no longer necessary to calibrate the usual heat clearance in absolute terms of flow by another method e.g. by the radioactive gas clearance. By this we will have a method for a continuous recording of local blood flow in absolute terms for acute and chronic animal experiments.

Reference

1. PERL, W.: Heat and matter distribution in body tissues and the determination of tissue blood flow by local clearance methods. J. theor. Biol. 2, 201 (1962).

Regional Brain Blood Flow Estimated by Using Particle Distribution and Isotope Clearance Methods

M. Meyer, T. Tschetter, A. Klassen, and J. Resch

University of Minnesota, Minneapolis

With the development of radio-active particles, investigators [1, 2, 4, 5, 6, 8, 10] have been re-examining regional distribution of the cardiac output (C.O.) to organs and tissues, including the brain of experimental animals. Carbonized microspheres appeared to be distributed in direct proportion to blood flow (BF) and by appropriate size selection, essentially all can be removed in one circulation. Preliminary studies [9] were performed to examine regional BF and total brain flow (TBF) in the dog brain. These have been extended and the effect of breathing 5% CO_2 on BF has been studied [10]. External isotope monitoring to detect the washout of residual (r) radioactivity of an inert gas in a tissue has become an important tecnique to study regional BF. Many investigators [3, 4, 7 and others] have used this approach to examine cerebral blood flow (CBF). This study proposed to determine regional and TBF by particle distribution method and also analyze cortical blood flow by the rCBF approach using ^{133}Xe to provide a comparative study.

Methods

Carbonized ^{169}Yb labeled microspheres with a mean diameter of about 25 ± 5 μ (SD) and a density of about 1.5 g/ml were suspended in dextran to give a concentration from 0.5 to 1.5×10^6 spheres/ml depending upon the desired activity to be injected. A known activity (1 ml) was washed slowly (80–100 sec) from a chamber and via a catheter into the left heart of lightly anesthetized dogs (5–7 months, 7–15 kg) breathing air (Group I) or air with 5% CO_2 (Group II) without a respirator. The C.O. was determined by isotope dilution using ^{42}K or ^{86}Rb. After sacrificing the dogs, samples of the frozen brain were taken from the cerebral cortex – Cc; cerebellum – Ce; white matter – Wm; caudate – Ca; and thalamus – Th. The fractional activity/gram was determined and the BF calculated in ml/min · g. TBF was estimated from activity measured in a mixed sample of the remaining tissues. Arterial blood samples were taken to determine the CO_2 content. In another group of dogs of similar age, and weight and also breathing air, 5–10 mc of ^{133}Xe in isotonic saline was injected in the left carotid artery. The left motor cortex was monitored with a collimated β detector through a craniotomy covered with a polyvinyl film. The residual washout curves were recorded with a ratemeter and also digitally over 0.1 min intervals. Graphical, and/or analog and digital computer techniques were used to resolve all clearance curves into a sum of exponentials.

Supported in part by USPHS Grant Nos. NB-03364, DE-02212 and DE-5269. Microspheres purchased from Nuclear Products, 3M Center, St. Paul, Minnesota, 55101.

Technique

Averages and ranges of regional and TBF using the particle distribution approach for Gp. I and Gp. II dogs are shown in Table 1. The percent of C.O. for TBF was calculated and is also shown. The average arterial CO_2 was 42.8 volumes percent for Gp. I; whereas, for Gp. II it averaged 50.2. Regional and total brain blood flows and arterial CO_2 were significantly correlated except for the white matter.

Table 1. *Averages and ranges for regional and total brain blood fow (TBF) in ml/min · g in young dogs using a particle distribution method. Group I : Breathing air, Group II : Breathing 5 % CO_2 in air*

						TBF		
	Cc	Ce	Wm	Ca	Th	ml/min·g	ml/min	% C.O.
Mean Gp. I (10)	1.15	1.23	0.66	1.34	1.08	0.91	71.8	2.3
Range	0.52–2.05	0.62–2.66	0.39–1.18	0.93–1.77	0.66–1.53	0.51–1.45	39.7–112.4	1.1–3.7
Mean Gp. II (9)	3.75	3.52	1.13	4.38	4.33	3.32	267.3	9.1
Range	1.58–5.71	2.06–5.07	0.59–1.67	2.09–7.00	1.20–8.15	1.46–5.54	121.3–489.2	3.5–16.3

See text for abbreviations.

Cortical blood flows from the rCBF method were estimated in several ways as shown in Table II. The ^{133}Xe partition coefficient for grey matter was taken from VEALL and MALLETT [10]. Curves were usually fitted by a sum of 2 exponentials. Cortical blood flows calculated from the larger exponential constant corresponded closer to those values determined by the particle distribution method. Flows estimated by using the "initial slope", i. e., taking the first derivative of the equation providing the best fit, as $t \to 0$ (beginning of the clearance curve) or by the ratio of the peak activity and area under the desaturation curve averaged much less than the average cortical flow determined by the particle distribution method.

Table 2. *Averages and ranges of blood flow (ml/min · g) in the cerebral cortex, of young dogs breathing air calculated from ^{133}Xe clearance curves*

	A	B	C
Mean	0.96	0.58	0.31
Range	0.35–1.45	0.31–0.97	0.17–0.55

A) Calculated from largest exponential constant
B) From initial slope
C) Ratio of initial amount and area under the curve

Discussion

It is assumed that spheres trapped initially have little or no effect on blood flow before subsequent microspheres are caught. The average value of 0.91 ml/min · g for TBF is within the range reported by others [2, 6] using the particle distribution approach on rabbits and monkeys. The %C.O. value of 2.3 for TBF is less than that reported by KAIHARA et al. [5]; however, in our concurrent study [10] on older dogs, the %C.O. for the TBF was comparable.

4*

On the other hand, effect of CO_2 on regional brain flow is much more dramatic than observed by others. It is possible that initial microsphere embolism causes metabolic changes which lead to a marked decrease in resistance to flow. However, since relatively large numbers of particles were injected, trapping would cause an increase in resistance and tend to offset the suggested metabolic effects. Moreover, if vasodilatation occurred, then the probability that spheres are 100% trapped could be diminished and the calculated flows would be underestimated. In allied studies [10], differences in brain blood flow were noted between young and old dogs. Thus, age may play some role in the marked increase in CBF to inhaling 5% CO_2 by young dogs. Furthermore, this increase in flow can be accounted for by a decrease in blood flow to skeletal muscle (unpublished data).

In this study the particle distribution method to determine cortical blood flow was not combined with the rCBF approach to analyze cortical flow in the same animal. Perhaps with the rCBF approach, this flow can be best estimated by considering a parallel flow model where the larger exponential constant of a multiexponential equation providing the best fit is related to cortical blood flow.

It appears that a wide variation in blood flow to different regions of the brain exists. Comparing the results of different approaches is difficult. The arterial (CO_2), anesthetic agent if one is used, and differences in techniques must be considered. Conceivably by combining the rCBF and particle distribution methods in the same animal, more enlightening knowledge might be obtained about both techniques.

References

1. Flohr, H. W.: Methode zur Messung regionaler Durchblutungsgrößen mit radioaktiv markierten Partikeln. Europ. J. Physiol. **302**, 268 (1968).
2. Forsyth, R. P., A. S. Nies, J. Neutze, and K. L. Melmon: Normal distribution of cardiac output in the unanesthetized, restrained Rhesus monkey. J. appl. Physiol. **25**, 736 (1968).
3. Häggendal, E., and B. Johansson: Effects of arterial carbon dioxide tension and oxygen saturation on cerebral blood flow autoregulation. Acta Physiol. scand. **66**, suppl. **258**, 25 (1965).
4. Høedt-Rasmussen, K., E. Sveinsdottir, and N. A. Lassen: Regional cerebral blood flow in man determined by intra-arterial injection of radioactive inert gas. Circulat. Res. **18**, 237 (1966).
5. Kaihara, S., P. D. Van Heerlen, T. Migita, and H. N. Wagner, Jr.: Measurement of distribution of cardiac output. J. appl. Physiol. **25**, 696 (1968).
6. Neutze, J. M., F. Wyler, and A. M. Rudolph: Use of radioactive microspheres to assess distribution of cardiac output in rabbits. Amer. J. Physiol. **215**, 486 (1968).
7. Obrist, W. D., H. K. Thompson, C. H. King, and H. S. Wang: Determination of regional cerebral blood flow by inhalation of Xe[133]. Circulat. Res. **20**, 124 (1967).
8. Rudolph, A. M., and M. A. Heyman: The circulation of the fetus in utero. Circulat. Res. **21**, 163 (1967).
9. Tschetter, T., M. Meyer, A. Klassen, and J. Resch: Regional cerebral blood flow using labeled microspheres. Proc. Int. Union of Physiol. Sc. **7**, 441 (1968).
10. —, A. Klassen, J. Resch, and M. Meyer: In preparation.
11. Veall, N., and B. L. Mallet: Regional cerebral blood flow determination by ^{133}Xe inhalation and external recording: the effect of arterial recirculation. Clin. Sci. **30**, 353 (1966).

Absolute Measurement of Brain Blood Flow Using Non-Diffusible Isotopes

W. H. OLDENDORF

UCLA School of Medicine, Los Angeles, California

We have pursued non-diffusible intravenous tracer techniques because of their safety and simplicity of administration and interpretation. If sufficiently safe and simple to administer, such a technique could be used in population surveys where repeated studies of flow might be desirable. This would allow, for example, prospective studies of stroke-proneness.

The mode transit time (turnover time) is easily obtained by following an intravenous bolus of non-diffusible indicator as it enters and leaves the brain blood pool. This turnover time of the pool would be more useful if it were a more absolute index of brain blood flow. We could assign an absolute value to total brain blood flow if we knew the pool volume and its turnover time.

The brain blood pool is capable of expansion and contraction. For example, we have studied changes in pool volume in response to neck vein compression and find the pool capable of expanding from its resting volume by 10–15%. The CSF pressure quickly rises in response to CO_2 and drops with hyperventilation, probably due largely to pool volume changes secondary to changes in flow rate.

In 1962 we sought to measure the pool volume by means of intravenous injection of a non-diffusible tracer (I^{131} RISA) and relating the head count to the peripheral blood count. We measured the entire head rostral to the floor of the cranial cavity. Our measured values were so much greater (50–100%) than Hedlund's value for brain blood pool of about 130 cc that we discarded the technique without understanding the reasons for our inordinately high values.

We now recognize several factors probably account for this discrepancy. First, some high-angle Compton scatter from the trunk was included. This can be excluded by high pulse-height threshold settings. Secondly, our measurements were made using a plasma label comparing cranial with peripheral whole blood drawn from a large vein. Because the hematocrit in brain is about 15% lower than in large vessels, this alone would cause an approximately 15% increase in apparent cranial pool volume.

A more important source of artifact is the blood present in scalp and skull. We have been at a loss for a means to measure the volumes of human scalp and skull blood pools.

To obtain some crude estimate we performed the following experiment in 5 rabbits. Under pentobarbital anesthesia we injected 5 microcuries of ^{125}I-RISA intravenously 5 min before death. We injected 5 microcuries of ^{131}I-iodoantipyrine 45 sec before death. Death was produced by sudden crushing of the neck to prevent blood loss. The neck was held in this condition for several minutes until all circulation had stopped.

The ^{125}I and ^{131}I concentrations were determined for brain, skull, and scalp overlying the cerebral hemispheres. Blood plasma counts were also obtained in order to calculate the albumin space of each tissue as an index of blood volume. The blood flows in scalp and skull relative to brain were obtained by comparing the antipyrine content.

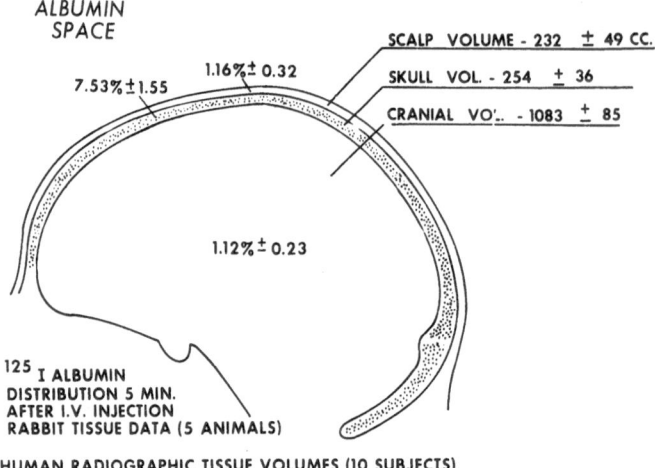

ALBUMIN
SPACE

7.53% ± 1.55   1.16% ± 0.32

SCALP VOLUME - 232 ± 49 CC.

SKULL VOL. - 254 ± 36

CRANIAL VOL. - 1083 ± 85

1.12% ± 0.23

125 I ALBUMIN
DISTRIBUTION 5 MIN.
AFTER I.V. INJECTION
RABBIT TISSUE DATA (5 ANIMALS)

HUMAN RADIOGRAPHIC TISSUE VOLUMES (10 SUBJECTS)

Fig. 1

The results of these studies are shown in Table 1. With an albumin space of 7.53% (Fig. 1), assuming a hematocrit of 41, the whole blood content of skull must be about 1.69 times as great or 12.69%. Correcting for the specific gravity of bone results in a skull whole blood pool volume of 22% of skull volume.

Table 1

Rabbit Tissues	Albumin Space*	Relative Blood Flow**
Brain	1.12% (0.23)	100%
Scalp	1.16% (0.32)	61% (16)
Skull	7.53% (1.55)	42% (10.2)

* (S.D.) Five Animals
** Eleven Animals

If these values are transposed to human skull volumes (determined radiographically) this suggests there is about 58 cc of blood in the average skull. This is 44% of the brain pool volume described by Hedlund.

Our data for scalp pool volume, similarly calculated, indicates a much smaller blood content for human scalp covering the cerebral hemispheres of 4.5 cc. This probably is artificially low since the relative flow rate per unit weight is 61% of brain. Most likely blood is lost during our experimental animal tissue excision despite our attempts to retain it. This would also explain our low calculated human brain pool volume of 20.5 cc. Using a mean brain flow approximation of 55 cc/min/100 gms the flow rate for 232 grams of scalp is 78 cc per minute. If this scalp flow is passing through a pool volume of 4.5 cc it would indicate a human scalp mean transit time of 3.5 sec which seems low.

In conclusion, these rabbit data, if transposed to human skull and scalp, suggest that there is something of the order of 60 cc of blood in the scalp and skull. The flow rates through skin probably vary by at least an order of magnitude in response to temperature and emotion. This analysis suggests the pool volume artifact created by scalp and particularly skull are so formidable as to preclude meaningful correction of count rates determined by total cranial counting. Scalp pool volume could be corrected for by using a soft gamma-emitter such as ^{125}I-albumin which is largely absorbed by the electron-dense skull preventing radiation from within the cranial cavity affecting the external count. The count from the skull pool could not be similarly corrected. I know of no way of substantiating the accuracy of any scalp of skull pool volume corrections.

The transit time alone is a semi-quantitative index of flow non-linearly related to flow. Combining this simple technique with attempts at measuring brain pool volume would not seem worthwhile.

A Method of Employing External Counting Techniques to Obtain the Mean Transit Time and Distribution of Transit Times of Isotope Passage Through the Head

D. L. Ciffone, J. W. Gamel, W. H. Marshall, and L. A. Sapirstein

Ames Research Center, NASA, Moffett Field, California, and Stanford University Medical Center, Palo Alto, California

A diagnostic tool is being developed which will allow the measurement of the distribution of transit times and the mean transit time of isotope passage through the head. The technique, applicable to any organ, is discussed herein only in terms of the brain. The procedure, a combination of clinical measurements and fluid-dynamic analysis, is almost entirely non-traumatic and total body radiation exposure is minimal (0.2 rad.).

Technique

A 10 millicurie bolus of Technetium (99 m Tc) in the pertechnetate form is introduced into the antecubital vein by a rapid cuff release while three external scintillometers laterally view each half of the cerebral hemisphere and the vertex of the aortic arch.

Since Technetium does not cross the brain-blood barrier, its rate of accumulation in the brain (measured by the head scintillometers) is the difference between arterial delivery and venous removal (i.e., the law of conservation of mass). The fraction of isotope leaving the head per unit of time as a result of isotope entering a given number of time units earlier is a time-dependent frequency function which, when taken over all time, defines the distribution of transit times of isotope passage through the head. Hence, this frequency function relates venous removal to arterial delivery statistically. The product of time and frequency function summed over all time then yields the mean transit time. If the venous-arterial flow could be measured at the site of the organ of interest, the desired distribution of transit times and mean transit time could be evaluated directly. However, utilizing conservation of mass and the fact that for the first second or two the head venous drainage is essentially zero, a scaling factor (the ratio of rate of accumulation within the field of view of the head counter to an arterial signal) can be evaluated, which allows (after a time-scale shift) an arterial measurement made elsewhere in the system to be used as the arterial delivery to the head. Venous removal from the head is then the difference between the scaled arterial flow and the head accumulation rate.

A physical dynamic model has been constructed which utilizes dyes and solar cells to test the analytic approach and to assess its sensitivity to physical changes within the system. Clinical measurements are being made in an attempt to establish normal mean transit times and head perfusion patterns as well as deviations from these in disease. Measurements are also being made to verify that the shape of the aortic isotope history, as viewed externally, is truly the shape of the arterial isotope concentration.

Fig. 1 shows a direct clinical comparison of the shape of the isotope concentration time history using an external scintillometer and using a catheter inserted via the femoral artery into the aortic arch. The curves have the same shape, which substantiates the external measurement. Fig. 2 shows the distribution of cerebral transit times as obtained from 2 patients at the Stanford Medical Center. The mean transit times are about the same ($4\frac{1}{2}$ sec); and the curve forms are similar. Patient I (a healthy, young male) was the subject in the blood measurement of Fig. 1. Patient II (a 52-year-old female) showed no recognizable dissimilarities in left-right hemispherical perfusion comparisons, although she displayed clinical evidence of severe arteriosclerosis.

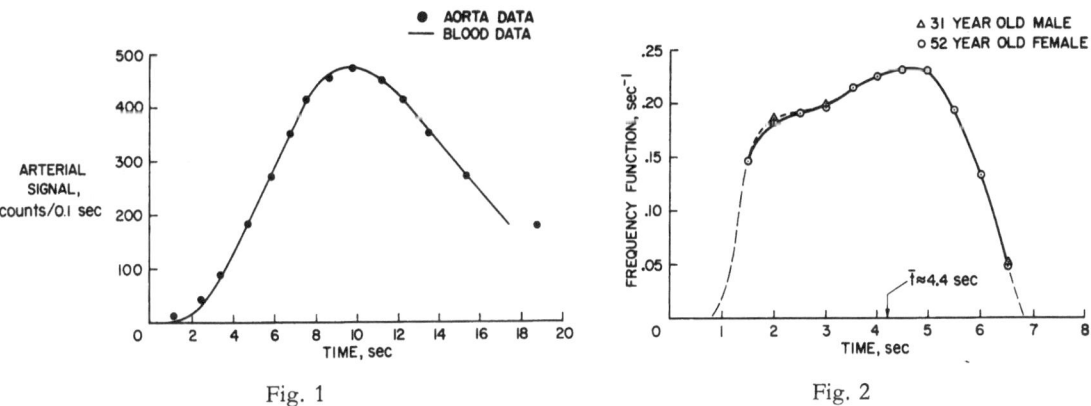

Fig. 1   Fig. 2

The distribution of transit times of dye particles, passing though one-half of a "brain box" used in the fluid-dynamic model, is shown in Fig. 3. The data measured directly at the entrance to and exit from the "brain box", and simultaneously measured 14 inches upstream of the box and within the box, yield mean transit times which agree reasonably well, but curve forms which do not. This failure lies largely in the data-analysis system.

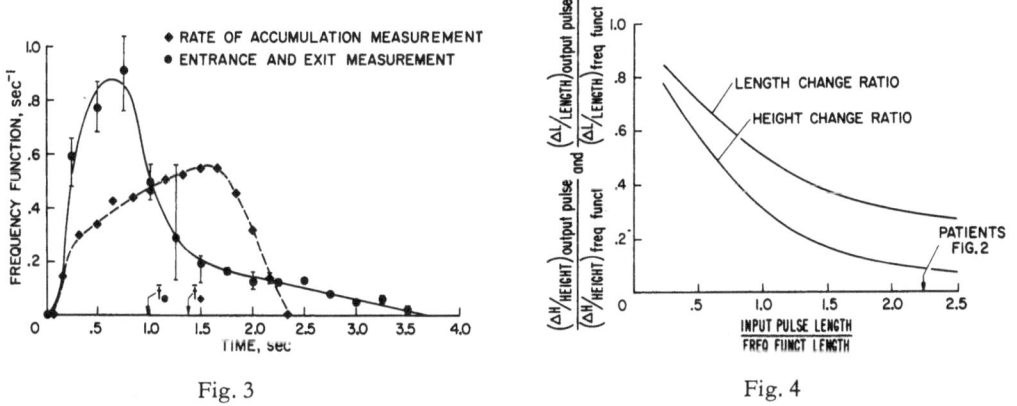

Fig. 3   Fig. 4

Fig. 4 presents the result of a sensitivity study which illustrates the need for improving the quality of the measured data and the data-processing techniques as the ratio of input pulse length to frequency-function length increases (where lengths are measured at the pulse half heights). For the clinical approach which is being developed, the bolus of isotope is dispersed by passage through the heart and lungs to the extent that this ratio is on the order of $2–2\frac{1}{2}$.

Hence a 10 percent increase in the length of the frequency function will be realized as a 3 percent change in the length of the output pulse. This insensitivity introduces limits to the diagnostic value of the shape of the distribution of transit times as obtained by the above proposed nontraumatic technique.

These preliminary results indicate that head perfusion patterns and rates can be measured nontraumatically. Work is in progress to improve both the data collection and reduction techniques in order to enhance the diagnostic value of the measurements.

Effects of Various Agents on Cerebral Circulation Studied by Ultrasonic Doppler Technique in Man

M. Miyazaki

Department of Internal Medicine, Osaka Municipal Kosaiin Hospital, Suita City, Osaka

The measurement of cerebral blood flow by ultrasonic Doppler technique allows an instantaneous and continuous study of dynamic changes of the cerebral circulation resulting from administration of various agents. In addition, by this technique blood flow can be measured individually for each vessel, e. g., the internal, external and common carotid artery or the internal and external jugular vein. In this paper, several clinical applications of this technique are presented.

Principle and Method

When ultrasonic waves impinge upon the blood stream, the reflected waves are subjected to a frequency change, due to the Doppler effect, caused by moving blood particles and turbulent flow. A sound can be obtained by comparing the reflected waves with the direct waves and by demodulating them. The frequencies of this sound are proportional to blood velocity.

When ultrasonic waves impinge upon the moving medium, the frequencies of reflected waves (f') are obtained as follows:

$$f' = \frac{c + u \cdot \cos \theta}{c - u \cdot \cos \theta} f$$

f: frequency of direct wave
c: sound velocity
u: velocity of moving subject
θ: angle between the direction of ultrasonic wave and the direction of moving subject
λ: wave length

The frequencies of Doppler beat (fd) are obtained as follows:

$$fd = f' - f = \frac{c + u \cdot \cos \theta}{c - u \cdot \cos \theta} f = \frac{2 u \cdot \cos \theta}{c} f = \frac{2 u}{\lambda} \cos \theta \qquad (c \gg u)$$

λ: wave length.

From the above formula it is postulated that the waves from the moving medium are converted into an audible sound, since the frequencies of the Doppler beat are proportional to the velocity of the medium and that the Doppler effect is theoretically minimal at $\theta = 90°$ and maximum at $\theta = 0°$ [1, 4].

Cerebral blood flow (ΔCBF), cerebral vascular resistance (ΔCVR) and cerebral oxygen consumption (ΔCMRO$_2$) are used as a measure of cerebral circulation and metabolism. Fig. 1 shows the diagram of the equipment.

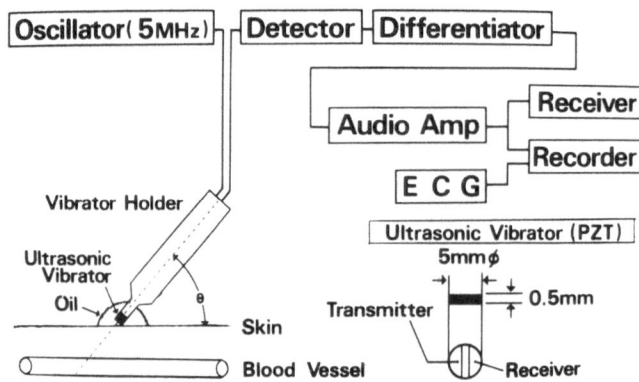

Fig. 1. Diagram of the employed equipment

Results and Discussion

1. Hemodynamic changes in the internal carotid artery were investigated before, during and after intravenous administration of 250 mg theophylline ethylenediamine dissolved in 20 ml of 5% glucose solution (Fig. 2 and 3). The effect upon cerebral circulation was a cerebral vasoconstriction during the drug administration and a cerebral vasodilation following it [2]. Theophylline ethylenediamine has been widely used for the treatment of cerebral

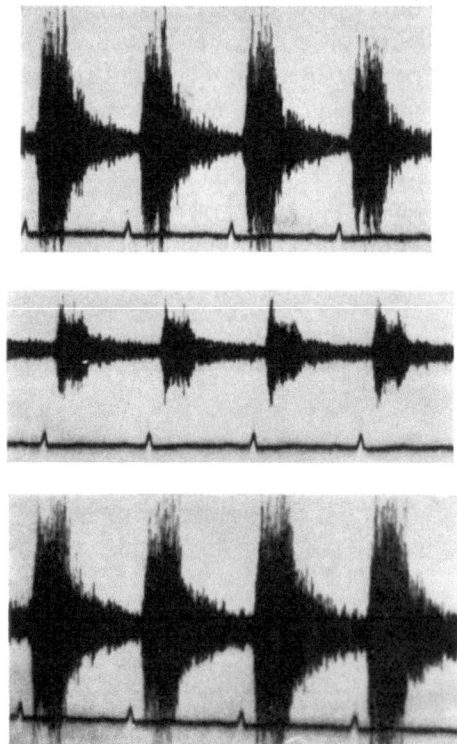

Fig. 2. Effect of theophylline ethylenediamine on internal carotid blood flow in a 72 years-old patient. Recordings made before (upper), during (middle) and after (lower) administration

vascular disease. The heterogeneous opinions as to its effect upon cerebral circulation may be partially derived from the above mentioned pharmacological characteristics of the drug.

Hemodynamic changes between the external carotid artery and the internal carotid artery before, during and after intravenous administration of 20 mg of nicotine acid dissolved in 20 ml of 5% glucose solution showed an increase in blood flow and a decrease in vascular resistance in the external and internal carotid arteries, mainly in the former, when facial blushing was evident. It is generally said that nicotinic acid only increases extracranial blood flow. Our results support the above concept.

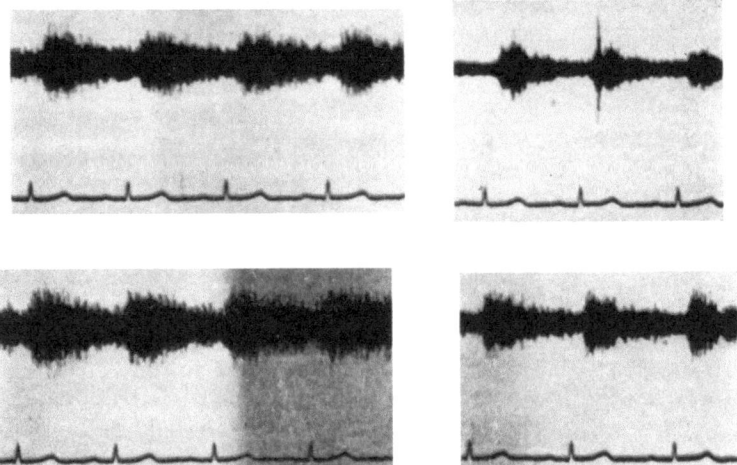

Fig. 3. Effect of nicotinic acid on blood flow in the internal (left recordings) and external (right recordings) carotid arteries. The upper recordings represent the controls

2. Hemodynamic changes in the internal carotid artery were also studied before, during and after cigarette smoking in young and elderly smokers [3]. The patients were instructed to smoke cigarettes of usual type with ordinary and rapid inhaling speed. Blood pressure and pulse rate were also measured throughout the smoking (Fig. 4). During ordinary smoking,

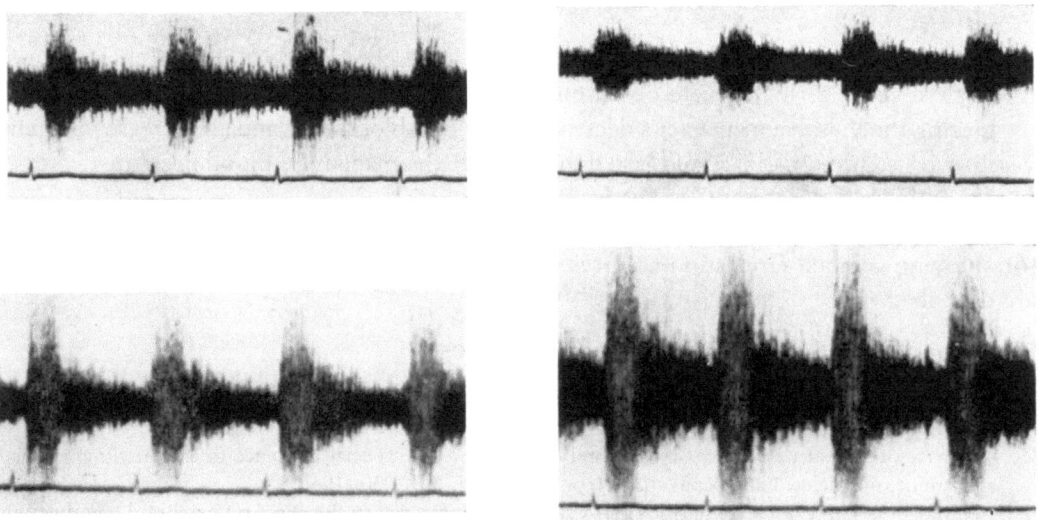

Fig. 4. Changes in internal carotid blood flow caused by ordinary (left recordings) and fast (right recordings) cigarette smoking. The upper recordings represent the controls

an increase in cerebral blood flow and a decrease in cerebral vascular resistance were observed in all subjects following one to three inhalations and lasting for about 10–20 min after smoking had been discontinued. On the other hand, blood pressure and pulse rate were either unchanged or slightly increased. It is suggested that the effect of tobacco smoking on CBF is mainly due to a direct effect of nicotine, since no significant correlation was observed between cerebral hemodynamics and blood pressure. During rapid smoking, the changes of CBF, blood pressure and pulse rate were more conspicuous than during ordinary smoking.

3. Hemodynamic changes before, during, and after voluntary hyperventilation and the Valsalva maneuver have also been studied (Fig. 5). The duration of voluntary hyperventilation ranged from 2–3 min, while the Valsalva maneuver lasted 30–40 sec. Blood pressure, pulse rate and blood gas content (pCO_2, pO_2, pH) were also determined during these studies [4].

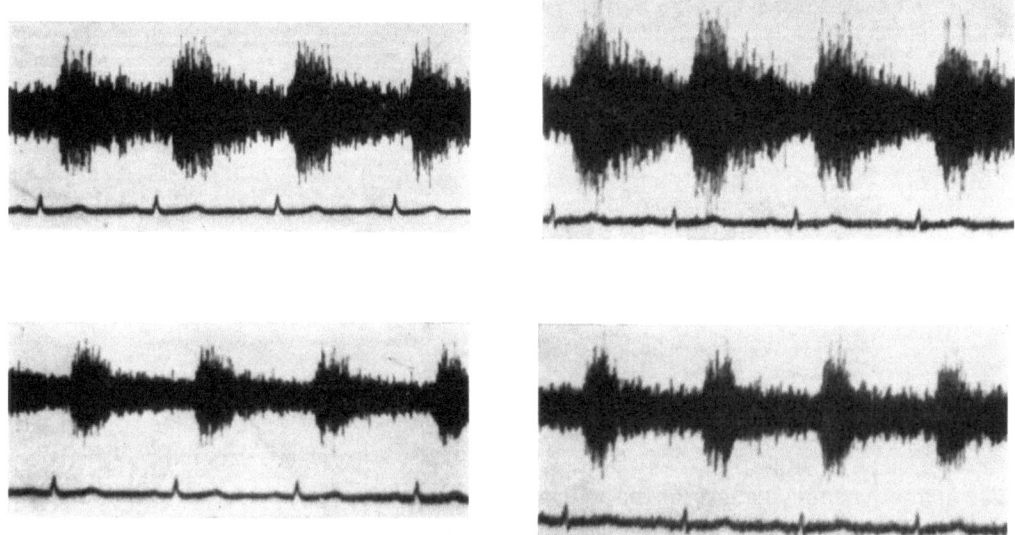

Fig. 5. Effect of hyperventilation (left) and of Valsalva maneuver (right) on cerebral circulation. The upper recordings represent the controls

During voluntary hyperventilation a decrease in cerebral blood flow and an increase in cerebral vascular resistance were observed.

During the Valsalva maneuver a decrease in cerebral blood flow and an increase in cerebral vascular resistance took place and lasted throughout the maneuver. Blood pressure decreased, pO_2 and pH increased and pCO_2 decreased.

In conclusion, the principle and clinical applications of the ultrasonic Doppler technique for studying cerebral circulation are presented, and the useful characteristics of the method are described.

References

1. MIYAZAKI, M.: Measurement of cerebral blood flow by ultrasonic Doppler technique: theory. Jap. Circulat J. **29**, 375 (1965).
2. — Effects of several vasodilators on cerebral circulation, with special reference to theophylline ethylenediamine, nicotic acid and papaverine. Jap. Circulat. J. **30**, 1230 (1966).
3. — Circulatory effect of cigarette smoking, with special reference to the effect on cerebral hemodynamics. Jap. Circulat. J., in press.
4. — Cerebral hemodynamics during voluntary hyperventilation and the Valsalva maneuver. Jap. Circulat. J. **32**, 315 (1968).

The Use of a New Cyclotron Produced Isotope Krypton-85m, for the Measurement of Cerebral Blood Flow

H. I. Glass, R. N. Arnot, J. C. Clark, and R. N. Allan

Medical Physics Department, Medical Research Council, Cyclotron Unit, Department of Clinical Pharmacology, Royal Postgraduate Medical School, Hammersmith Hospital, London

Introduction

Xenon-133 has been widely used for the measurement of regional cerebral blood flow by both the carotid injection and inhalation methods. The principal disadvantage of Xenon-133 is the low energy of the radiation which is very readily scattered. The energy of the radiation scattered at 180° (the backscatter peak) depends on the energy of the primary beam. The backscatter peak of the 81 keV of ^{133}Xe occurs at 69 keV. Since the resolution of a sodium iodide detector at 80 keV is usually about 25 keV, counting over the full photopeak includes a substantial amount of scattered radiation and there will also be a contribution from lead X-rays at 73 keV, if the collimators contain lead. It is not possible to eliminate the scattered energy without a substantial loss in the sensitivity even by accepting only the upper portion of the photopeak. Wilkinson (personal communication) has indicated that if the lower discriminator gate is set at 77 keV approximately 23% of the counts recorded by the detector come from outside the "optical" field of view of the collimator. Most of the papers on local blood flow do not specifically indicate that counting conditions were such as to exclude scattered radiation. The results ascribed to regional areas as measurements of local blood flow are therefore of questionable accuracy.

If the inhalation technique is used then since Krypton is less soluble in fat than Xenon there will be less recirculation of Krypton than Xenon. We would like to draw attention to the use of Krypton-85m which emits a gamma ray of 150 keV.

Production and Physical Properties of Krypton-85m

The physical characteristics and radiation dosimetry of ^{85m}Kr and its impurities are shown in Table 1. It has a 4.4 h half life and emits 2 gamma rays, one at 150 keV (78%) and the other at 305 keV (13%). At the present time ^{85m}Kr is produced by the ^{84}Kr(d, p)^{85m}Kr reaction (^{84}Kr–56.9% abundant). 3 l of Krypton are irradiated in the external beam of the M.R.C. cyclotron at Hammersmith with 15 MeV deuterons at a beam current of 35 μA. The yield of ^{85m}Kr is 790 μCi/μA h, and a typical specific activity for a 2.5 h irradiation is 3.5 μCi/ml at time of use (14 h after end of bombardment). ^{87}Kr and ^{79}Kr are produced concurrently by the ^{86}Kr (17.37%) (d, p)^{87}Kr and ^{78}Kr (0.35%) (d, p)^{79}Kr reactions. The irradiation is normally carried out the evening before use and the amount of ^{87}Kr, which has a half life of 78 min, has reduced to 0.6% by the time the gas is used. This is important since the radiation dose and scatter radiation of ^{87}Kr would otherwise be excessive. The relative amount of ^{79}Kr (1% at the time of administration) gradually increases, however, and makes a small contribution to the radiation dose and the scatter radiation.

Table 1. *Physical characteristics and dosimetry of Krypton-85 m and impurities*

	^{85m}Kr	^{87}Kr	^{79}Kr
Half life	4.4 h	1.3 h	34.9 h
Principal γ rays (MeV)	.150 (78%)	.403 (87%)	.261 (9%)
	.305 (13%)	.847 (13%)	.398 (10%)
		2.57 (22%)	.511 (15%)
			.616 (10%)
k-factor	1.13	11.0	1.3
Average β energy (MeV)	0.25	1.40	0.08
Percent present at administration	100	0.6	1.0
Radiation Dose* Mucosal Tissue (mrads)	187	6	0.7
Radiation Dose* Lung (mrads)	34	1	0.2

* For inhalation of ^{85m}Kr at a concentration of 0.5 mCi/l for 2 min

Because of the method of production the ^{85m}Kr is only available at present in gaseous form, which is also the case if the ^{85m}Kr is produced in a reactor. ^{85m}Kr can be obtained in a pure carrier free form by α-particle irradiation of enriched Selenium-82. This introduces the possibility of obtaining the gas dissolved in saline which would allow estimation of cerebral blood flow by carotid injection. Due to the complete absence of ^{87}Kr, interference from scatter radiation due to this isotope is eliminated. The radiation dose to the mucosal tissue following inhalation of 0.5 mc/l of pure ^{85m}Kr for 2 min is 187 mrad. With 0.6% ^{87}Kr and 1% ^{79}Kr present the radiation dose is 194 mrad.

Method

Since the isotope is only available at present in gaseous form, we have been using the inhalation technique for the measurement of cerebral blood flow. The subjects inhaled oxygen containing ^{85m}Kr at a concentration of 0.5 mc/l, for 2 min. The activity in the head and expired air were monitored. The data were analysed using the Obrist deconvolution program, the parameters for the air curve fit being obtained using the Marquardt programme. Duplicate measurements have been carried out on 10 normal volunteers. These results will be reported elsewhere. Simultaneously the spectrum of the radiation reaching one of the head detectors was monitored on a multichannel analyser.

Results

Fig. 1 shows the scatter spectrum of ^{85m}Kr obtained using a 6 mm thick NaI detector from the head of a patient. The unscattered spectrum is also shown normalized to the height of the 150 keV photopeak. By subtracting the unscattered spectrum the contribution due to the scattered energy was assessed. The scatter peak due to the 305 keV energy interferes with the 150 keV photopeak. In order to carry out reliable regional flow measurements the lower gate should be set at 140 keV, which still permits 60% of the photopeak energy to be detected while completely excluding scatter radiation. There is a small contribution (5%) to the scatter peak at 130 keV due to the ^{87}Kr impurity which will be absent if ^{85m}Kr is obtained from enriched Selenium-82.

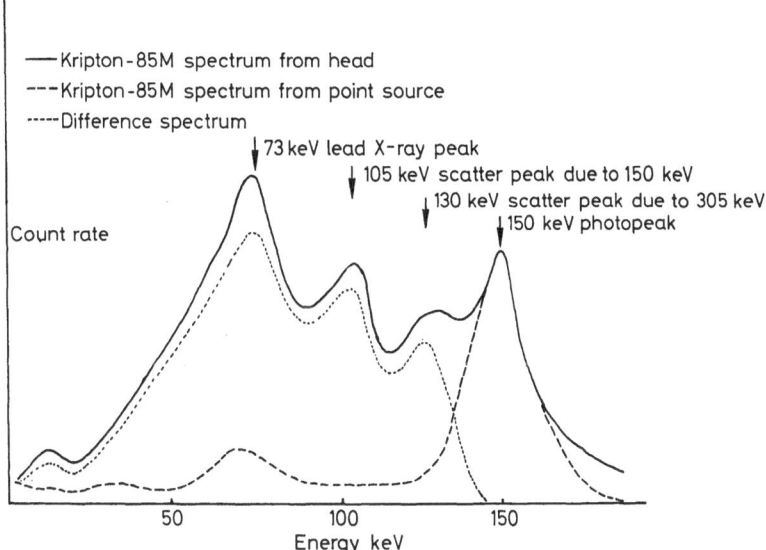

Fig. 1. Scatter spectrum of Krypton-85 *m*

Discussion

Krypton-85 m can be obtained from a reactor by the irradiation of natural Krypton. A flux of 10^{14} n/cm²/sec for 24 h would be required to produce 20 mCi in 3 ml of gas at the end of bombardment. ^{87}Kr and ^{79}Kr would still be present as impurities. Although the ^{87}Kr could be allowed to decay, the ^{79}Kr content would probably be higher as the irradiation time would be longer. 100 mg of enriched selenium-82 would cost about £ 150 and this could produce many millicuries of krypton-85 m. The selenium could be recovered and reused. This production process is now being investigated. Until the Krypton-85 m becomes available dissolved in saline we are continuing to use the inhalation method of measuring cerebral blood flow. The short half life is a disadvantage but since we normally do not use it until 14 h after production this is not a major disadvantage for hospitals remote from a cyclotron. Since 6 cyclotrons have recently been or are currently being installed in hospitals around the world, this isotope should become increasingly available. It seems likely that more precise and diagnostically significant information on regional cerebral blood flow will be obtained with Krypton-85 m.

The Simultaneous Measure In Vivo of Regional Cerebral Blood Flow and Regional Cerebral Oxygen Utilization by Means of Oxyhemoglobin Labelled with Radioactive Oxygen[15]

M. M. Ter-Pogossian, J. O. Eichling, D. O. Davis, and M. J. Welch

Edward Mallinckrodt Institute of Radiology, Washington University School of Medicine, St. Louis, Missouri

Regional cerebral blood flow and oxygen utilization are measured simultaneously *in vivo* with a single radioactive label by the following method.

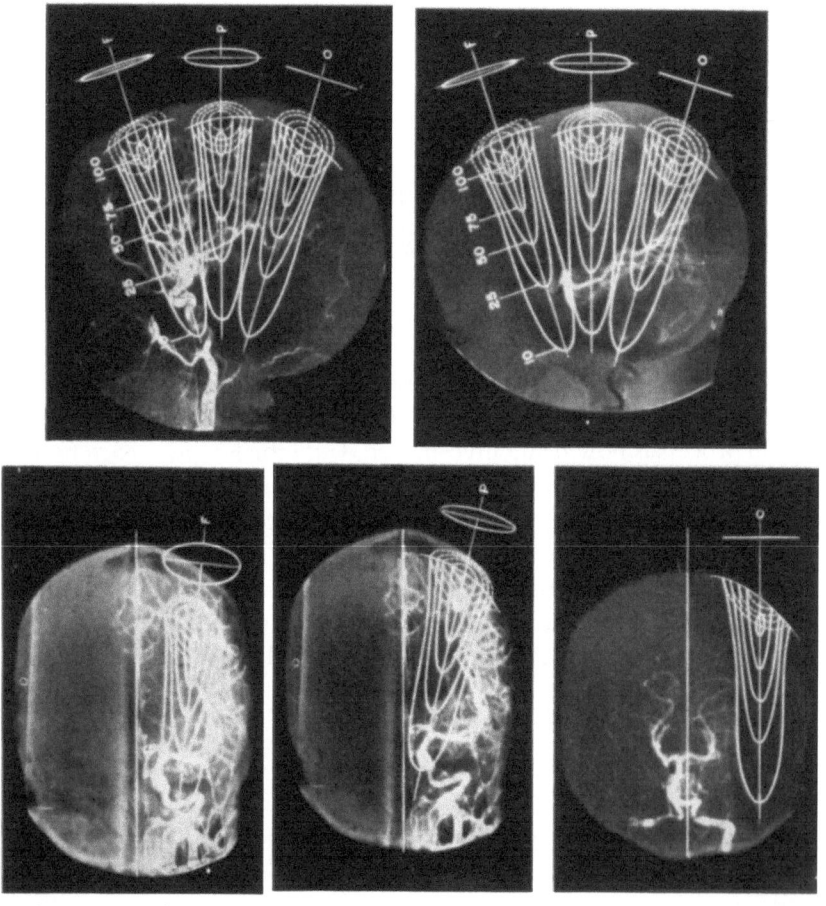

Fig. 1. Isosensitivity curves for the annihilation radiation for three of the scintillation probes used in this study. Each probe is fitted with a sodium iodide crystal 3 inches in diameter and 2 inches thick. Radiation collimation is achieved by means of a 19-hole "focusing" collimator. 3 probes are placed over each cerebral hemisphere

Supported by Public Health Service Grants No. GM 14889-06 and No. NB-HE-06833-01.

A small volume of blood (2–3 cc) with radioactive oxygen-15 tagged hemoglobin is rapidly injected into the internal carotid artery of a patient. The distribution of the radioactive label in the brain is measured, as a function of time, by six collimated scintillation probes placed over the subject's head (Fig. 1). This recording (Fig. 2) reflects (1) the arrival of the labelled oxygen into the tissues, (2) its partial conversion into water of metabolism, and (3) the washout of labelled water from the brain. (The contribution of these components was determined by comparing recordings obtained with successive injections of blood labelled with (1) ¹⁵O-oxyhemoglobin, (2) ¹⁵O-carboxyhemoglobin, and (3) ¹⁵O-water (Fig. 3)). The ratio of the amount of oxygen perfusing the tissues to the amount of labelled water formed is

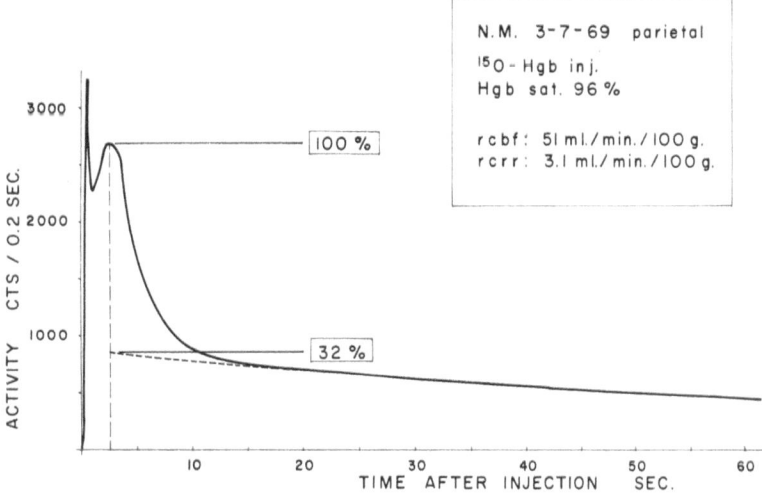

Fig. 2. (Patient N. M.) Recording over the parietal area subsequent to the injection of 2 cc of blood with the hemoglobin labelled with about 1 mCi of ¹⁵O. The maximum counting rate reached before corrections was about 10.000 counts per second and the sampling time was 0.2 sec. The first sharp peak is interpreted as "non nutritional" flow. The ratio of the second peak to the extrapolated wash-out curve provides the fraction of O₂ utilized

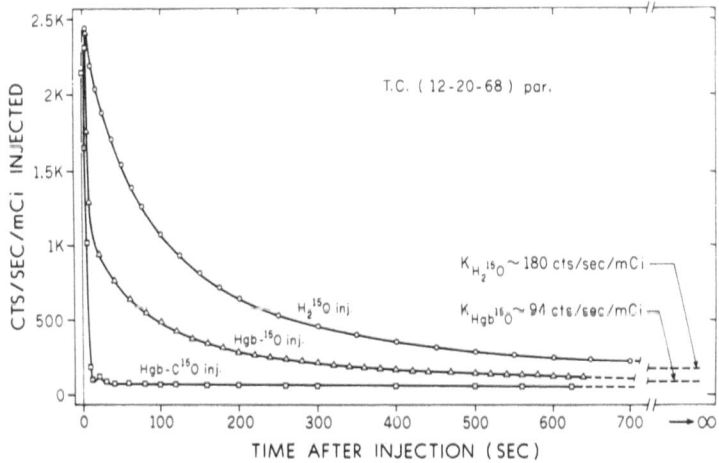

Fig. 3. (Patient T. C.) Curves obtained subsequently to the injection of blood labelled with (1) H₂¹⁵O, (2) ¹⁵O-Hgb and (3) C¹⁵O-Hgb. Note: 1. The near complete retention of activity in curve (1) which means a nearly complete equilibration of water with one passage of the bolus. 2. The partial retention of activity in curve (2) and 3. The minimal recirculation of activity in curve (3)

5*

a relative measure of oxygen utilization (Fig. 2), and the rate of washout of the labelled water provides a measure of blood flow (Fig. 4). The product: oxygen utilization × blood flow × arterial oxygen content expresses tissue oxygen consumption rate in absolute units. Regional cerebral respiration rates for a series of patients with cerebral pathology are shown in Table 1.

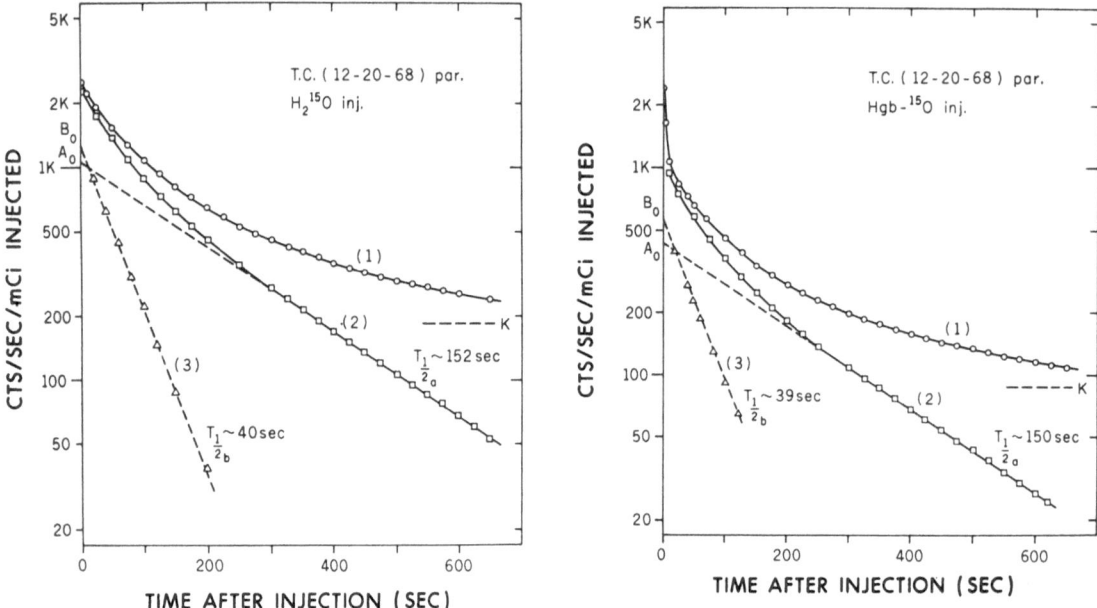

Fig. 4. (Patient T. C.) Analysis of the ^{15}O-Hgb (a) and H$_2$^{15}O (b) curves in Fig. 3. Curve 1: Semi-logarithmic plot of recording. Curve 2: Is obtained by subtracting the equilibrium value "K" (Fig. 3) from curve 1. Curve 3: (and 2): Analysis of curve 2 into two exponential components. Note that "a" and "b" show nearly identical wash-out components, which is interpreted as further indication that ^{15}O in "a" is washed out as H$_2$^{15}O. If it is assumed the "fast" and "slow" components of the wash out of labelled water occur in white and grey matter, absolute blood flow in these compartments can be derived from this data by conventional calculations, and by using the water partition coefficients for grey and white matter

Table 1. *Mean rCBF, O$_2$ retention and rCMRO$_2$ (or tissue respiration rate) for 3 patients with angiographically demonstrated cerebral pathology (W.T., A.D. and T.C.) and 1 patient (N.M.) with negative angiograms.*

	Frontal			Parietal			Occipital		
Subject	rCBF ml/min/ 100 g	O$_2$ Re-tention %	rCMRO$_2$ ml/min/ 100 g	rCBF ml/min/ 100 g	O$_2$ Re-tention %	rCMRO$_2$ ml/min/ 100 g	rCBF ml/min/ 100 g	O$_2$ Re-tention %	rCMRO$_2$ ml/min/ 100 g
W.T. 11-13-68	48	45	4.2	43	44	3.7	55	39	4.2
A.D. 11-21-68	39	46	3.6	36	41	3.0	36	39	2.8
T.C. 12-20-68	42	48	4.1	43	45	4.0	52	46	4.8
N.M. 3-7-69	55*	29	3.2	51*	33	3.4	61*	25	3.1

* The mean rCBF is calculated from "grey" and "white" matter water wash-out curves, weighted for the relative size of the washed-out compartments. Oxygen retention is obtained as shown in Fig. 2

Comments to Chapter I

Methodology

The first group of papers in this chapter refers to the problems of evaluation of inert gas clearance curves. The main interest concentrated on the possibilities of clinical application of the evaluation methods discussed, while purely theoretical aspects of this problem deserved less attention. Analysis of the reported material leads to 2 conclusions: 1. despite certain limitations, the isotope clearance method has become such a useful clinical tool and such a valuable complement to other methods that its place in clinical practice is assured. 2. The regional circulation of the brain, taken as a measurable parameter, bears such a clinical importance in itself that almost no attempt has been made to gain additional information from the recorded clearance curves. An example of this fact is the multicompartmental analysis, which gives no satisfactory answer to the question of local inhomogeneity of tissue flow. Thus, interpretation of the measured flow values also remains an open question whenever an inhomogeneous flow takes place and a functionally significant and well defined subdivision in various compartments is not possible. The interesting attempt of REIVICH et al. could be a first step towards the understanding of such inhomogeneities.

The papers of SVEINSDOTTIR et al. and HEISS et al. illustrate how much the answer to certain differentiated questions, as for instance the exact location of focal rCBF disturbances, may depend of technological sophistication and equipment improvement. It is beyond doubt that the time and work consuming task of evaluation of isotope clearance curves will be increasingly delegated to on-line working computers. In such cases the computer could also be employed to survey other vital parameters as blood pressure, blood gas tensions, acid-base status, etc.

A further group of authors (LÜBBERS et al., METZGER et al., SEM JACOBSEN et al.) reports rCBF measurements by means of H_2 clearance. This method offers the advantages 1. of allowing long-term studies by means of chronically implanted electrodes and 2. of further reducing the volume of tissue recorded from. However, it must be taken into account that, according to the position of the electrode tip in relation to a capillary or to a larger vessel, clearance curves of different shape may be obtained, the interpretation of which would presuppose an exact knowledge of the electrode tip position.

It has further been shown in this chapter that the use of heat clearance (MÜLLER-SCHAUENBURG, and BETZ), radioactive microspheres (MEYER et al.), non-diffusible indicators (OLDENDORF, CIFFONE et al.), as well as ultra-sound (MIYAZAKI) constitutes a valuable complement in the study of certain aspects of cerebral circulation.

Finally, the report by TER-POGOSSIAN et al. will certainly exert a substantial influence upon future research on regional brain blood flow and energy metabolism. By using radioactive ^{15}O-oxyhemoglobin, ^{15}O-carboxyhemoglobin and ^{15}O-water it now seems to have become possible to measure regional tissue oxygen consumption directly in absolute units.

H. HUTTEN
W. OLDENDORF

Influence of Hypoxia on the Response of CBF to Hypocapnia

E. Häggendal and I. Winsö

Departments of Clinical Physiology and Anesthesiology, University of Göteborg,
Sahlgrenska Sjukhuset, Göteborg

Already in 1944 Noell and Schneider [3] proposed that the vasocontricting influence of hypocapnia on the cerebral blood flow (CBF) was limited by secondary tissue hypoxia. They found that when the venous oxygen tension reached a value just below 20 mmHg there was no further influence on the cerebral arterio-venous oxygen difference in spite of more pronounced hyperventilation.

This presumed hypoxic influence generally occurs at an arterial carbon dioxide tension of about 20 mmHg and this value has later often been considered as the limiting value for the influence of hypocapnia on the CBF [1]. A hypoxic influence on the vascular reactivity to the arterial carbon dioxide tension would, however, occur at different carbon dioxide tensions depending on the arterial oxygen content. Also all metabolic factors which influence the oxygen availability to the tissue, as the hydrogen ion concentration (the Bohr effect), the temperature and so on, would affect the vascular response to hypocarbia.

The purpose of the present study was to investigate the existence and nature of an hypoxic counteraction on the cerebrovascular response to the hypocapnia during hyperventilation.

Methods

The study was performed on mongrel dogs during nitrous oxide anesthesia, muscle relaxation and artificial ventilation, which was gradually increased up to maximally 2 l/(kg body weight · min). The body temperature was kept at about 38°. After every change of the ventilation 20–30 min elapsed before measurements were performed. Both arterial and venous blood (superior sagittal sinus) was analysed for oxygen saturation (with a spectrophotometrical method) and for pH and carbon dioxide tension. Arterial samples were analysed also for the haemoglobin concentration. The oxygen tension was calculated from the oxygen saturation and the pH by use of the oxygen dissociation curve of human blood [4]. Intra-arterial pressure was recorded continuously and mean blood pressure was obtained by electric integration.

Cerebral blood flow was measured with the inert gas elimination technique using short injections of small volumes of radioactive krypton (^{85}Kr) solution into one of the vertebral arteries. The γ-elimination curve was recorded through the intact skull for at least 25 min. The fast component of the composite elimination curve was taken as representative of blood flow in the cerebral grey matter [2].

Supported by the Swedish Medical Research Council (Project K 69-14 X-2726-01 A) and Göteborgs Läkarsällskap.

Arterial hypoxia was induced by lowering the oxygen content of the inspired gas mixture. Metabolic alkalosis was induced by infusion of sodium bicarbonate. A decrease of the hydrogen ion concentration of about 0.1–0.2 pH units was generally produced.

Results and Comments

Fig. 1 shows the mean values of the blood flow of the cerebral grey matter (CBF) and the arterio-venous oxygen difference. The values are collected within limited ranges of arterial carbon dioxide tensions, namely one almost normocapnic group, one with moderate hypocapnia and 2 groups of pronounced hypocapnia with an arterial carbon dioxide tension of about 15 and 8 mmHg.

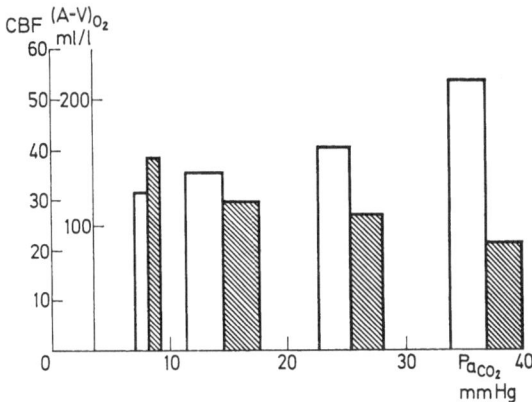

Fig. 1. Mean values of the cerebral blood flow (CBF) and the arterio-venous oxygen differences (A-V)$_{O_2}$ collected within limited ranges of arterial carbon dioxide tensions (PaCO$_2$)

The mean values show a marked decrease of the CBF from the first to the second group, then a smaller decrease between the following groups. A corresponding increase of the arterio-venous oxygen difference is seen. The resulting cerebral oxygen consumption showed no significant change. Due to the great spread of the flow and the arterio-venous oxygen difference there was in fact no significant difference between the two most hyperventilated groups.

In Fig. 2 the cerebral venous oxygen tension is correlated to the arterial carbon dioxide tension. The different values of venous oxygen tension at different arterial carbon dioxide tensions from each experimental animal are connected with each other. The figure shows a pronounced spread of the venous oxygen tension values within the "normoventilated" range. The lines are, however, convergent until an oxygen tension of just below 20 mmHg is reached. There are a few lines which at the "normoventilated" range start at a venous oxygen tension of just below 20 mmHg. These particular lines are almost horizontal. Other lines, which start at higher tension values, show a tendency to become horizontal at the same oxygen tension.

In accordance with the earlier suggestion and findings [3] the present results suggest that the cerebrovascular constriction due to very pronounced hyperventilation is counteracted by "tissue hypoxia".

In order to prove this counteraction in a more definite manner we introduced a lowered arterial oxygen content in one series of experiments by diminishing the arterial oxygen saturation to a level which at normoventilation did not interfere with the cerebral blood flow. At pronounced hyperventilation the same lowered oxygen content always resulted in an increase of the cerebral blood flow. This flow increase was generally of such a magnitude that the venous oxygen tension remained at the low oxygen tension level (15–20 mmHg). The lower the arterial oxygen content was, the less pronounced the hyperventilation was before the hypoxic counteraction of the vascular reactivity to hypocapnia occurred. Of course a corresponding effect on CBF at extreme hyperventilation is obtained by a corresponding diminishing of the oxygen content by a decreased haemoglobin concentration.

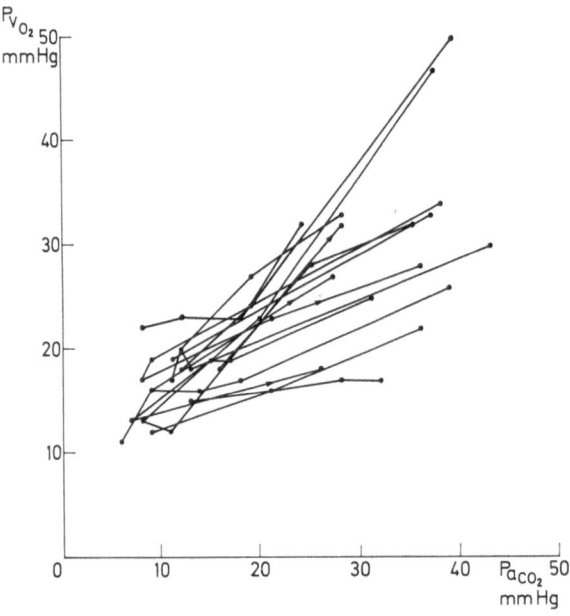

Fig. 2. Cerebral venous oxygen tension ($PvCO_2$) correlated to the arterial carbon dioxide tension ($PaCO_2$). Values connected with each other refer to the same experimental animal

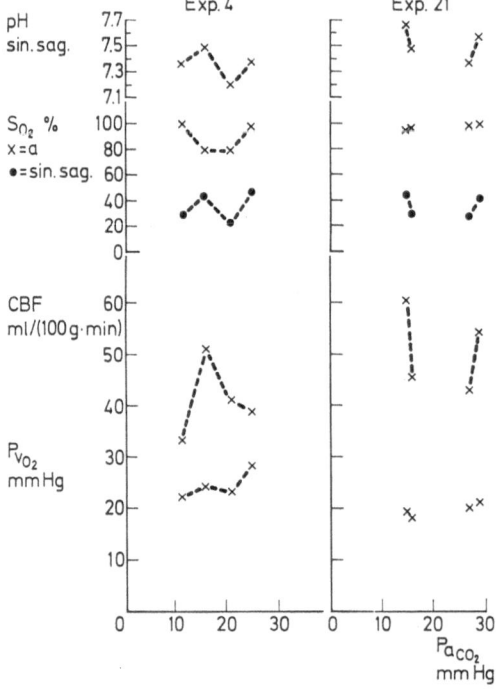

Fig. 3. Cerebral blood flow response to hypocapnia modified by metabolic alcalosis and/or arterial hypoxia

The effect of changing the oxygen availability by inducing metabolic alkalosis is illustrated in Fig. 3, where the data from two typical experiments are presented. One of these animals (Exp. 4) shows in addition the effect of a lowered arterial oxygen saturation. The hypoxia hardly influenced the CBF at a metabolically acidotic situation, measured in the cerebral venous blood, but elicited a marked flow increase, when a metabolic alkalosis was added. During these changes there was hardly any difference observed between the cerebral venous oxygen tension values. Exp. 21 (Fig. 3) illustrates the influence of metabolic pH changes. Both at $aPCO_2$ ' 15 and $aPCO_2$ ' 27 mmHg, respectively, an increase of pH of about 0,2 pH units induced an increase of the CBF of about 20 per cent with the cerebral venous oxygen tension maintained around 20 mmHg.

The effect on CBF of the induced metabolic alkalosis is summarized in Fig. 4. The correlation is given between the increase of CBF and the induced metabolic alkalosis, when cerebral venous oxygen tension was equal to or below 20 mmHg. The values are widely scattered. Part of this spread can, however, be explained by alterations of the haemoglobin concentration between the experiments, for which no corrections were performed.

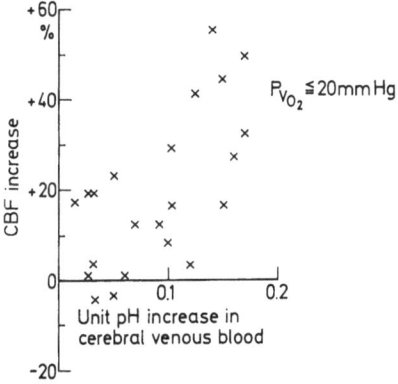

Fig. 4. Increase of CBF correlated to the increase of pH of cerebral venous blood in hyperventilated dogs with cerebral venous oxygen tension equal to or below 20 mmHg

Concluding Remarks

The present support results strongly the existence of some kind of "hypoxic" influence on the cerebrovascular response to excessive hypocapnia. This interference with a further vasoconstriction due to hypocapnia seems generally to occur independent of the actual arterial carbon dioxide tension value when the tissue hypoxia has reached a level which is represented by a cerebral venous oxygen tension around or below 20 mmHg. In the present study the CBF response to hypocapnia has been modified by introducing metabolic alkalosis and/or arterial hypoxia, thus eliciting the hypoxic counteraction.

Our results are in accordance with the findings of WOLLMAN et al. [5], who reported an increased CBF in man during excessive hyperventilation, when a metabolic alkalosis was induced. We conclude that the most likely explanation as they did of the flow increase is "hypoxic" vasodilation due to the induced change of the oxygen availability to the brain.

References

1. HARPER, A. M.: The inter-relationship between Pa CO_2 and blood pressure in the regulation of blood flow through the cerebral cortex. Acta neurol. scand. Suppl. **14**, 94 (1965).

2. Häggendal, E., N. J. Nilsson, and B. Norbäck: On the components of ^{85}Kr clearance curves from the brain of the dog. Acta physiol. scand. **66**, 5 (1965).

3. Noell, W., u. M. Schneider: Über die Durchblutung und die Sauerstoffversorgung des Gehirns. Pflügers Arch. ges. Physiol. **247**, 514 (1944).

4. Severinghaus, J. W.: Oxyhemoglobin dissociation curve correction for temperature and pH variation in human blood. J. appl. Physiol. **12**, 485 (1958).

5. Wollman, H., Th. C. Smith, G. W. Stephen, E. T. Colton, III, H. G. Gleaton, and S. C. Alexander: Effects of extremes of respiratory and metabolic alkalosis on cerebral blood flow in man. J. appl. Physiol. **24**, 60 (1968).

The Influence of Oxygen Affinity of Blood and Cerebral Blood Flow on Cerebral Oxygen Supply

J. GROTE

Physiological Institute, University of Mainz

The quantity of oxygen transported, per unit of time, by the blood to the brain, is determined by the blood flow, the oxygen capacity, and the oxygen affinity of the blood. The O_2-exchange between the blood and the tissue cells depends mainly on the oxygen transport characteristics of the blood and the O_2 diffusion conditions in the blood and tissue.

In order to be able to answer the question regarding the effect of the oxygen affinity of the blood on the cerebral tissue oxygen supply conditions, we first took a group of 50 adolescent subjects of both sexes, examined the oxygen affinity (O_2 dissociation curve) of their blood and its dependence on the CO_2 partial pressure, or pH (Bohr-effect), as well as the effect of the oxygen saturation of their blood on the acid base status (Haldane-effect), at temperatures of 28° C, 32° C, 37° C, and 40° C. The results could be combined in line graphs, which represent the correlation of the values C_{CO_2}, P_{CO_2}, pH, P_{O_2}, and S_{O_2}.

The line graph for the various respiratory gas values, shown in Fig. 1, is valid at a temperature of 37° C, for the blood of healthy subjects. If the exact values of 2 of the 5 interdependent blood specimen values are known, the relevant data for the other parameters can be derived directly from the nomogram.

If one wants to estimate the oxygen exchange between the capillary blood and the cerebral tissue, and the extent to which it is affected by the oxygen affinity of the blood, one must begin by inquiring about the oxygen pressure in the brain capillaries. The degree and form of the oxygen pressure decrease in the cerebral blood, during the passage of the capillary, are determined by the arterial O_2 pressure, the O_2 consumption of the tissue, the cerebral blood flow, as well as the O_2 capacity and O_2 affinity of the blood.

If the O_2 consumption, the CO_2 output, the blood flow value, and the respiratory gas partial pressures in the arterial blood are known, it is possible, with the aid of the nomograms, to calculate the O_2 pressure decrease in an assumed average brain capillary, and to analyse the effect of the oxygen affinity of the blood on the oxygen supply conditions of cerebral tissue.

According to our calculations, the mean O_2 pressure profiles, shown in Fig. 2, are present in the cerebral cortex capillaries, under normal conditions and conditions of respiratory alkalosis and acidosis.

Curve 1 shows the O_2 pressure decrease in a cerebral cortex capillary under conditions of $Pa_{CO_2} = 94$ mmHg, and $Pa_{CO_2} = 40$ mmHg. The oxygen consumption and the blood flow were assumed to be normal.

Curve 3 represents the O_2 pressure changes to be expected in the blood, during its passage through the cerebral cortex capillaries, under arterial CO_2 pressure conditions of 55 mmHg, if the changes in the blood flow [1, 2, 3], and the oxygen affinity of the blood, expected under the assumed conditions, are taken into consideration in the calculations. If only the changes in the oxygen affinity of the blood are taken into account, we then get the O_2 pressure course shown by curve 2. The figure clearly shows that under non-compensated respiratory acidosis the conditions for the oxygen supply of the cerebral tissue can be improved through the change in the oxygen affinity of the blood.

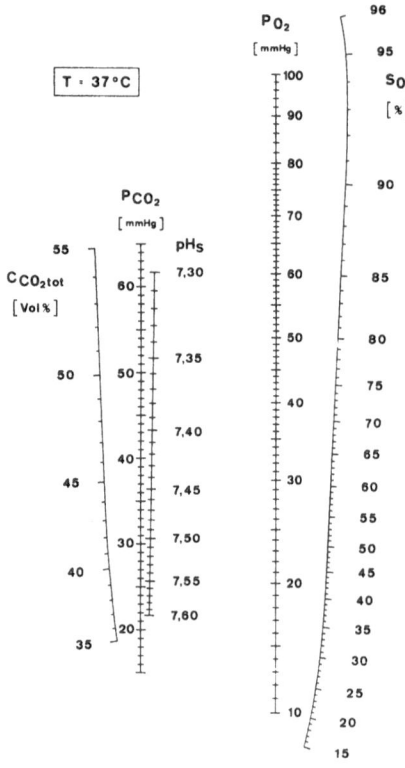

Fig. 1. Nomogram for the respiratory gas values and the acid base status for the blood of healthy subjects. $C_{CO_2 tot}$ = total CO_2 concentration, P_{CO_2} = CO_2 pressure, pH_s = pH value, P_{O_2} = O_2 pressure, S_{O_2} = percentage O_2 saturation, t = 37° C. The points of intersection of any straight line give the correlated values

If the O_2 pressure decrease in the cerebral cortex capillaries is calculated under non-compensated alkalosis conditions, we then get for an assumed CO_2 pressure of 30 mmHg in the arterial blood, the O_2 pressure decrease within the cerebral cortex, as shown by the dotted line (curve 5).

Under these conditions the decrease in the cerebral blood flow and the increase of the blood oxygen affinity lead to such an oxygen pressure decrease in the cerebral cortex capillaries that a deficiency in the oxygen supply of the cortical tissue must be expected, as well as the counter regulation which it triggers. The O_2- [3, 4, 5, 6, 7] and CO_2-dependent blood flow regulation under these conditions is taken into account.

The O_2 partial pressure profile that would be expected in the cerebral cortex capillaries, if only the changes in the oxygen affinity of the blood are taken into account is shown by curve 4.

Because a comparison of the theoretically obtained values for specific gas exchange situations in the human cerebral cortex capillaries with directly determined values was, at first, only possible in a few cases, comparative examinations were carried out on dogs. The respiratory gas partial pressures measured in the cerebral venous blood were compared with partial pressure values calculated for the same conditions. As in the comparative tests on humans, the theoretically-determined values and the directly measured data closely agreed in all the experiments.

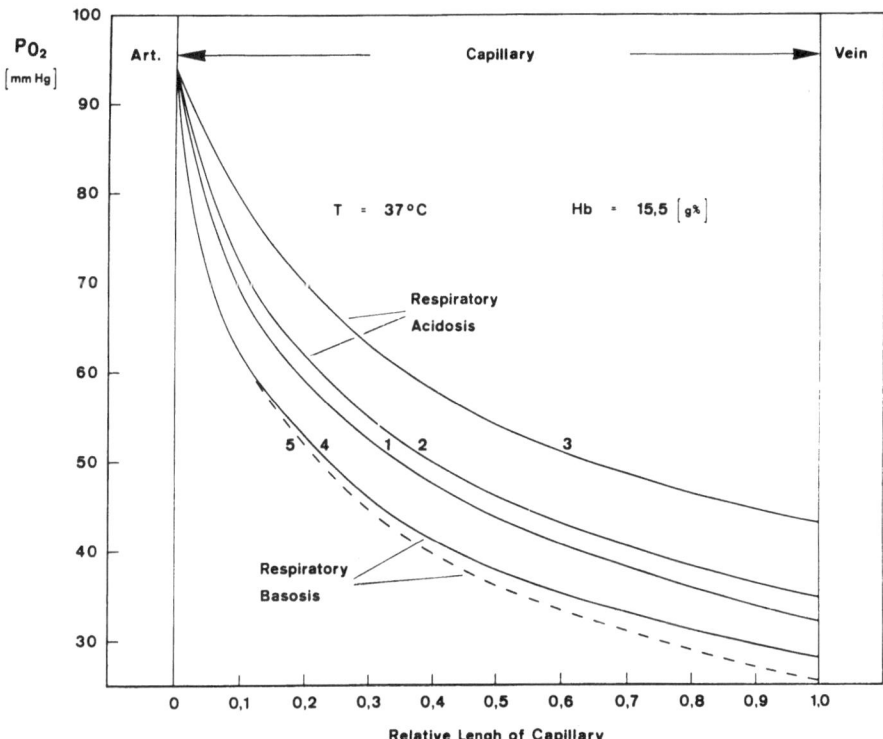

Fig. 2. Mean O_2 partial pressure decrease in the cerebral cortex capillaries, calculated for the following conditions: Curve 1, normal conditions, (Pa_{CO_2} = 40 mmHg). Curve 2, non-compensated respiratory acidosis, (Pa_{CO_2} = 55 mmHg), assuming a constant cerebral blood flow of normal value. Curve 3, non-compensated respiratory acidosis, (Pa_{CO_2} = 55 mmHg), taking into account the CO_2 and O_2 pressure-dependant cerebral blood flow regulation. Curve 4, non-compensated respiratory alkalosis, (Pa_{CO_2} = 30 mm Hg), assuming a constant cerebral blood flow of normal value. Curve 5, non-compensated respiratory alkalosis (Pa_{CO_2} = 30 mmHg), taking into account the CO_2 and O_2 pressure-dependent cerebral blood flow regulation. Abscissa: relative capillary length. Ordinate: oxygen pressure in mmHg

Because we are well acquainted with the boundary values of the O_2 pressure in the mixed cerebral venous blood, which, when fallen short of, cause an oxygen supply deficiency to occur, a calculation like the present one, of the mean O_2 pressure changes in the cerebral blood can provide important information for estimating the oxygen supply of the brain.

References

1. ALEXANDER, S. C., H. WOLLMAN, P. J. COHEN, P. E. CHASE, and M. BEHAR: Cerebrovascular response to Pa_{CO_2} during halothane anesthesia in man. J. appl. Physiol. 19, 561 (1964).
2. HARPER, A. M.: The inter-relationship between Pa_{CO_2} and blood pressure in the regulation of blood flow through the cerebral cortex. Acta neurol. scand. Suppl. 14, 94 (1965).
3. KREUSCHER, H., and J. GROTE: Effects of hyper- and hypoventilation on CBF under anesthesia. Presented to the International CBF Symposium, Mainz, April 1969 (this volume, p. 244).

4. Lambertsen, C. J.: In: V. A. Drill: Pharmacology in medicine, p. 820. New York: McGraw-Hill 1958.

5. McDowall, D. G.: Interrelationships between blood oxygen tensions and cerebral blood flow. In: A symposium on oxygen measurements in blood and tissues and their significance. London: Churchill 1966.

6. Noell, W., u. M. Schneider: Über die Durchblutung und die Sauerstoffversorgung des Gehirns im akuten Sauerstoffmangel. III. Die arteriovenöse Sauerstoff- und Kohlensäuredifferenz. Pflügers Arch. ges. Physiol. 246, 207 (1942).

7. — — Über die Durchblutung und Sauerstoffversorgung des Gehirns. IV. Die Rolle der Kohlensäure. Pflügers Arch. ges. Physiol. 250, 514 (1944).

Lack of Adaptation of CBF and CSF pH in Hypoxic Hypercapnia

A. Agnoli, N. Battistini, M. Nardini, S. Passero, and C. Fieschi

Department of Neurology and Psychiatry, University of Genoa

Previously [1] we reported that during chronic normoxic respiratory acidosis, lasting 8–11 h, the CSF pH tends to return toward normal values despite the persistence of acidosis in the blood. At the same time the CBF, which increased at the beginning, returned to the initial values. This tendency of the CBF and of the CSF pH to adapt has been repeatedly observed in chronic experimental conditions [3, 4, 5, 6, 8].

Acclimatization, however, is not present in severe cases of respiratory acidosis with diffuse encephalopathy observed in human pathology [7]. Only one factor differentiates experimental chronic hypercapnia from human pathology, namely the presence in the latter of hypoxia. We therefore decided to test the hypothesis that hypoxia is the responsible factor for the lack of the adaptation of CSF pH and CBF. In the present report we present a study of the acid-base status of the CSF and of arterial blood and to the flow of the caudate nucleus during hypoxic respiratory acidosis.

Experimental Procedure and Methods

Experiments were carried-out on 6 adult cats weighing 1.8–2.5 kg, immobilized with a single dose of succinylcholine, rapidly tracheostomized, and mechanically ventilated (24 breaths/ min; respiratory volume 1.5–2.5 l/min) with an anesthetic mixture (N_2/O_2 60/40 per cent). Muscular relaxation was maintained with Gallamine.

The femoral artery and vein were cannulated, and a thin needle was inserted into the suboccipital cisterna. Electrodes for measuring CBF with the H_2 clearance technique were implanted into the caudate nucleus by stereotaxic technique. 15–20 min after the completion of the surgical procedure, the acid-base status of arterial blood and of CSF was assessed. The CBF was also measured. At this time the hypoxic hypercapnia started reaching values of $aPCO_2$ and aPO_2 of between 40–45 mmHg.

Samples of arterial blood, CSF and rCBF determination were taken at 30′, 180′, 360′, 480′, 660′ after the beginning of the hypoxic-hypercapnia. The EEG, MABP, end-tidal CO_2 and rectal temperature were continuously monitored.

Results

The MABP remained constant around the control value; rectal temperature was kept at the 37° level. The EEG pattern showed slight slowing of dominant frequencies 6–8 h after the induction of respiratory acidosis and returned to normal at the end of the experiment.

Supported by US-PHS Grant No. 05017.

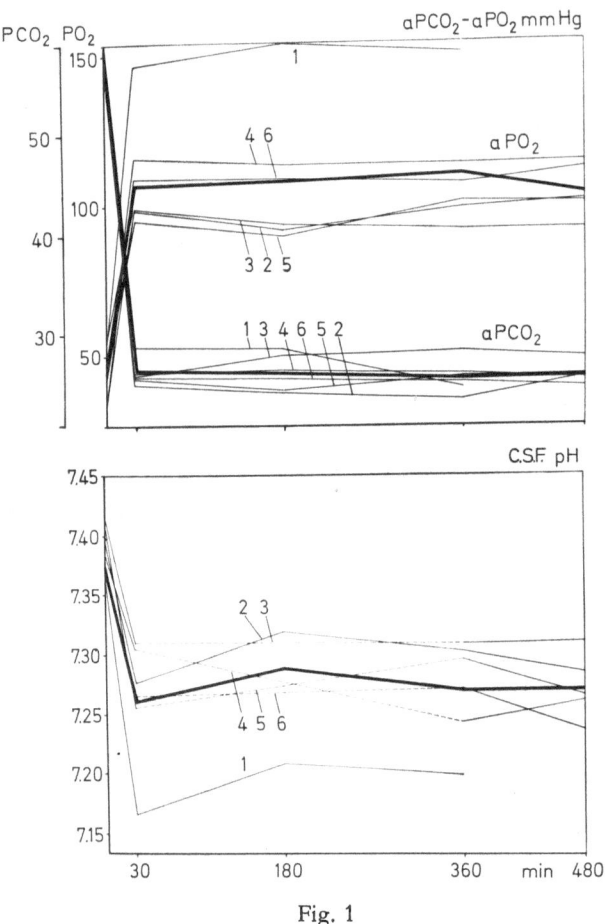

Fig. 1

The aPCO$_2$ and aPO$_2$ were quite stable around 45 mmHg. The CSF pH remained stable at the acidotic level (Fig. 1). The CSF bicarbonate, after a slight increase during the first hours, was practically constant. The CBF was constant at high levels throughout the experiment, showing a mean increase of 160% over the resting value (Fig. 2).

Comment

The evidence reported so far suggests that CSF pH does not adapt to chronic hypoxic hypercapnia, and also that the CBF does not show adaptation.

Our data confirm that, in this experiment, a strict correlation between CSF pH and CBF is present, stressing the same correlation observed in the normoxic hypercapnia.

The responsible factor for the absence of acclimatization seems to be the impairment of the transfer of [HCO$_3^-$] through the blood brain barrier.

Concerning the role of hypoxia as the interfering factor in the process of adaptation, it has been shown that neither hypoxia nor hypercapnia alone modifies the cerebral uptake of either sodium (^{24}Na) or phosphate (^{32}P), while the combination of the 2 factors increases these uptakes and modifies their distribution [2]. If we consider the derangement in the compartmental distribution of sodium and phosphate, such an effect can perhaps be extrapolated to the [HCO$_3^-$] as well, explaining the lack of adaptation of CSF pH and consequently of the CBF.

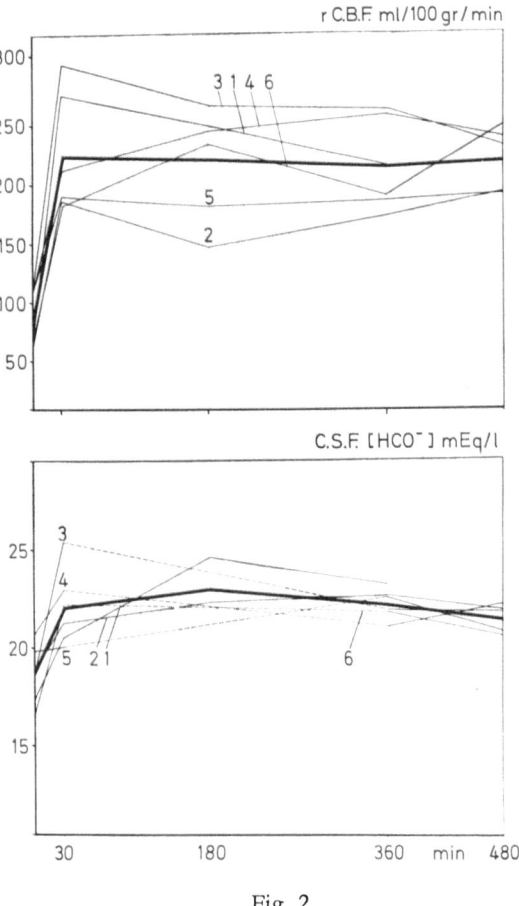

Fig. 2

References

1. AGNOLI, A.: Adaptation of CBF during induced chronic normoxic respiratory acidosis. Scand. J. lab. clin. Invest. Suppl. **102**, p. VIII:D (1968).

2. BAKAY, L., and C. LEE: The effect of acute hypoxia and of hypercapnia on the ultrastructure of the Central Nervous System. Brain **91**, 697 (1968).

3. BETZ, E., and D. HEUSER: Cerebral cortical blood flow during changes of acid-base equilibrium of the brain. J. appl. Physiol. **23**, 726 (1967).

4. BLEICH, G. L., P. M. BERKMANN, and W. B. SCHWARTZ: The response of cerebrospinal fluid composition to sustained hypercapnia. J. clin. Invest. **43**, 11 (1964).

5. MARSHALL, B. E., H. E. GLEATON, M. HEDDEN, A. AGNOLI, and S. C. ALEXANDER: The adaptation of cerebral blood flow during sustained hypocarbia in goats: the role of CSF pH and cerebral metabolism. Circulat. Res., In preparation.

6. PONTEN, U.: Consecutive acid-base change in blood brain tissue and cerebrospinal fluid during respiratory acidosis and baseosis. Acta neurol. Scand. **42**, 455 (1966).

7. POSNER, J. B., A. G. SWANSON, and F. PLUM: Acid-base balance in cerebrospinal fluid. Arch. Neurol. **12**, 479 (1965).

8. SWANSON, A. G., and A. ROSENGREEN: Cerebrospinal fluid buffering during acute experimental respiratory acidosis. J. appl. Physiol. **17**, 812 (1962).

Cerebral Blood Flow in Patients with Chronic Respiratory Insufficiency, with Special Regard to Induced Acute Changes of the Blood Gas Situation

B. Ekström-Jodal and E. Häggendal

University Lung Clinic, Renströmska Sjukhuset, Göteborg, and the Department
of Clinical Physiology, University of Göteborg

Convincing evidence of a tendency to normalization of the cerebral blood flow (CBF) in patients with chronic hypercapnia and simultaneous normalization of the pH of the cerebrospinal fluid (CSF) were presented by Skinhøj [10] at the International Symposium on "CSF and CBF" in Lund and Copenhagen in 1968. If the resistance of the cerebral blood vessels were mainly regulated by the CSF pH it seems natural that CBF is normalized parallel to the adaptation of the CSF (or more likely the extracellular fluid) pH to the altered arterial carbon dioxide tension [2, 3, 8, 9].

The purpose of our study was to see whether there was a normalization of the CBF in patients with chronic respiratory insufficiency, and if any correlation could be found between the CBF and CSF pH. Furthermore we studied the effect of acute spontaneous or induced changes of the blood gas situation.

Methods

Subjects : 18 patients of both sexes with an average age of 61 years and with chronic respiratory insufficiency were investigated. No signs of mental disturbances attributable to the hyperventilation were noted. Most of the patients were examined in a "steady state" after at least 1 week of practically unchanged blood gases. Two of them were studied during several consecutive days after an acute exacerbation of the disease. Furthermore 1 patient with normal blood gases was added to the material.

Experimental Design

After the catheterization of the internal jugular vein and of a peripheral artery the patient rested for 15 min before the first flow measurement was performed, before and during which simultaneous arterial and cerebral venous blood samples were collected for blood gas analyses. Then a lumbar puncture was made and a sample of CSF was taken together with arterial and venous blood samples. Afterwards new blood samples were taken after 15 min of changed ventilation. During this period the patient breathed a gas mixture containing five per cent carbon dioxide or hyperventilated. In some cases it was controlled that the blood gases were

Supported by the Faculty of Medicine, University of Göteborg and by the Swedish Medical Research Council (Project K 69-14 X 2726-01 A).

constant during the last 5–7 min of the period. Between the different procedures there was always an interval of at least half an hour.

Measurements

CBF was measured with the inert gas inhalation technique using slight modifications of the original KETY technique. The sampling period was 13 min. In order to keep the blood loss within acceptable limits (each measurement required about 120 ml) and also because it was hard to keep the changed blood gas situation constant during the indicator inhalation, CBF was then calculated from the arteriovenous oxygen difference. The cerebral oxygen consumption ($CMRO_2$) has been demonstrated to be constant in spite of great variations of the arterial carbon dioxide tension [5]. Oxygen saturation and hemoglobin concentration were measured spectrophotometrically. Blood and CSF carbon dioxide tension were determined from pH measurements with a microelectrode technique. The CSF pH measurements and bicarbonate content calculations were performed according to SEVERINGHAUS [7]. Intra-arterial blood pressures were recorded during the examination.

Results

Determination of CBF was performed in 16 patients. The difference between the measured CBF values and values calculated from the normal $CMRO_2$ in this age group, 3.15 ml/ (100 g min), amounted to 1.3 ± 5.14 ml/(100 g.min). This normal value (3.15) for $CMRO_2$ was used in the following results for calculation of CBF from the measured arteriovenous oxygen differences. Fig. 1 consists of a correlation between the calculated CBF of all the subjects and the arterial carbon dioxide tension. All values obtained from 1 patient during the

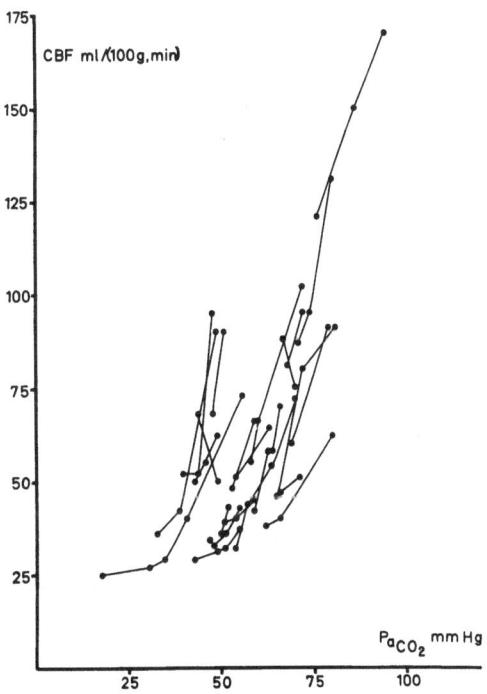

Fig. 1. Correlation lines between arterial carbon dioxide tension and CBF in all patients where the arterial carbon dioxide tension was changed

same day are connected. There is a clear tendency to 2 groups of curves, where 1 group
has considerably lower flow values at about the same values of arterial carbon dioxide tension
in the range from 45–55 mmHg than the other. The former group was studied at very pronounced hypercapnia where some remarkably high flow values were obtained. The figure
shows that the 2 groups of correlation lines seem to be parallel to each other.

The arterial blood pressure was maintained almost constant in each patient throughout the
study. In a few cases, where small changes occurred, arteriovenous oxygen differences were
measured at the different blood pressures without signs of impaired cerebral blood flow autoregulation.

At all levels of arterial carbon dioxide tension there were patients with arterial hypoxia
but in no case was the cerebral venous oxygen tension below 27 mmHg. In a separate study
on the reaction to lowered oxygen tension in patients with high arterial carbon dioxide
tension CBF was not increased until the arterial oxygen tension was below 45 mmHg.

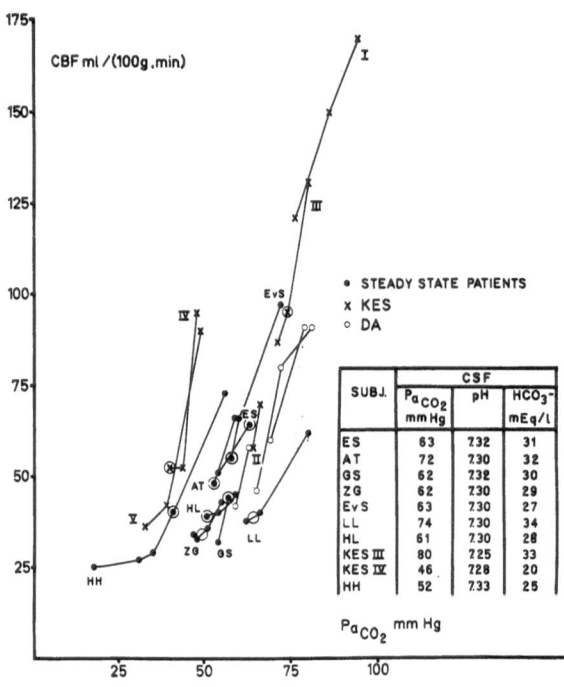

Fig. 2. Correlation lines between arterial carbon dioxide tension and CBF in 2 patients studied during
several days (KES and DA) and some "steady state" patients (see text). The carbon dioxide tension, pH
and bicarbonate content of CSF are listed. Large circles indicate where the CSF samples were taken

Fig. 2 illustrates the corresponding curves from 2 patients, in whom consecutive measurements were performed during several days. In addition some typical patients in a steady state
concerning the blood gas situation are presented. The CSF pH, carbon dioxide tension, and
bicarbonate values are listed. One patient (KES) was investigated 3 times during the first 4 days
after admission to hospital. At the beginning he had a partly compensated respiratory acidosis,
which on the 4th day was fully compensated. The CSF pH was then low (pH = 7.25) with
high bicarbonate content. No shift of the curve between arterial carbon dioxide tension and
CBF was noted during those days (KES I–III), where some very high flow values were obtained. Gradually he developed an alveolar hyperventilation and during some days before a new
measurement was made (KES IV) the arterial carbon dioxide tension had been around

30 mmHg. This is probably the reason why the CSF pH then was somewhat low whereas the arterial carbon dioxide tension was normal. One week later another CBF measurement was performed (KES V), when a somewhat lower CBF value was found. Between the correlation lines KES I–III and KES IV–V there is noted a clear shift to the left i.e. higher flow at lower carbon dioxide tension.

In the other patient (DA), who was studied during the first 3 days after arrival, the CBF values were also fairly high without significant shift of the relation curve between arterial carbon dioxide tension and CBF. He had a normal arterial pH during the whole study, except during the induced changes of carbon dioxide tension. No CSF samples were obtained.

Discussion and General Conclusion

Our results confirm earlier findings of a normalization of CBF in conditions with long-standing moderate hypercapnia and, parallel to that, a normalization of the pH of the CSF [10]. At pronounced hypercapnia there was a clear tendency to supranormal CBF although the scatter of the calculated values was large. 1 patient (KES) had extremely high CBF in combination with low CSF pH after at least 4 days of pronounced hypercapnia during which time his blood pH had normalized completely. No data regarding the time required for adaptation of extracellular pH to hypercapnia are available in man, but results from animal experiments indicate that it should have been completed in the time mentioned [1]. Also from studies in man with prolonged hypocapnia this could be expected [8]. The present results speak then in favor of an incomplete capacity for adaptation to very high arterial carbon dioxide tensions, as has been demonstrated earlier [4, 10].

The CBF reaction to an acute change of the arterial carbon dioxide tension was of the same magnitude as that in normal subjects. This means that the CBF closely follows the alterations of the blood gas situation that may occur in a short time in patients with chronic respiratory insufficiency.

References

1. AGNOLI, A.: Adaptation of CBF during induced chronic normoxic respiratory acidosis. Scand. J. clin. Lab. Invest. **22**, Suppl. 102, p. VIII:D (1968).
2. BETZ, E., u. R. KOZAK: Der Einfluß der Wasserstoffionenkonzentration der Gehirnrinde auf die Regulation der cortikalen Durchblutung. Pflügers Arch. ges. Physiol. **293**, 56 (1967).
3. FENCL, V. J., R. VALE, and J. R. BROCH: Cerebral blood flow and pulmonary ventilation in metabolic acidosis and alkalosis. Scand. J. clin. Lab. Invest. **22**, Suppl. **102**, p. VIII:B (1968).
4. HUANG, C. T., and H. A. LYONS: The maintenance of acid-base balance between cerebrospinal fluid and arterial blood in patients with chronic respiratory disorders. Clin. Sci. **31**, 273 (1966).
5. KETY, S. S., and C. F. SCHMIDT: The effects of altered arterial tensions of carbon dioxide and oxygen on cerebral blood flow and cerebral oxygen consumption of normal young men. J. clin. Invest. **27**, 484 (1948).
6. LASSEN, N. A.: Cerebral blood flow and oxygen consumption in man determined by the inert gas diffusion method. Christtreus bogtrykkeri, 1958.
7. SEVERINGHAUS, J. W., R. A. MITCHELL, B. W. RICHARDSON, and M. M. SINGER: Respiratory control at high altitude suggesting active transport regulation of CSF pH. J. appl. Physiol. **18**, 1155 (1963).
8. —, H. CHIODI, E. J. EGER, II, B. BRANDSTATER, and T. F. HORNBEIN: Cerebral blood flow in man at high altitude. Role of cerebrospinal fluid pH in normalization of flow in chronic hypocapnia. Circulat. Res. **19**, 274 (1966).
9. SKINHØJ, E.: Regulation of cerebral blood flow as a single function of the interstitial pH in the brain. Acta neurol. Scand. **42**, 604 (1966).
10. — CBF adaptation to chronic hypo- and hypercapnia and its relation to CSF pH. Scand. J. clin. Lab. Invest. **22**, Suppl. **102**, p. VIII:A (1968).

Arterial pCO$_2$ and Blood Flow in Different Parts of the Central Nervous System of the Anesthetized Cat

H.-W. Flohr, M. Brock, R. Christ, R. Heipertz, and W. Pöll

Physiological Institute, University of Bonn, and Neurosurgical Department,
University of Mainz

It cannot be regarded as proven that the mechanisms involved in the regulation of local circulation are identical in all parts of the central nervous system. Evidence has been presented that carbon dioxide exerts differential effects in different parts of the CNS [2, 3, 7, 10].

This problem has been studied in a series of cats under nembutal anesthesia using the particle distribution method. This technique permits simultaneous quantitative determinations of flow rates in various parts of the CNS. Details of the method have been described elsewhere (Flohr, Neutze et al.). Its application for the measurement of CNS blood flow was studied by Tschetter et al. and Flohr et al [6, 12].

Methods

The present study was performed in 15 cats anesthetized with nembutal (35 mg/kg body weight intraperitoneally). After tracheotomy the animals were curarized and ventilated with a Starling pump. Different arterial pCO$_2$ levels between 20–40 mmHg were induced by varying the ventilation rate and tidal volume. In the hypoventilated group arterial oxygen tensions were kept above 85 mmHg by adding oxygen. Arterial blood pressures (Statham pressure-gauge transducer) and end-expiratory CO$_2$ (infrared analyser) were continuously recorded. Arterial pO$_2$, pCO$_2$ and pH were determined according to the micromethod of Astrup. Cardiac output was measured by the thermodilution technique.

Blood flow determination was performed at steady state. A calibrated amount of ^{131}I-labelled macroaggregated albumin (50 μC/kg body weight) was injected by direct puncture of the left ventricle through the incised thoracic wall. The particles used had a diameter of 10–50 μ; recirculation of the indicator was less than 5%. In the amounts used no hemodynamic effects were detected.

4 min after the injection the animals were sacrificed by a intravenous injection of KCl. The CNS was removed and the following parts were dissected after removal of the dura: prosencephalon, cerebellum, brain stem, cervical cord, thoracic cord, lumbosacral cord.

The radioactivity found in each segment was estimated using a well-type scintillation counter. The amount of indicator found in each part is an index of the fraction of the cardiac output passing to that region. The absolute flow to each region is obtained from local indicator concentration, the total amount of the isotope injected and cardiac output at the time of injection:

$$Fi = \frac{CO}{Ao} \cdot Ac \cdot 100$$

Fi = regional flow (ml/100 g min)
CO = cardiac output (ml/min)
Ao = total amount of activity injected (μC)
Ac = regional activity (μC/g tissue)

Results

1. At normal arterial pCO_2, pO_2 and pH levels for the cat [1, 4, 11], blood flow through prosencephalon, cerebellum, brain stem, cervical, thoracic and lumbosacral cord were 51.3, 57.8, 39.3, 20.3, 16.5 and 23.7 ml/100 g/min, respectively.

2. For all parts studied a significant correlation between flow and arterial pCO_2 was found within the range studied, as had been previously suggested by others [7].

3. A plot of pCO_2 versus blood flow (Fig. 1) suggested that these 2 parameters showed a linear correlation within the range studied. This assumption was borne out by computer analysis.

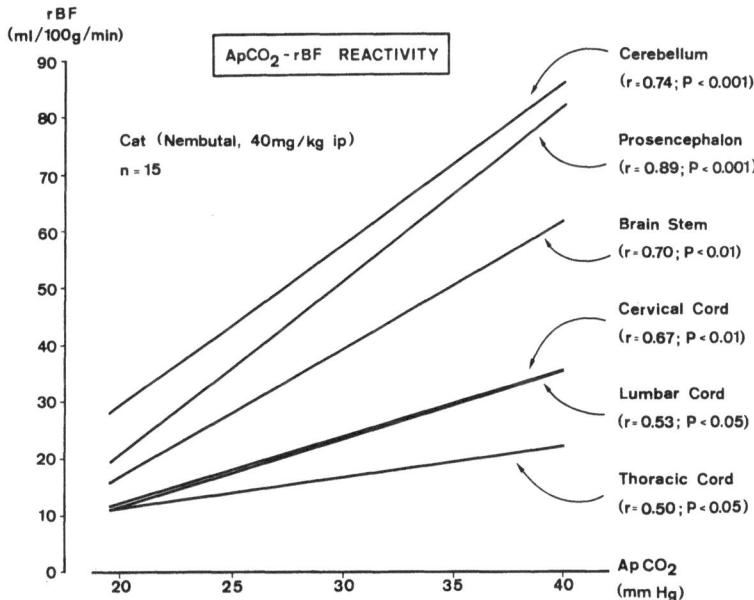

Fig. 1. ApCO₂-reactivity of CBF in various parts of the CNS of the cat

The formulae describing the regressions were:

$y = 3.05 \quad x - 40.1$ (prosencephalon)
$y = 2.84 \quad x - 27.3$ (cerebellum)
$y = 2.26 \quad x - 28.6$ (brain stem)
$y = 0.91 \quad x - \ 6.9$ (cervical cord)
$y = 0.54 \quad x + \ 0.3$ (thoracic cord)
$y = 1.17 \quad x - 11.5$ (lumbosacral cord)

4. The regression coefficients were significantly different between prosencephalon and brain stem, prosencephalon and spinal cord, cerebellum and brain stem, cerebellum and spinal cord, brain stem and spinal cord.

5. The absolute sensitivity of the various parts of the CNS in ml/100 g. min/mmHg CO_2 tension were 3.05 for the prosencephalon, 2.84 for the cerebellum, 2.26 for the brain stem, 1.17 for the eumbosacral cord, 0.91 for the cervical cord and 0.54 for the thoracic cord. These values seem to be directly related to the blood flow values of the corresponding parts.

Summary and Conclusions

Our data indicate that carbon dioxide is an essential factor in circulatory control at all levels of the CNS. This is in accordance with the findings of Kety and Schmidt, and Palleske [9]. There are however quantitative differences in the effects exerted by carbon dioxide on absolute flow at various CNS levels. Apparently $dF/dpCO_2$ art. of a given structure is related to its blood flow and $CMRO_2$ under normal conditions.

References

1. Brock, M.: Unpublished data, 1968.
2. Capon, A.: Les régulations vasculaires dans la moelle épinière. Acta neurol. belg. 61, 227 (1961).
3. Field, E. J., J. Grayson, and A. F. Rogers: Observations on the blood flow in the spinal cord of the rabbit. J. Physiol. 114, 56 (1951).
4. Fink, B. R., and M. Schoolman: Arterial blood acid-base balance in unrestrained waking cats. Fed. Proc. 21, 440 (1962).
5. Flohr, H. W.: Methode zur Messung regionaler Durchblutungsgrößen mit radioaktiv markierten Partikeln. Europ. J. Physiol. 302, 268 (1968).
6. —, M. Brock, R. Heipertz, R. Christ u. A. Hadjidimos: Quantitative Bestimmung regionaler Durchblutungsgrößen des Zentralnervensystems mit radioaktiv markierten Partikeln. In: Radioisotope in Pharmakokinetik und klinischer Biochemie. In Press, 1969.
7. Freygang, H. W., Jr., and L. Sokoloff: Quantitative measurement of regional circulation in the central nervous system by the use of radioactive inert gas. In: Advances in Biological and Medical Physics, vol. VI. p. 263. New York: Academic Press 1968.
8. Neutze, J. M., F. Wyler, and A. M. Rudolph: Use of radioactive microspheres to assess distribution of cardiac output in rabbits. Amer. J. Physiol. 214, 486 (1968).
9. Palleske, H., and H. D. Herrmann: Experimental investigations on the regulation of the blood flow of the spinal cord. Acta neurochir. 19, 73 (1968).
10. Schmidt, C. F.: The intrinsic regulation of the circulation in the hypothalamus of the cat. Amer. J. Physiol. 110, 137 (1934).
11. Sørensen, S. C.: Arterial PCO_2 in cats calculated from gas tensions in subcutaneous pockets. Resp. Physiol. 3, 261 (1967).
12. Tschetter, T., M. Meyer, A. Klassen, and J. Resch: Regional cerebral blood flow using labelled microspheres. Proc. Internat. Union Physiol. Sc. Washington DC 1968. Vol. VII, Abstr. 1321, 441 (1968).

Changes of the Transmural Pressure – the Probable Stimulus to Cerebral Blood Flow Autoregulation

B. Ekström-Jodal, E. Häggendal, N. J. Nilsson, and B. Norbäck

Departments of Clinical Physiology I and II, Anesthesiology II, and Surgery III,
University of Göteborg

At the International Symposium on "CSF and CBF" in Lund and Copenhagen in 1968 one of the main topics concerned the mechanisms underlying the cerebral blood flow autoregulation. The results and discussions were considered to a large extent to support the opinion that metabolic mechanisms play an important role in the normal cerebral blood flow autoregulation [2].

The metabolic theory about the autoregulation, so attractive at first sight, meets, however, with considerable difficulties when analysed in detail. The constancy of flow at marked changes of blood pressure on different flow levels, e. g. during hypo- or hypercapnia [3] represents an important obstacle. It implies that the regulation would have an extraordinarily high sensitivity to metabolic changes retained at different levels. Furthermore previous results have demonstrated cerebrovascular reactivity to carbon dioxide during abolished autoregulation [3] which seems to exclude carbon dioxide or secondary chemical factors (pH) as the main cause of the flow autoregulation.

This paper presents some experimental approaches to the understanding of the mechanisms behind the normal cerebral flow autoregulation.

Methods

The experiments were performed in mongrel dogs anesthetized with pentobarbital and artificially ventilated. Arterial and cerebral venous blood from the superior sagittal sinus was at every flow measurement analysed for pH, carbon dioxide tension and oxygen saturation. The hemoglobin concentration was measured on the arterial samples.

Intraluminal blood pressures were recorded during the flow measurements and mean blood pressures were obtained by electrical integration.

Cerebral blood flow was determined according to the inert gas elimination principle using short injections of small amounts of radioactive Krypton into the intact vertebral artery and external recording of the γ-activity. The flow values given are those for cerebral grey matter. Methodological details are given in an earlier paper [4].

Supported by the Medical Faculty of the University of Göteborg and by the Swedish Medical Research Council (Projects B 69-14 X-2683-601 and K 69-14 X-2726-01 A).

Studies on the Vascular Reactivity to Altered Blood Gas Situation

By ventilation with different gas mixtures, low in oxygen and/or high in carbon dioxide content, different blood gas situations were established. The arterial blood pressure was changed by bleeding and reinfusion of the blood. No measurements were performed until 20 min after any change of the gas situation or 5 min after alterations of the blood pressure.

Results and Comments

1. Reactivity to Carbon Dioxide

In a previous report various cerebrovascular resistance values were found corresponding to the carbon dioxide tension, when the autoregulation was abolished by severe arterial hypoxia [3]. Those results were obtained from different animals. The present experiments in which animals were made hypotensive beyond the autoregulatory range show that the change in the cerebrovascular resistance was about the same when the arterial carbon dioxide tension was altered as in the normotensive autoregulated states.

2. Reactivity to Induced Arterial Hypoxia

It has been demonstrated several times that arterial hypoxia abolishes the cerebral blood flow autoregulation and it is also well known that arterial hypoxia increases the blood flow. In some experiments we have studied the pressure-flow relationship at different hypoxic levels in order to determine if the flow increase starts before autoregulation is impaired. The general result from these studies was that the flow increase occurred earlier at a gradually increased hypoxia than the abolishment of autoregulation. Thus in one experiment it was possible during marked arterial hypoxia (40 per cent arterial oxygen saturation) to reduce the perfusion pressure from 133 to 82 mmHg with measured flow values of 77 and 73 ml/(100 g min), respectively.

The effect of arterial hypoxia was also studied during pronounced hypotension, i.e. passive pressure-flow relationship. In accordance with the corresponding experiments on the carbon dioxide effect, there was a clear vasodilatation induced by the hypoxia, during hypotension and this was of the same order of magnitude as during normotension and functioning autoregulation.

Studies on Reactive Hyperemia after Previous Arterial Hypotension

Also the reactive hyperemia, i.e. high flow values due to impaired autoregulation combined with normal or high perfusion pressures have generally been considered as caused by metabolic factors [6] and recently even a specific metabolic factor, lactic acid, has been described as a probable cause of the reactive hyperemia [7]. In an earlier paper [5] we have noted that reactive hyperemia could be elicited without a previous period with low cerebral venous oxygen saturation during the perfusion pressure reduction.

Results and Comments

In the present experiments we have been able to produce reactive hyperemia in hypercapnic animals where the flow during the pressure decrease was never reduced below the normal flow value. In these experiments it seems unlikely that "hypoxia" or other metabolic factors could elicit the hyperemia. A hypothetic objection that carbon dioxide might lower the hypoxic threshold was tested by another type of experiment, where the flow was increased with papaverine infusions. In a typical experiment (Fig. 1) we found a reactive hyperemia after a period of flow decrease from about 300–188 ml/(100 g min).

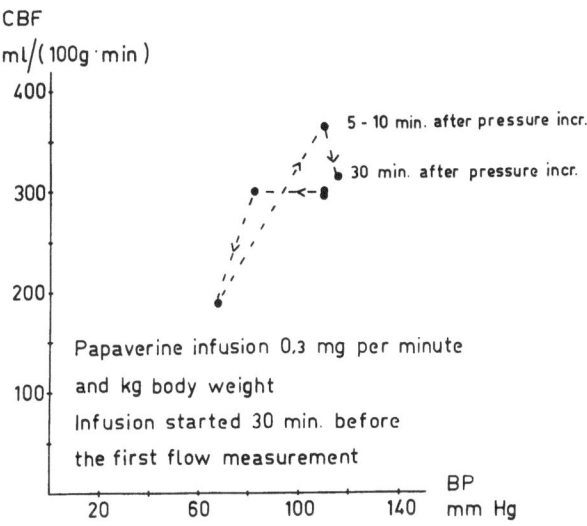

Fig. 1. The figure illustrates an experiment where reactive hyperemia was produced during papaverine infusion and the previous hypotension (67 mmHg) had reduced CBF to 188 ml/(100 g min). The figure also exemplifies autoregulation on a very high flow level

From these results it could be suggested that this type of reactive hyperemia or impaired autoregulation is due to the smooth muscle cells responsible for the hypotensive vasodilatation being incapable of very fast constriction at pressure increments after a period of maximum dilatation.

Differentiation Between Transmural and Perfusion Pressure Changes

Besides the metabolic explanation of autoregulation the myogenic theory is often discussed in relation to the situation where the autoregulating stimulus is a change of transmural pressure in the resistance vessels. An experimental dissociation between changes in transmural and perfusion pressures would therefore be expected to help clarifying the mechanisms. With an unchanged transmural pressure and an induced change of the perfusion pressure a mainly metabolic mechanism would establish autoregulation while a myogenic one would fail. The experimental approach was to increase the venous pressure with the brain surface exposed permitting expansion when the venous pressure was raised thereby minimizing the increase of tissue tension. The skull and dura were removed over most parts of both hemispheres. The brain was protected by a plastic dome. The venous pressure was increased by inflating a balloon in the right atrium and the pressure measured through a catheter inserted in the exposed superior sagittal sinus. The usual γ-technique was used for flow measurements.

Results and Comments

The result from a typical experiment is presented in Fig. 2. After the operative procedure the flow was measured at normal arterial and venous pressures. The next step involved flow measurement during the period of increased venous pressure. Thereafter the inflated balloon was released and the flow measured at about the same perfusion pressure as during the period of increased venous pressure and later at even lower arterial pressure. The figure shows that complete autoregulation did not occur when the perfusion pressure was reduced by the venous pressure increase, whereas it was functioning at even lower perfusion pressure due to arterial hypotension.

Our present results from these experiments are summarized in Fig. 3 and speak strongly against a metabolic mechanism as the main cause of the normal cerebral flow autoregulation.

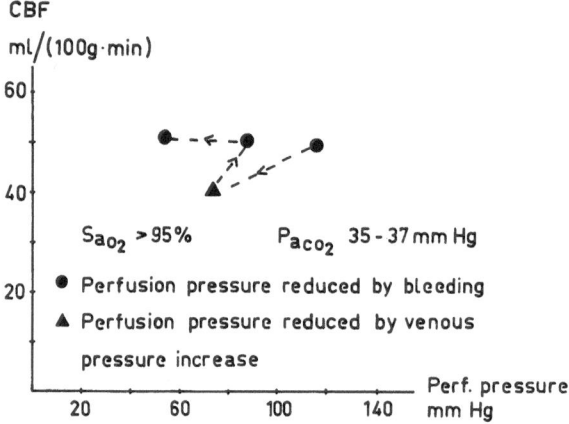

Fig. 2. The figure demonstrates the absence of normal autoregulation when the perfusion pressure was reduced by increased venous pressure whereas the autoregulation during arterial hypotension is preserved

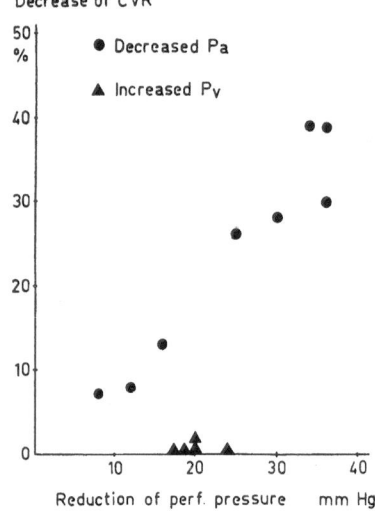

Fig. 3. The figure shows the change in cerebrovascular resistance (CVR) during reduction of the perfusion pressure established by bleeding or increased venous pressure. As reference value of the perfusion was taken the arterial blood pressure value during the period of increased venous pressure minus the measured normal value of the venous pressure

Concluding Remarks

From the present results it can be concluded that neither the normal autoregulation nor the reactive hyperemia after preceding periods of hypotension are mainly due to metabolic mechanisms. Our results speak in favor of a different, efficient autoregulatory mechanism – probably of myogenic origin. However, when the normal autoregulation is exhausted or abolished due to e. g. severe arterial hypotension or hypoxia, metabolic factors are still capable to influence the cerebral vessels in an autoregulatory sense. This conception is also supported by the findings that the passive pressure-flow relationship, which usually can be found when the normal autoregulation is lost seems to be remarkably linear which a true passive pressure flow relation curve should not be [1]. The linearity found may then be caused by metabolic factors.

References

1. FOLKOW, B , and B. LÖFVING: The distensibility of the systemic resistance blood vessels. Acta physiol. scand. **38**, 37 (1956).
2. HARPER, A. M., and E. HÄGGENDAL: Discussion and comments on autoregulation. Scand. J. clin. Lab. Invest. **22**, Suppl. **102**, p. V:K (1968).
3. HÄGGENDAL, E., and B. JOHANSSON: Effects of arterial carbon dioxide tension and oxygen saturation on cerebral blood flow autoregulation in dogs. Acta physiol. scand. **66**, Suppl. **258**, 27 (1966).
4. –, N. J. NILSSON, and B. NORBÄCK: On the components of ^{85}Kr clearance curves from the brain of the dog. Acta physiol. scand. **66**, Suppl. **268**, 5 (1966).
5. –, J. LÖFGREN, N. J. NILSSON, and N. ZWETNOW: The influence of induced changes in cerebrospinal fluid pressure on the cerebral blood flow in dogs. III. International Symposium on Cerebral Circulation, Salzburg, 1966.
6. LASSEN, N. A.: The luxury perfusion syndrome of the brain: a condition of relative cerebral hyperemia occuring in a variety of acute brain disorders. Lancet II, 1133 (1966).
7. SIESJÖ, B. K., A. E. KAASIK, L. NILSSON, and U. PONTÉN: Biochemical basis of tissue acidosis. Scand. J. clin. Lab. Invest. **22**, Suppl. **102**, p. III:A (1968).

CBF in Non-Pulsatile Perfusion

K. Held, U. Gottstein, and W. Niedermayer

Department of Internal Medicine, University of Kiel

The mechanism of cerebral autoregulation is still open to discussion [4]. One important factor involved is described by the myogenic theory. From our recent experimental work dealing with the influence of extracorporeal circulation upon CBF and cerebral metabolism we obtained results which, we think, support the myogenic theory. The term autoregulation in its restricted sense describes the intrinsic ability of cerebral circulation to maintain a constant blood flow at varying perfusion pressures. This fact is usually demonstrated in the familiar pressure-flow diagram, where CBF is plotted against an arbitrary mean pressure. This procedure however neglects the fact that normal blood pressure is pulsatile. If one agrees to the concept of transmural pressure as the important factor in myogenic autoregulation a different response to normal pulsatile and to a constant non-pulsatile perfusion pressure can be expected.

To test this hypothesis we have used, in 10 dogs, a bypass that permitted a pulsatile and non-pulsatile perfusion of the animals head only. CBF was measured electromagnetically using the cerebral venous outflow of the superior sagittal sinus.

Our results are shown in Fig. 1. The lower curve of CBF in pulsatile perfusion demonstrates the familiar finding of cerebral autoregulation above a threshold of approximately 70 mmHg. In non-pulsatile perfusion (upper curve) we find a very similar pressure-flow relationship, which proves that the ability to autoregulation is maintained also in this special hemodynamic situation. Autoregulation now is, however, taking place on a higher flow level. The difference between the 2 curves is significant at the 1% level.

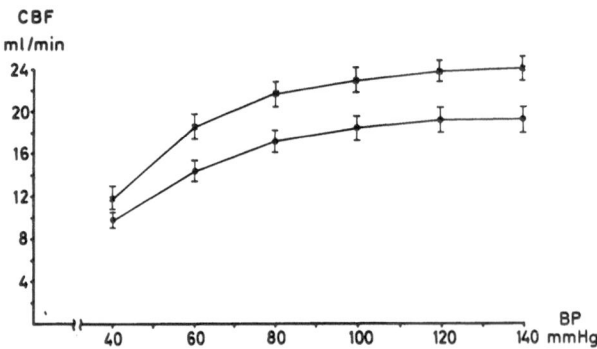

Fig. 1. CBF in pulsatile (lower curve) and in non-pulsatile perfusion (upper curve)

From these findings it is concluded that cerebral autoregulation responds to both qualities of pressure: its pulsations and its constant component. Pulsatile pressure seems, however to be more effective as a stimulus evoking a regulatory response. Such a different reaction to constant and pulsatile pressure is a well-known quality of mechanoreceptors, as e.g. the baroreceptors of the carotid sinus.

This response of cerebral circulation to changes in the pressure form is further evidenced by the following experiment. When a constant perfusion pressure is turned acutely to pulsatile pressure the mean perfusion pressure remains constant at first for a few seconds, while CBF rises temporarily 25%, thereby indicating pulsatile pressure to be more "effective". Within 5–10 sec a regulatory response occurs: CBF returns almost to the initial level whereas mean pulsatile pressure rises about 20%. In the following steady state cerebral vascular resistance is increased approximately 20%.

The only change in this experiment evoking an autoregulatory response is the alternation of the adequate stimulus for a myogenic regulation, the form of the perfusion pressure, while other important factors as arterial and cerebral venous pH, pO_2 and pCO_2 remained constant. In our opinion these results are further evidence of the myogenic mechanism of cerebral autoregulation.

Our findings are also of clinical interest. In bypass-procedures of cardiac and vascular surgery cerebral circulation is exposed to a virtually non-pulsatile perfusion. There are only very few and contradictory observations about the regulation of CBF in this situation. While HALLEY et al. [2] and ANKENEY et al. [1] demonstrated a linear pressure-flow relationship over the whole pressure range tested, WOLLMAN et al. [3] believed perfusion pressure to be unimportant as a factor in extracorporeal circulation. From our findings it is however evident that an effective cerebral autoregulation is maintained also during bypass-procedures – provided the perfusion pressure is kept above the critical threshold.

References

1. ANKENEY, J. L., and P. H. VILES: The effect of total body perfusion on carotid blood flow. Surgery **49**, 209 (1961).
2. HALLEY, M. M., K. REENTSMA, and O. CREECH JR.: Cerebral blood flow, metabolism and brain volume in extracorporeal circulation. J. thorac. Surg. **36**, 506 (1958).
3. WOLLMAN, H., G. W. STEPHEN, A. J. CLEMENT, and G. K. DANIELSON: Cerebral blood flow in man during extracorporeal circulation. J. thorac. Surg. **52**, 558 (1966).
4. ZWETNOW, N.: CBF Autoregulation to Blood Pressure and Intracranial Pressure Variations. Scand. J. clin. Lab. Invest., Suppl. **102**, p. V–A (1968).

On the Location of the Vascular Resistance
in the Cerebral Circulation

E. Kanzow and D. Dieckhoff

Physiological Institute, University of Göttingen

From the investigations of M. Fog, H. S. Forbes, G. I. Mchedlishvili, H. G. Wolff [1, 2, 3, 4, 5, 6] and others it is known, that the pial arteries change their diameter with any change of cerebrovascular resistance. But, it was unknown, to what extent the arterial vessels in the pia contribute to the blood flow resistance and to what degree their diameter changes give rise to alterations of the total cerebrovascular resistance. The experiments to be described were carried out to study the possible role of the pial arteries in the flow resistance by determinations of the pressure drop from the aorta to large and small pial arteries.

The experiments were performed on cats anesthetized with chloralose-urethane. Pressure measurements were made in the small pial arteries using micropipettes with sharpened tips of about 15–30 micron in outside diameter and the counter-pressure method of Landis. In large pial arteries and in the aorta the pressures were continuously measured with Statham pressure transducers. The relative pressure drop is calculated as difference between aortic and pial artery pressures in percent of the aortic blood pressure.

The relationship of the relative pressure drop to the outside diameter of the punctured vessels illustrates the diagram. Solid circles represent values at aortic pressures from 65 to 120 mmHg, open triangles values at more than 120 mmHg. At normal aortic pressures the relative pressure drop from the aorta to the middle cerebral artery near the circle of Willis and its large branches is about 20%. In the smaller branches the relative pressure drop increases to about 30% up to arteries 80–120 microns in diameter, and to about 40% to arteries 30–41 microns in diameter.

As the relative pressure drop corresponds to the percentage of total flow resistance in the circulation of the area supplied from the punctured vessels, the data demonstrate that about 20% of the total flow resistance are located between the aorta and the middle cerebral artery. In the branches of the middle cerebral artery up to vessels 30–40 microns in outside diameter are located further 20%, mostly in vessels smaller than 100 micron in diameter. From the aorta to pial arteries 30–40 microns in diameter, a mean of 39.4% of the total flow resistance was found at a mean of 88 mmHg aortic pressure. At 100 mmHg aortic pressure this percentage rises to about 42% as a result of the relationship between the relative pressure drop and the aortic pressure.

These data demonstrate the importance of the flow resistance of the vessels between the aorta and the small pial arteries for the adjustment of the vascular resistance in the pial circulation. It is evident that the resistance of these vessels must also be changed with greater changes of the total cerebrovascular resistance. This view is supported by preliminary results of other experiments in which the relative pressure drop was nearly unchanged following vasodilatation induced by CO_2 and following increased blood pressure by norepinephrine infusion.

Since the pial arteries decrease to diameters smaller than 30 microns and the pressure drop increases more and more in the smaller vessels, and since also in the venous system there exists a pressure drop, it is to be expected that much more than 50% of the blood flow resistance in the circulation of the cerebral cortex are not enclosed in the cerebral tissue. Therefore the major part of the flow resistance cannot be adjusted by a direct influence of metabolites on the vascular smooth muscle.

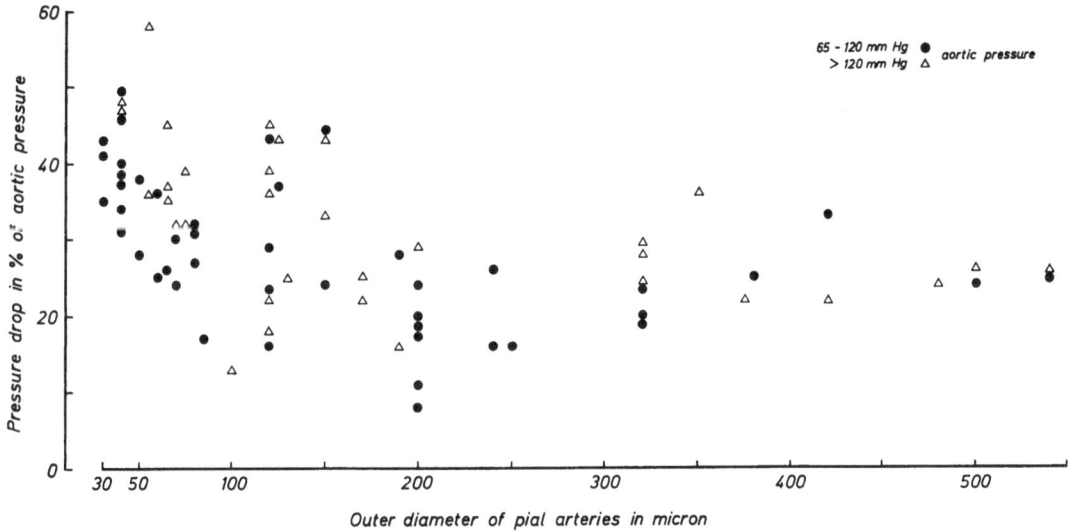

Fig. 1. Relationship of the relative pressure drop to outside diameter of pial arteries

References

1. Fog, M.: Cerebral circulation. Reaction of pial arteries to increase in blood pressure. Arch. Neurol. Psychiat. **41**, 260 (1939).
2. — Cerebral circulation. The reaction of the pial arteries to a fall in blood pressure. Arch. Neurol. Psychiat. **37**, 351 (1937).
3. Forbes, H. S.: The cerebral circulation. Observation and measurement of pial vessels. Arch. Neurol. Psychiat. **19**, 751 (1928).
4. Mchedlishvili, G. I., u. S. Nikolaishvili: Zum nervösen Mechanismus der funktionellen Dilatation der Piaarterien. Pflügers Arch. ges. Physiol. **296**, 14 (1967).
5. —, D. G. Baramidze, and L. S. Nikolaishvili: Functional behaviour of pial and cortical arteries in conditions of increased metabolic demand from the cerebral cortex. Nature **213**, 506 (1967).
6. Wolff, H. G.: The cerebral circulation. Physiol. Rev. **16**, 545 (1936).

The Spasm of the Internal Carotid Arteries

G. I. Mchedlishvili

Institute of Physiology, Georgian Academy of Sciences, Tbilisi

Until the 1960's almost nobody was interested in the role of the major arteries of the brain (internal carotids and vertebrals) in the control of blood supply to the brain, since it was generally believed that the regional circulation is controlled everywhere in the body only by the smallest arteries, i.e. arterioles, and there were no suitable techniques to investigate the functional behaviour of the major arteries of the brain (localized mainly inside the bones on the base of the skull).

Vascular spasm, i.e. a continuous and strong vasoconstriction disturbing the normal circulation, may certainly occur in any cerebral blood vessel possessing a smooth muscle layer. However, high sensitivity to vasoconstrictor substances and nerve impulses, as well as high vasoconstricting tendency of vascular responses are necessary for the appearance of spasms under natural conditions. At present there is little doubt that it is the larger arteries that possess these features more pronounced than the other cerebral blood vessels.

In a recently published monograph by the present author on the *"Functional Behaviour of the Vascular Mechanisms of the Brain"* (1968) present information on the responses of the different parts of the vascular system of the brain were summarized: a) The internal carotid arteries possess a well developed muscular layer and are able to constrict up to almost a full closure of the vascular lumen; b) an abundant nerve plexus emanating from different sources surrounds the wall of the major arteries of the brain, while the pial arteries are more poorly innervated, and the intracerebral arteries and arterioles are supplied only with 1–2 nerve fibres, if with any at all; c) Vasoconstrictor responses of the internal carotid arteries have been observed under a number of various conditions, while the typical vascular response of the pial arteries in vasodilatation which appears regularly under conditions of increased metabolic demands of the cerebral tissue of any kind; d) Adreno- and cholinoreceptors were found in abundance is the vascular walls of the internal carotid arteries. Their vasoconstriction caused by sympathetic stimulation, as well as brought about by the reflex from the baroceptors of the venous sinuses were also demonstrated; e) continuous and strong constriction of the internal carotid arteries followed by a decreased blood supply to the brain and by death of the animal has been observed. All these experimental findings with animals furnished the basis of the hypothesis that the spasm of the internal carotid arteries should play an important role in the disturbances of the blood supply of the human brain. Recent studies of Dr. V. Gabashvili with neurological patients support this assumption.

To investigate the pathophysiological mechanism of the spasm of the major arteries of the brain a new technique of the isolated internal carotid artery in dogs was developed (in collaboration with Dr. L. Ormotsadze). All anastomoses connecting the artery with the extracerebral blood vessels were occluded and inside the skull a thin polythene catheter was inserted into the artery to carry blood directly into the jugular vein by-passing the other cere-

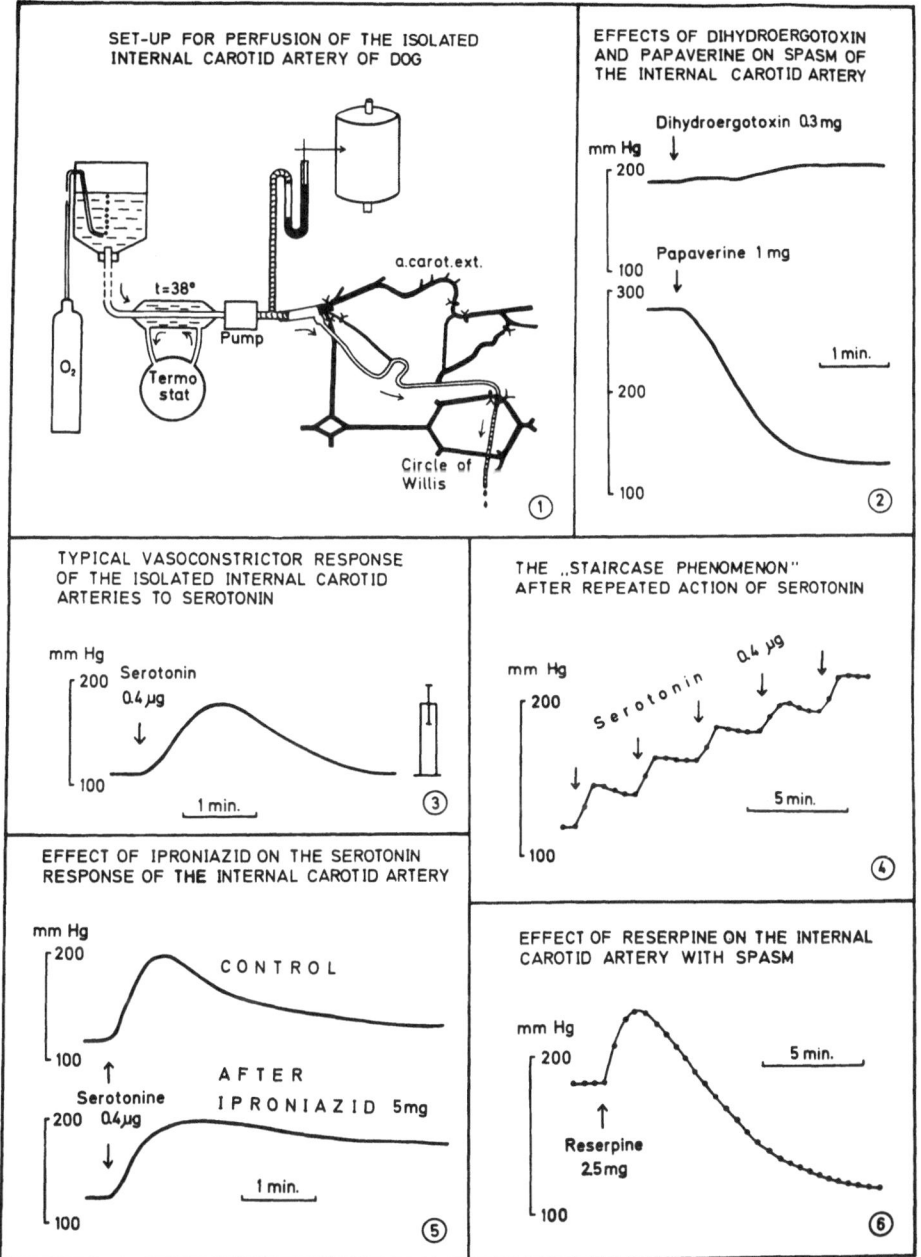

Fig. 1. Set-up for perfusion of the isolated internal carotid artery of dogs (see the text) with oxygenated Ringer-Krebs solution

Fig. 2. Persistance of spasm of the internal carotid artery after administration of dihydroergotoxin (α-adrenergic blocker) and relaxation of the spastic arterial wall after administration of papaverine

Fig. 3. Typical vasoconstrictor effect of serotonine on the isolated internal carotid artery of the dog

Fig. 4. "Staircase phenomenon" in some experiments where the contraction-relaxation cycle was disturbed

Fig. 5. Prolongation of the serotonine response of the isolated internal carotoid artery after preliminary administration of iproniazid (inhibitor of monoamine oxidase)

Fig. 6. Effect of reserpine (causing release of serotonine) on a spastic internal carotid artery

bral vessels. The artery thus remains in the body with all its nervous connections intact. The resistance in the artery, being dependent of the tension on its muscular layer, was measured directly using a perfusion pump with a constant stroke volume. The artery was initially perfused with blood (from the common carotid artery), and then for a better analysis of the vascular responses – with an oxygenated Ringer-Krebs solution (Fig. 1).

The experiments showed that the resistance in the isolated internal carotid artery varied within the range of about 4.8–31.8 mmHg/ml/min (when a spasm occurred). Sometimes the resistance was even higher, but the equipment used could not measure it.

The following experiments helped to clarify whether the arterial spasm already developed was dependent of the nervous or muscular mechanism: the intraarterial administration of alpha adrenergic blocking agents (dihydroergotoxin and Regitine Ciba) did not cause relaxation of the vascular wall tension, while Papaverine resulted in its significant relaxation (Fig. 1 to 2). Thus, the arterial spasm, when it is already in evidence, is not brought about by vasoconstrictor nerve impulses, but results from disturbances in the contraction-relaxation cycle in the smooth muscle fibres of the arterial wall.

Amongst endogenous vasoconstrictor substances tested, serotonine was found to be the most potent one. When the artery was perfused with blood its effect was on the average 3 times greater than that of norepinephrine, while under conditions of perfusion with the Ringer-Krebs solution it was 50–100 times greater. The inhibition of the serotonine responses by reserpine and chlorpromazine proves that they were specific.

Under normal conditions the vasoconstrictor response of the artery to serotonine lasted about 3 min, and then the vascular wall tension regained its initial value (Fig. 1–3). However, under conditions of some disturbances in the contraction-relaxation cycle the repeated serotonine intraarterial administration resulted in a "staircase phenomenon", i.e. the tension of the arterial wall increased gradually, and hence the contracture of the muscular layer developed (Fig. 1–4).

The inhibitors of monoamine oxidase, e.g. iproniazid, caused a significant prolongation of the serotonine induced vasoconstriction (Fig. 1–5); nialamid also resulted in an increase (85 ± 28 percent) and prolongation (102 ± 37 percent) of the serotonine response.

When a spasm of the internal carotid artery was in evidence reserpine caused a temporary increase in tension with the subsequent relaxation of the vascular wall (Fig. 1–6). Since reserpine releases serotonine, our findings seem to provide evidence that serotonine is responsible for the contracture of the muscular layer of the arterial wall.

We are in no position yet to draw any general conclusions about the pathophysiological mechanism of a spasm of the internal carotid artery. But we may state that a) spasm of the artery may occur, b) that it is caused by some abnormality in its muscular layer and c) that an important role in its appearance may be played by serotonine circulating in the blood.

Acknowledgement. The author is indebted to Prof. P. KOMETIANI for helpful suggestions.

Regional Cerebral Blood Volume During Paradoxical Sleep

J. Risberg, L. Gustavsson, and D. H. Ingvar

Departments of Clinical Neurophysiology and Psychiatry I, University Hospital,
Lund

Introduction

Sleep, accompanied by slow waves in the EEG (orthosleep), does not lead to any marked alterations of the cerebral oxygen uptake or blood flow in man [7]. In dogs, however, a slight reduction of the cortical blood flow and oxygen uptake has been demonstrated during the same form of sleep [6]. Paradoxical sleep (parasleep) accompanied by a fast EEG rhythm and periods of rapid eye movements (REMs) leads, however, to an increase of the cerebral circulation as demonstrated with a thermoelectric technique [1, 5]. This finding has recently been confirmed by REIVICH [8], with a quantitative autoradiographic rCBF-technique. COOPER and HULME [3], found an increase of the intracranial pressure during REM sleep, probably caused by an increase of cerebral blood flow and blood volume (see below).

In the present study we have measured regional cerebral blood volume (rCBV) during paradoxical sleep in normals. The rCBV-technique [10] is untraumatic and offers a continuous measure of regional cerebral circulatory changes.

Materials and Methods

19 normal men were studied during 1 night following sleep deprivation. The rCBV-measurements were carried out in a specially constructed bed which enabled fixation of the subject's head in relation to the 8 scintillation detectors mounted within the head rest in lead collimators. Sleep periods of 4–8 h were recorded. 14 records had to be discarded due to the lack of parasleep or movement artefacts. 2 of the 5 final cases woke up following paradoxical sleep and reported vivid dreaming.

Continuous recordings were made of the EEG, the oculogram and the EMG of neck muscles with conventional techniques. For the rCBV-measurements, 100 micro Curie RISA (^{131}I) was administered intravenously and the recordings started 30 min later. Counting rates of about 1000 cpm per detector were obtained.

Results

In Fig. 1 it is seen that there was an increase of the mean rCBV (from all 8 detectors) which showed a fairly clear-cut relation to the occurrence of REMs. The limited material does, in fact, suggest some proportionality between the frequency of the eye-movements and the height of the rCBV increase. Some comparison is also possible between orthosleep and

Supported by the Swedish Medical Research Council (Contracts No.: B 6721 X 8403 and B 6821 X 8404) and the Wallenberg Foundation, Stockholm, Sweden.

paradoxical sleep but unfortunately due to movement artefacts, such transitions were rarely recorded in a satisfactory fashion. Cases 2, 3, and 5 do, however, show that the rCBV was lower during ortho- than during parasleep.

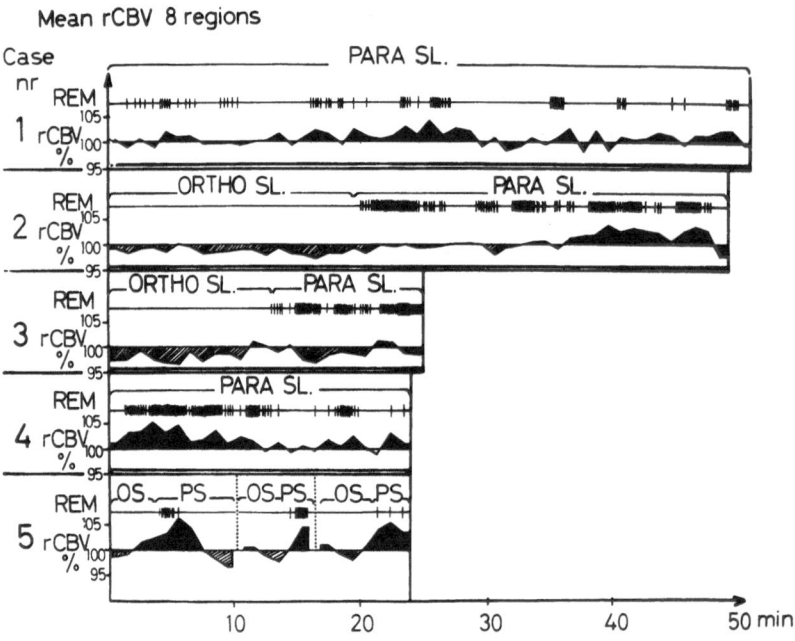

Fig. 1. Changes of mean regional cerebral blood volume (rCBV, mean of 8 regions) during sleep in 5 normal subjects. Note the relationship between the REM-frequency and the height of the rCBV increase

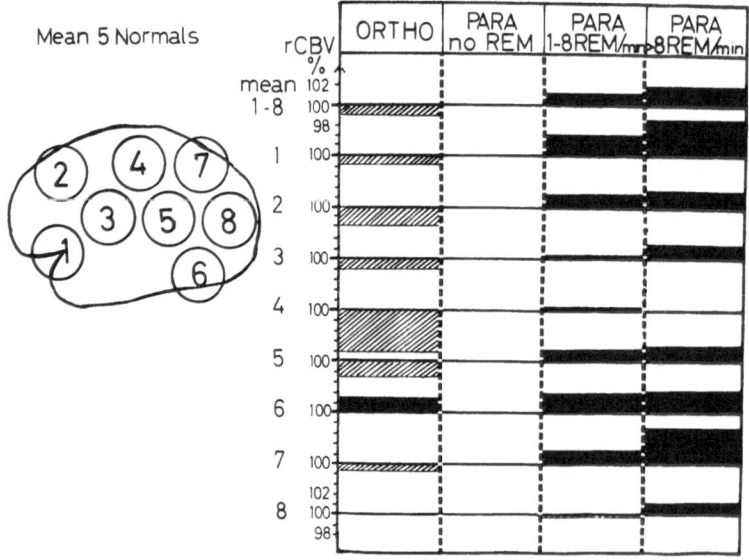

Fig. 2. Changes of regional cerebral blood volume (rCBV) during sleep in 5 normals. The increase of rCBV in region 7 during paradoxical sleep (parasleep) with more than 8 REMs per min is significant at the 5 percent level (*t*-test)

In Fig. 2 a regional analysis of the rCBV-changes is shown. Means for all 5 subjects are given for the 8 regions studied. Periods with parasleep and REMs were parted into 2 categories, those with few (1–0) and those with many (more than 8) REMs per minute. It is seen that the most pronounced rCBV increase was recorded at high REM frequency in regions 1 (low frontally) and 7 (parietally). Only the change in region 7 was, however, statistically significant (*t*-test: $p < 0.05$).

Discussion

We have recently in animal experiments demonstrated a linear relationship between changes of rCBV and rCBF [9]. The rCBV changes demonstrated during parasleep thus most likely reflect changes of rCBF.

Indeed, respiratory and systemic circulatory changes may cause alterations of the cerebral blood flow and volume, but such factors do not seem to have been of importance for the rCBV changes in parasleep. In this state there is, in fact, a slight hypocapnia [2], which should lead to a *general* and not a regional reduction of rCBF and rCBV. The increase of blood pressure and heart rate [4] during parasleep should also, due to vasoconstriction, cause general changes in rCBV opposite to those found.

The results confirm the findings of KANZOW et al. [5], BAUST [1] and REIVICH et al. [8], and they provide support for the theory that the cerebral metabolic rate is increased during paradoxical sleep, and, furthermore, as shown by REIVICH et al. [8], that this increase may show a regional pattern. The most significant increase of rCBV during parasleep with REM ("phasic" paradoxical sleep) in our subjects occurred in the parietal region, which is commonly assumed to be of a special importance for spatial orientation and imagery. The increase in the lower frontal region might to some extent be an artefact due to eye movements, but it could perhaps be related to rCBV-increase in this region during mental activity of abstract type [10].

Summary

Regional cerebral blood volume (rCBV) was recorded during sleep in 5 normals. During the phasic part of paradoxical sleep accompanied by rapid eye movements (REM) a slight increase in rCBV was recorded. The parietal region appeared to be the one in which the largest rCBV changes took place. There was also a general proportionality between the height of the rCBV increase and the REM frequency.

References

1. BÜLOW, K.: Respiration and wakefulness in man. Acta physiol. scand. **59**, Suppl. 209 (1963).
2. COOPER, R., and A. HULME: Intracranial pressure and related phenomena during sleep. J. Neurol. Neurosurg. Psychiat. **29**, 564 (1966).
3. HARTMANN, E.: The Biology of Dreaming. Springfield: Charles C. Thomas, (1967).
4. KANZOW, E., D. KRAUSE u. H. KÜHNEL: Die Vasomotorik der Hirnrinde in den Phasen desynchronisierter EEG-Aktivität im natürlichen Schlaf der Katze. Pflügers Arch. ges. Physiol. **274**, 593 (1962).
5. LÜBBERS, D. W., D. H. INGVAR, E. BETZ, H. FABEL, M. KESSLER u. F. W. SCHMAHL: Sauerstoffverbrauch der Großhirnrinde in Schlaf- und Wachzustand beim Hund. Pflügers Arch. ges. Physiol. **281** (1964).
6. MANGOLD, R., L. SOKOLOFF, E. CONNER, J. KLEINERMAN, P. THERMAN, and S. S. KETY: The effects of sleep and lack of sleep on the cerebral circulation and metabolism in normal young men. J. clin. Invest. **34**, 1092 (1955).
7. REIVICH, M., G. ISAACS, E. EVARTS, and S. S. KETY: The effect of slow wave sleep and REM sleep on regional cerebral blood flow in cats. J. Neurochem. **15**, 301 (1968).
8. RISBERG, J., D. ANCRI, and D. H. INGVAR: Correlation between cerebral blood volume and cerebral blood flow in the cat. Exp. Brain Res., in press 1969.
9. —, and D. H. INGVAR: Regional changes in cerebral blood volume during mental activity. Exp. Brain Res. **5**, 72 (1968).

The Effect of Hemodilution
Caused by Low Molecular Weight Dextran on Human Cerebral Blood Flow and Metabolism

U. Gottstein and K. Held

Department of Internal Medicine, University of Kiel

It is generally accepted that a *cerebral vasodilatation* and thereby an augmentation of CBF cannot be achieved signficantly by intravenous infusions of *vasodilating substances* [3]. The vasodilating effect of CO_2-*inhalation* has some disadvantages: 1. blood flow in ischemic areas is not increased; it may even lead to an intracerebral steal as has been shown by Hoedt-Rasmussen et al. [5]; 2. the acidosis due to CO_2 is undesirable.

Because of the impaired autoregulation, CBF can be increased in ischemic [2] and injured areas [6] by *pressure-agents*. An augmentation of CBF can be achieved, furthermore, if initially there was a pathological low blood pressure.

According to Hagen-Poisseuille's law, blood flow depends not only on the pressure and the vascular cross-section but also on the *viscosity*: the lower the viscosity the higher the flow values. It could therefore be expected that by lowering the viscosity blood flow should improve regardless of whether the vessels are stenosed or if myogenic autoregulation is effective or impaired. The viscosity of whole blood is directly dependent on the hematocrit as has been demonstrated by Bollinger et al. [1]. There is a linear correlation between a hematocrit of 35% and 52% as proved at 3 different shear rates. The variations of CBF in anemia and polyglobulia [4] are probably caused by the different viscosities in these states.

Bollinger [1] has shown that hematocrit and whole blood viscosity can be lowered by intravenous infusions of low molecular weight dextrans. The plasma viscosity which is unimportant for the blood flow increases slightly.

We have therefore measured CBF before and after intravenous infusions of 500 ml Rheomacrodex and Macrodex in order to test whether a decrease of hematocrit and of whole blood viscosity does also augment CBF. As shown in Table 1 the blood flow increased significantly with both solutions both in the group with normal and with initially reduced values. Cerebral oxygen and glucose uptakes remained unchanged. This constancy of metabolism in spite of an augmented blood flow is not remarkable, because this reaction is known from increases of blood flow brought about otherwise, e.g. by intracarotid infusions of AMP or Papaverin [3].

During CO_2-inhalation cerebral oxygen uptake also remains constant in normal persons and in patients with slight or severe arteriosclerosis, although CBF is considerably increased [3].

In summary, we have been able to demonstrate, that the CBF in man can be augmented significantly by lowering the total blood viscosity with low molecular weight dextran infusions. Whether this result is of therapeutic value will have to be tested under double-blind conditions.

Table 1. *The effect of intravenous infusions of Rheomacrodex, Macrodex and 0,9% NaCl on cerebral blood flow and metabolism*

i. v. infusion 500 ml	No.	cerebral blood flow ml/ 100 g · min		cerebral O_2-up-take ml/ 100 g · min		cerebral glucose uptake mg/ 100 g · min		cerebral Lac-tatedeliv. mg/ 100 g · min		cerebral Pyru-vatedeliv. mg/ 100 g · min	
		pre	post	pre	post	pre	post	pre	post	pre	post
Rheo-macrodex	19	47.5	68.1[c]	3.18	3.54[b]	3.81	3.67[a]	0.7	0.9	0.02	0.06
Macrodex	16	54.1	66.8[c]	3.42	3.73[a]	4.52	5.80[a]	0.5	0.8	0.05	0.07
0.9% NaCl	5	61.9	56.2[a]	3.97	3.67[a]	4.68	4.69	0.1	0.4	0.06	0.04

i. v.	hemato-krit %		art. O_2 content Vol. %		mean art. pres-sure mmHg		"cerebral vasc. resistance" mmHg/ml/ 100 g · min		pH art.		PCO_2 art. mmHg	
	pre	post	pre	post	pre	post	pre	post	pre	post	pre	post
Rheomacrodex	43	36[c]	16.6	13.9[c]	116	119	2.5	1.8[c]	7.44	7.42	35	35
Macrodex	42	37[c]	16.2	14.1[c]	117	117	2.3	1.8[c]	7.45	7.42	35	37
0.9% NaCl	46	44	17.5	16.7	117	117	1.9	2.3	7.40	7.39	38	37

[a] $P = 0,10$

[b] $P > 0.01 < 0.05$

[c] $P < 0.001$

References

1. BOLLINGER, A., E. LÜTHY u. E. JENNY: Vollblutviskosität bei verschiedenen Schergeschwindigkeiten und ihre Beeinflussung durch niedermolekulares Dextran. Klin. Wschr. **45**, 939 (1967).
2. FREEMAN, J., and D. H. INGVAR; Elimination by hypoxia of the cerebral blood flow autoregulation and EEG relationship. Exp. Brain Res. **5**, 61 (1968).
3. GOTTSTEIN, U.: Der Hirnkreislauf unter dem Einfluß vasoaktiver Substanzen. Heidelberg: Hüthig-Verlag 1962.
4. — Physiologie und Pathophysiologie des Hirnkreislaufes. Med. Welt **715** (1965).
5. HØEDT-RASMUSSEN, K., E. SKINHØJ, O. PAULSON, J. EWALD, J. K. BJERRUM, A. FAHRENKRUG, and N. A. LASSEN: Regional cerebral blood flow in acute apoplexy. Arch. Neurol. **17**, 271 (1967).
6. REIVICH, M.: Regional changes in autoregulation following cerebral trauma. Fourth Internat. Symposium of the Research Group on Cerebral Circulation, Salzburg 1968.

Cerebral Vascular Responses to Changes in the Ionic Osmolar Composition of Blood

C. E. RAPELA and J. P. BUYNISKI

Department of Physiology, Bowman Gray School of Medicine, Wake Forest University, Winston-Salem, North-Carolina

Previous studies on the effect of changes in blood osmolality on the cerebral vasculature have been made by observing the pial or cortical vessels, or by indirect methods of estimating flow. In these studies the hyperosmotic solutions were injected intravenously or intraperitoneally; the results obtained were at variance though a dilator effect followed by delayed constriction seemed to be the most frequent response reported [5, 8]. Interpretations of the mechanisms of the hyperosmotic response were also at variance; most workers considered the hyperosmolar response as secondary to cerebrospinal fluid or arterial blood pressure changes. Intravenous infusions of hyperosmolar solutions have been used in the treatment of increased cerebrospinal pressure [3], and induced hyperosmolality has been reported to reduce the histological damage resulting from a period of cerebral ischemia [2]. Until recently, the osmolality of the immediate environment surrounding vascular smooth muscle has not been considered as a variable likely to regulate the "tone" of blood vessels. MELLANDER [6], and JOHANSSON [4] have postulated, respectively, that hyperosmolality may play a role in exercise hyperemia of skeletal muscle and in the electrical and mechanical events of vascular smooth muscle *in vitro*. We have been studying the effects of hyperosmolar solutions [1] by means of a direct method to measure cerebral venous blood flow in the dog [7]. The confluence of the cerebral sinuses was cannulated and, with the lateral sinuses occluded, the cerebral venous outflow was passed through an electromagnetic flowmeter. Brain perfusion pressure was taken as the mean between common carotid and wedged vertebral artery pressures. Cerebral venous outflow osmolality and hematocrit were measured during induced changes of blood osmolality produced by rapid intracarotid injection of hyperosmolar solutions in volumes not producing marked changes in mean arterial pressure or systemic blood osmolality. Hyperosmolar solutions of Na, Li and K produced marked increases in cerebral vascular conductance (CVC) in contrast to the minimal changes in CVC produced by equally hyperosmolar dextrose, urea and mannitol. Hyperosmotically equivalent solutions of Na, Li, K, and dextrose produced similar changes in cerebral venous outflow osmolality and hematocrit. The effect of hyperosmolar solutions were not secondary to changes in cerebrospinal fluid pressure, since the changes in cerebrospinal fluid pressure that did take place passively followed the changes in cerebral blood flow. Besides the difference in the vascular effects exerted by ionic or non-ionic solutions, K had a greater effect on conductance (375 ± 41.4, $n = 6$) than Na and Li (263.5 ± 12.9, $n = 6$) at similar injected osmolalities. Cerebral vascular responses to Na and Li were identical.

The effect of intracarotid injections of hyperosmotically similar solutions of Na, dextrose and K is shown in Fig. 1. Dextrose produced a small increase in cerebral blood flow. Similar hyperosmotic amounts of Na and K produced larger increases in cerebral flow and were preceded by periods of vasoconstriction. This period of decreased flow could be due to cell agglutination as shown by the fall in hematocrit, or to the specific ionic stimulus of Na and K on cerebral smooth muscle. Since injection of hyperosmotically similar amounts of dextrose produced a fall in hematocrit but did not decrease the flow, this could mean that cell agglutination might not be the major factor causing the periods of decreased flow seen with Na and K. Equal volumes of isotonic NaCl produced only minimal changes in venous outflow hematocrit and cerebral blood flow, and indicated that hemodilution did not play a major role in the increased blood flow response observed with intracarotid injection of hyperosmolar Na and

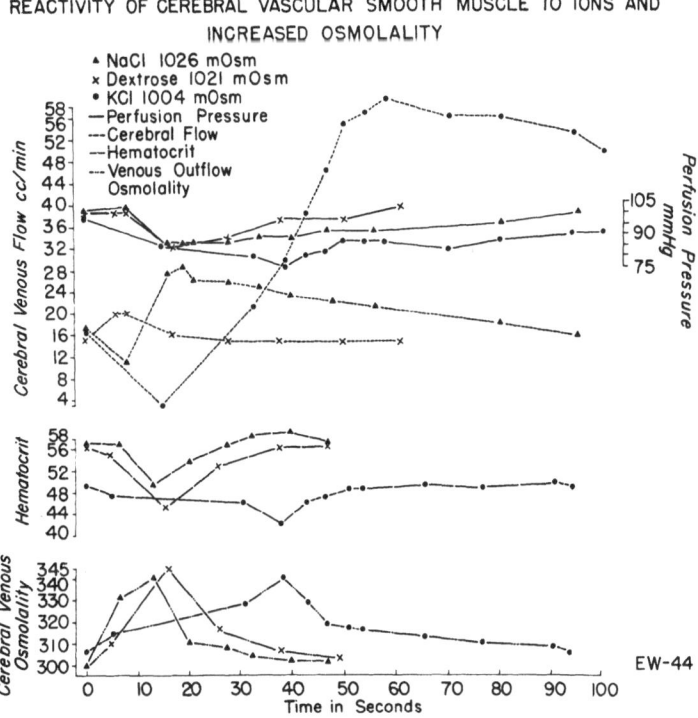

Fig. 1. Effect of rapid intracarotid injection of hyperosmolar Na, K, and dextrose on cerebral venous osmolality (mOsm/kg), % hematocrit, cerebral venous flow (cc/min) and perfusion pressure (mmHg). The volume injected was 1.5 cc in each common carotid just before zero time. This shift in the K measurements was probably due to the longer period of vasoconstriction this substance caused

K. The peak change in cerebral venous osmolality corresponded with the peak fall in hematocrit and this could mean that the fall in hematocrit was due to the transcapillary flux of water out of brain tissue and cerebral smooth muscle. Lack of similarity in the magnitude of the increased flow response observed with injections of hyperosmotically similar Na, K, and dextrose does not appear to be due to marked differences between plasma and interstitial space osmolality, since venous outflow osmolality for Na, K, and dextrose is approximately the same (Fig. 1). These results indicate that probably specific characteristics of ions as well as osmolality play a role in the responses of the cerebral vasculature to blood hyperosmolality.

References

1. Buyinski, J. P., and C. E. Rapela: Proc. Int. Un. Phys. Sci. **7**, 61 (1968).
2. Cantu, R. C., and A. Ames: Distribution of cerebrovascular obstruction following ischemia. Circulation, **4**, suppl. VI, 51 (1968).
3. Goluboff, B., H. A. Shenkin, and H. Haft: The effects of mannitol and urea on cerebral hemodynamics and cerebrospinal fluid pressure. Neurology **14**, 891 (1964).
4. Johansson, B., and O. Jonsson: Cell volume as a factor influencing electrical and mechanical activity of vascular smooth muscle. Acta physiol. scand. **72**, 456 (1968).
5. Lassen, N. A.: Cerebral blood flow and oxygen consumption in man. Physiol. Rev. **39**, 183 (1959).
6. Mellander, S., B. Johansson, S. Gray, O. Jonsson, J. Lundvall, and B. Ljung: The effects of hyperosmolarity on intact and isolated vascular smooth muscle. Possible role in exercise hyperemia. Angiologica, **4**, 310 (1967).
7. Rapela, C. E., and H. D. Green: Autoregulation of canine cerebral blood flow. Circulat. Res. **15**, 205 (1964).
8. Sokoloff, L.: The action of drugs on the cerebral circulation. Pharmacol. Rev. **11**, 1 (1959).

Comments to Chapter II

Regulation of CBF

The significance of various physiological and pathophysiological effects on CBF were discussed in this chapter.

The shift of the O_2 dissociation curve to the left decreases the O_2 supply of the cerebral tissue in respiratory alkalosis. With an additional metabolic alkalosis a further decrease in the O_2 supply can be expected because of the BOHR-effect. The low CBF in respiratory alkalosis is an additional cause for the low tissue pO_2. Adding a metabolic alkalosis to a respiratory alkalosis, however, causes an increase in CBF as shown by HÄGGENDAL and WINSÖ. The mechanism of this effect still remains unclear. The possible regulatory factor may be the extravascular H^+-ion concentration within the vascular smooth muscle cells and in the extra-cellular space. The pH of the extracellular space is different from that in the blood at least in a non-stationary state of metabolic alkalosis or acidosis as discussed by LASSEN, by FIESCHI and by BETZ. The O_2 affinity of the blood and its relation to the energy supply of the brain has been calculated by GROTE. From normograms of cerebral-venous O_2-values certain predictions on critical oxygen pressure-values within the cerebral tissue can be made. In the discussion it was pointed out that CBF mainly influences the O_2-capacity of the blood within the tissue. The clinically important problem of adaptation of CBF to hypoxia and hypercapnia was discussed in connection with the paper of AGNOLI et al. In hypoxia and in hypercapnia the initial increase of CBF returns to normal values in the course of adaptation, and normalization is accompanied by a normalization of the extracellular pH. Increasing PCO_2 in the arterial blood beyond 55 mmHg causes a loss of CBF adaptation as pointed out by EKSTRÖM-JODAL and HÄGGENDAL. The loss of adaptation of CBE is also seen in combined hypoxia and hypercapnia in anesthetized animals (AGNOLI et al.). In the discussion it was shown that in these experiments the cortical extracellular pH did not return to normal and that CBF was related to the reactions of the H^+ concentration in tissue. The reaction of blood flow to increased CO_2 is similar in the various segments of the central nervous system (including the spinal cord), but there are quantitative differences (FLOHR et al.). WÜLLENWEBER demonstrated that the reactions of the vessels in the spinal cord of patients qualitatively follow the same regulations as in the cortex of the brain.

In the discussion of myogenic reactions caused by changes of the intraluminar pressure EKSTRÖM-JODAL et al. showed that autoregulation was not a purely metabolic problem but a consequence of the smooth muscle tension. However, it is almost impossible to seperate metabolic effects (changes of pCO_2 or changes of the H^+ production) from variations of smooth muscle tensions *in situ*.

KETY discussed and confirmed the findings of HELD, GOTTSTEIN and NIEDERMEYER, who saw that non-pulsatile perfusion of the brain resulted in a decreased autoregulatory response, and a lower tissue perfusion, even when the perfusion pressure was the same as in pulsatile perfusion. These results were interpreted as further evidence in favour of the myogenic mechanism of CBF autoregulation.

KANZOW and DIECKHOFF, by measuring the pressure drop from the aorta to the pial arteries found that more than 50% of the so-called cerebrovascular resistance originate from blood vessels situated outside of the cerebral tissue proper. These authors pointed out that under such circumstances the major part of flow resistance within the brain can hardly be due to direct metabolic regulation of vascular tone.

The possible role of chemical induction of vasoconstrictor responses and of vascular spasm of extracerebral arteries was stressed by MCHEDLISHVILI.

Finally, a new aspect of regulation of CBF was discussed by RISBERG et al., who demonstrated physiological variations of regional cerebral blood volume (rCBV) during various stages of sleep.

GOTTSTEIN and HELD, who saw increases of CBF after infusion of 500 ml Rheomacrodex in man, regarded this as a sign of improvement in the state of the patient and ascribed the effect to the lower blood viscosity. As stressed by NILSSON, however, "the possible influence of the reduction in oxygen capacity was not considered, although this would seem to be rather important in their patients, who had an arterial oxygen content of only 16–17 vol. %, a value at which the oxygen capacity assumes a considerable importance".

RAPELA and BUYNISKI were able to show that specific characteristics of ions play a role in the response of cerebral vessels to hyperosmolarity.

E. BETZ

Regional Cerebral Blood Flow at Rest and During Functional Tests in Occlusive and Non-Occlusive Cerebrovascular Disease

O. B. PAULSON

Departments of Neurology and Clinical Physiology, Bispebjerg Hospital, Copenhagen

This paper presents the results obtained by measuring the regional cerebral blood flow (rCBF) in 2 groups of patients with cerebrovascular disease: One group with occlusion of the middle cerebral artery, and one group with apoplexy without angiographic signs of arterial occlusion. These results will be reported in more detail elsewhere [6, 7]. The results obtained in patients with transient ischemic attacks are reported separately at this symposium [8].

Method

The Xenon-133 injection method was used for the rCBF measurement [1, 5, 6]. About 2 mCi Xenon-133 dissolved in 2–4 ml of saline were injected as a slug into the internal carotid artery; the duration of the injection was 1–2 sec. The clearance of the isotope was followed by 16 small NaI (Tl) scintillation detectors placed externally over the lateral part of the hemisphere studied.

In all except 1 case the rCBF was measured 2–4 times at intervals of about 15 min. The first clearance curves were obtained in the "resting state". The next set of curves was obtained after the arterial pCO_2 had been changed (inhalation of 8% carbon dioxide or voluntary hyperventilation). The third set of clearance curves was recorded after increasing or decreasing the blood pressure by drug infusions [metaraminol bitartrate (Aramine) or angiotensin amide (Hypertensin) or pentolinium tartrate (Ansolysen)]. The final set of curves was obtained 5 min after an intravenous injection of 200 mg theophylline (Aminophylline). The arterial pCO_2 and the mean arterial blood pressure was measured after each Xenon-133 injection.

The rCBF values were calculated from the initial part (2 min) of the logarithmically recorded clearance curves as:

$$rCBF_{(initial)} = 2 \cdot 100 \cdot D_{(initial)} \ ml/100\,g/min$$

2 is the product of the λg value for gray matter taken to be 0.87 and of the factor 2.3 for converting natural to decade logarithm. 100 is a factor for calculating flow per 100 g tissue. $D_{(initial)}$ is the initial slope of the clearance curve in decade unit per minute [5, 6].

Normally the clearance curve is practically monoexponential for the first 2 min of clearance. But in pathological cases it may be multiexponential, i. e. initial peaks may be seen [5].

Results

41 patients were studied: 10 had an occlusion of the middle cerebral artery. The other 31 had no angiographically demonstrated arterial occlusion or stenosis considered of haemo-dynamic importance. All patients had acute hemiparesis of different severity and of different degree of remission. In all patients the symptoms persisted for at least 24 hrs.

The abnormalities of rCBF encountered were to some extent similar in the 2 groups. The most commonly encountered abnormalities were: *Ischemic foci*, *hyperemic foci* and *focal vasoparalysis* (i. e. focal loss of vasodilatation during hypercapnia, loss of vasoconstriction during hypocapnia, loss of autoregulation, and focal loss of theophylline induced vaso-constriction). In addition to these focal abnormalities an impaired autoregulation over the entire hemisphere and globally depressed flow values were observed in many cases. But, other abnormalities were also seen. All these abnormalities will be discussed below separately for the 2 groups.

1. Middle Cerebral Artery Occlusion

All 10 patients were studied within the first 3 days after the onset of symptoms. In 9 of the patients functional tests were performed. The rCBF abnormalities encountered may be summarized as:

Ischemic foci	in 9 of 10 patients
Perifocal hyperemic regions (very extensive in 1 case)	in 4 of 10 patients
Focal vasoparalysis	in 7 of 9 patients
Global impairment of autoregulation	in 0 of 5 patients
Globally depressed flow values	in 8 of 10 patients

Three of the cases were restudied once or twice, 7–30 days after the onset of the disease. In one of these cases the occlusion of the middle cerebral artery had disappeared at the time of the restudy; in this case there was a hyperemic focus with vasoparalysis, but the autoregula-tion was also affected outside the focus. In the other 2 patients the occlusion of the middle cerebral artery persisted when the restudies were performed. Three restudies were performed in these 2 cases. We observed:

Ischemic foci	in 3 of 3 studies
Perifocal hyperemic regions	in 1 of 3 studies
Focal vasoparalysis	in 3 of 3 studies
Global impaired autoregulation	in 3 of 3 studies
Globally depressed flow values	in 3 of 3 studies

2. Apoplexy Without Arterial Occlusion

Twenty of the 31 patients were studied within the first 14 days after the acute attack. One of these patients was restudied once. The other 11 patients were studied later after the onset of symptoms.

The rCBF abnormalities encountered were (see also Fig. 1).

Hyperemic foci disclosed during the resting state were seen in the first few days of the disease; focal vasoparalysis was also related to the acute phase of the disease although it was not confined to the very first days as the hyperemic foci; ischemic foci as disclosed during the resting state were seen without close correlation to any specific phase of the disease; global impairment of autoregulation, this was seen during the first 2 weeks after disease; globally depressed flow values were observed in the more severely affected cases manifesting drowsiness. The findings summarized here were the most frequent ones encountered. But, in addition, two other abnormalities were found in a few cases:

A "paradoxical" focal flow decrease during hypertension combined with a global loss of autoregulation was observed in three cases. Changes of the arterial pCO_2 showed in these cases a focal loss of vasodilatation during hypercapnia or of vasoconstriction during hypocapnia, but normal reactions outside the focal area. This finding will be discussed in further details in another contribution to this Symposium [3].

A focal loss of vasodilatation during hypercapnia but preserved vasoconstriction during hypocapnia and hypertension was observed in one case corresponding to an area between the regions supplied from the anterior and the middle cerebral artery. This finding will be discussed in further details in another contribution to this Symposium [3].

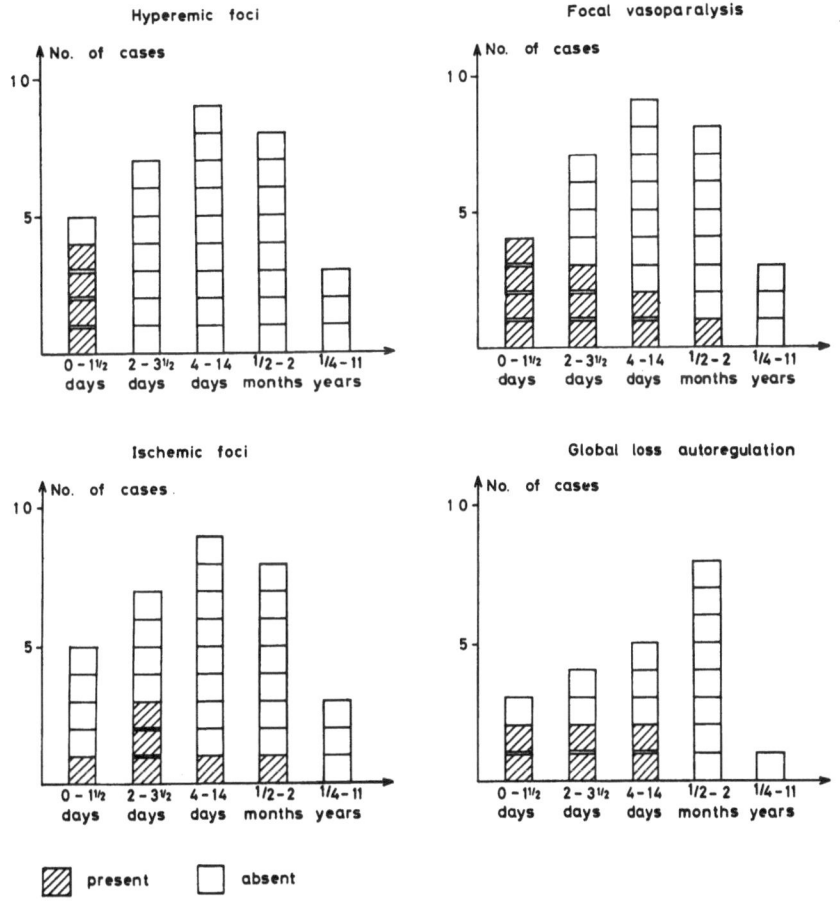

Fig. 1. Diagram showing the occurrence of hyperemic foci, of focal vasoparalysis, of ischemic foci, and of global loss of autoregulation as correlated to the time following the attack

Discussion

This study has demonstrated that focal flow disturbances with vasoparalysis and hyperemic or ischemic foci are frequently encountered in apoplexy with an arterial occlusion as well as apoplexy where no arterial occlusion could be demonstrated. These flow abnormalities have been suggested to be due to lactic acid and/or CO_2 tissue acidosis [2] and recently the acidosis has been confirmed in biopsy taken from patients with intracerebral haematomas [4].

In apoplexy without arterial occlusion, the flow abnormalities which were frequently encountered in the acute phase of the disease constitute evidence of fairly large tissue damage in the hemisphere and probably represent the sequels of a rather extensive ischemic lesion. Such an ischemic lesion can readily be explained as an acute thromboembolic arterial occlusion with subsequent rapid thrombolysis as will be discussed in another report at this symposium [8].

References

1. Høedt-Rasmussen, K., E. Skinhøj, O. Paulson, J. Ewald, J. K. Bjerrum, A. Fahrenkrug, and N. A. Lassen: Regional cerebral blood flow in acute apoplexy. The "luxury perfusion syndrome" of brain tissue. Arch. Neurol. 17, 271 (1967).
2. Lassen, N. A.: The luxury-perfusion syndrome and its possible relation to acute metabolic acidosis localized within the brain. Lancet II, 1113 (1966).
3. —, and O. B. Paulson: Partial cerebral vasoparalysis in patients with apoplexy: dissociation between carbon dioxide responsiveness and autoregulation. Presented at the International CBF Symposium, Mainz, April 1969 (this volume, p. 117).
4. Olesen, J.: Total CO_2, lactate and pyruvate in brain biopsies taken from patients with intracerebral tumors after freezing the tissue in situ. Presented at the International CBF-Symposium, Mainz, April 1969 (this volume, p. 181).
5. Paulson, O. B., S. Cronquist, J. Risberg, and F. I. Jeppesen: Regional cerebral blood flow: a comparison of 8-detector and 16-detector instrumentation. J. nucl. Med., 10, 164 (1969).
6. — in preparation, 1969.
7. —, N. A. Lassen, and E. Skinhøj: in preparation.
8. Skinhøj, E.: Regional cerebral blood flow at rest and during functional tests in transient ischemic attacks. Presented at the International CBF Symposium, Mainz, April 1969 (this volume, p. 115).

rCBF at Rest and During Functional Tests in Transient Ischemic Attacks

E. Skinhøj

Department of Neurology, Bispebjerg Hospital, Copenhagen

It is well known that in more than half of the cases with cerebral apoplexy, particularly, but not exclusively, in the transient attacks, it is impossible to detect cerebral arterial occlusion or hemorrhage.

A possible explanation for this is that part of the cerebral vascular system is impaired by arteriosclerosis or stenosis in some afferent vessel to such an extent that the blood supply is marginal or the normal autoregulation is abolished. If this is so, then a transient decrease in systemic blood pressure, an increase in venous pressure, a lack in oxygen saturation of the blood, a hypoglycemia or kinking and extracerebral compression of cerebral vessels can result in a focal cerebral hypoxia and transient or permanent neurological symptoms. This is the haemodynamic crisis of Denny-Brown and Meyer. In single cases such mechanisms have been proven, but do they play a role in the majority of cases?

To study this problem we examined 24 patients with strokes without angiographically detectable occlusion, half of them belonging to the so-called transient focal cerebral ischaemic group with repeated attacks.

We used the intracarotid injection method with ^{133}Xe with 16 extracranial recordings and with functional tests in repeated measurements after controlled changes in the systemic blood pressure and in the arterial pCO_2 [1].

The results appear in the table. The symbol $(+)$ indicates that neighbouring channel-curves differ significantly from the rest of the remaining curves during rest/or during the functional tests.

It appears from the table that focal abnormalities, i. e. hyperaemic or ischaemic foci, focally impaired reaction upon BP changes, or focally impaired reaction upon changes in the arterial pCO_2, were only seen shortly after an attack. These findings following an attack are presumably explained by a postischaemic focal acidosis [2, 3].

The lack of such findings in between, i. e. "before" an attack makes it unlikely that they should have played a pathogenetic role in our cases.

These investigations thus give evidence against the haemodynamic theory and they indirectly support the theory of emboli of fibrine and platelets from diseased endothelium as the origin of the majority of transient focal cerebral ischaemic attacks.

The more definite solution of this problem seems to be fundamental for any prophylactic approach.

Table 1. *Focal findings in patients with transient ischemic and minor apoplectic attacks*

Case no.	Time interval since last attack	Focal abnormalities at rest	Focal loss of reaction to changed arterial pCO_2	Focal loss of autoregulation
Transient ischemic attacks				
42	½ hour	hyperemic		+
43	2 days	no	no	no
44	2½ days	(ischemic)	no	no
45	5 days	no	no	no
46	7 days	no	no	no
47	7 days	no	no	no
48	8 days	no	no	no
49	8 days	no		no
50	17 days	no	no	no
51	1½ month	no	no	no
52	3 months	no	no	(+)
53	4 years	no	no	no
Minor apoplectic attacks				
54	½ day	ischemic		+
55	1 day	hyperemic	+	paradox
56	1 day	hyperemic	+	+
57	2 days	no	no	no
58	3½ days	hyperemic	no	no
59	4 days	no	+	+
60	21 days	no	no	no
61	1 month	no	no	no
62	2 months	ischemic	no	no
63	2 months	no	no	no
64	3½ months	no	no	
65	2 years	no	no	

References

1. Høedt-Rasmussen, K., E. Skinhøj, O. Paulson, J. Ewald, J. K. Bjerrum, A. Fahrenkrug, and N. A. Lassen: Regional cerebral blood flow in acute apoplexy. Arch. Neurol. **17**, 271 (1967).
2. Skinhøj, E.: CBF adaptation in man to chronic hypo- and hypercapnia and its relation to CSF pH. Scand. J. clin. Lab. Invest. Suppl. **102**, p. VIII:A (1968).
3. — Regulation of cerebral blood flow as a single function of the interstitial pH in the brain. Acta neurol. scand. **42**, 604 (1966).

Partial Cerebral Vasoparalysis in Patients With Apoplexy: Dissociation Between Carbon Dioxide Responsiveness and Autoregulation

N. A. LASSEN and O. B. PAULSON

Departments of Clinical Physiology and Neurology, Bispebjerg Hospital, Copenhagen

A dissociation of the cerebral vasomotor response in the form of a preserved vasoconstriction to hypocapnia and an abolished vasoconstriction to hypertension was reported in a patient with a parasagittal meningeoma by EASTON and PALVÖLGYI [1]. The present paper analyzes the frequency of such partial vasoparalysis in a material of 23 cases of apoplexy studied with the 16 channel rCBF technique. These 23 cases represent all cases of PAULSON's series of 41 apoplexies in which both carbon dioxide reactivity and autoregulation studies were per-

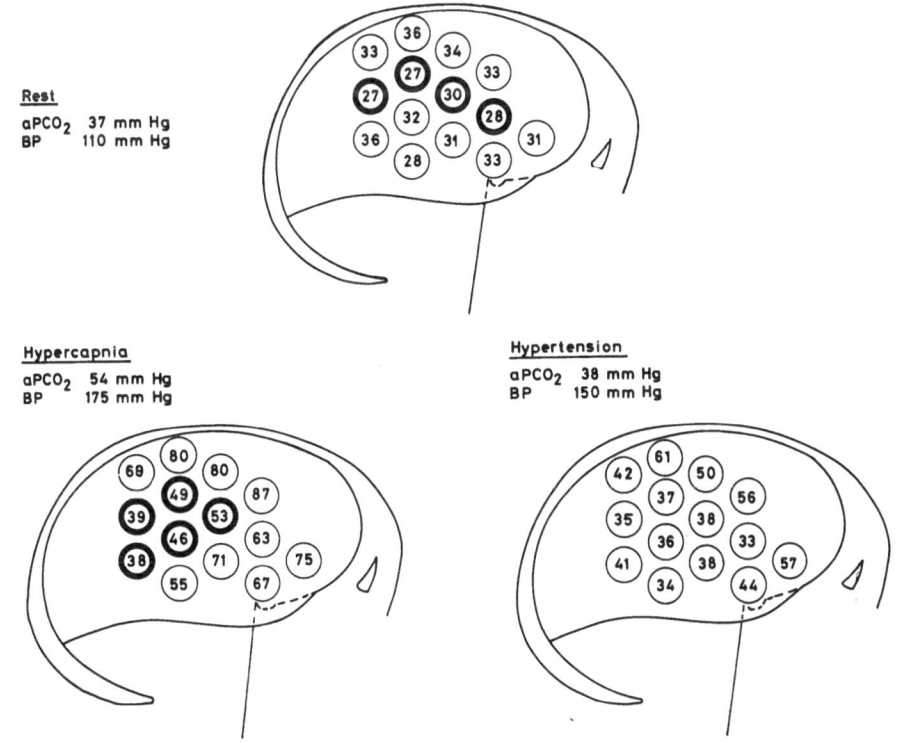

The rCBF initial values are indicated in the circels

Fig. 1. An ischemic focus in a patient with apoplexy (without arterial occlusion) studied 1 day after the onset of symptoms is illustrated. In the ischemic focus there is a markedly reduced reaction to hypercapnia. During hypertension, one observes loss of autoregulation in all regions studied; flow increase is more marked in the non-focal areas than in the focal ones

formed [5]. 4 of the patients (group O) had occlusion of the middle cerebral artery, while 19 (group NO) had no occlusions on the angiographic films.

A localized dissociation of response was seen in 1 patient (NO, 5 days after onset) where the focal area showed a preserved autoregulation as well as preserved vasoconstriction during hypocapnia, but a lack of vasodilatation during severe hypercapnia.

A more widespread dissociation of response (Fig. 1) was seen in 7 patients (O, 1 case, NO, 6 cases) out of a total of 15 studied within 2 weeks of the onset of symptoms, while no such dissociation was seen in the remaining 8 cases which were studied at a later time. All seven had in most or all regions a loss of autoregulation (in the form of flow increase during Aramine or Angiotensine infusion), but a preserved CO_2 responsiveness. Out of the 7 cases with lost autoregulation 3 had percentually a smaller flow increase in focal areas or even a regional flow decrease during induced hypertension.

Discussion

The observations reported here all deviate from the simple pattern regarding cerebro-vascular pathophysiology in apoplexy: one might have expected the focal area to be characterized by vasoparalysis affecting all normal vasomotor responses, and that non-focal areas had preserved responsiveness. This "reasonable" pattern was indeed typical in our series of acutely investigated patients with apoplexy – both with and without angiographically proven occlusion of cerebral arteries [2]. The 3 types of deviations will be separately commented on.

Localized vasoconstriction to blood pressure increase (autoregulation) and to hypocapnia but no vasodilatation during severe hypercapnia. This pattern was only seen in 1 single case (NO) that was studied 5 days after the acute onset. No simple explanation of this dissociation of vasomotor responses is apparent to us. However, the intensity of the hypercapnia (aPCO$_2$ = 73 mmHg) and the marked flow increase in non-focal areas suggest a marked rise in intra-cranial pressure which may well be of importance.

Widespread loss of autoregulation with preserved carbon dioxide responsiveness. This dissociation was found in not less than 6 out of 11 patients with apoplexy with no vascular occlusion on angiography, and studied within the first 2 weeks.

As a tentative hypothesis for the mechanism of this dissociation we may assume that from the focal lesion lactic acid has leaked out; after a while this cerebrospinal fluid acidosis also reaches the extracellular fluid surrounding the cerebral arterioles [3]; this in turn shifts the autoregulatory range towards a higher than normal pressure level [2]. These theoretical considerations were given since they allow us to predict two phenomena which can be tested: first, this type of dissociation should be associated with cisternal lactacidosis and the autoregulation should reappear if tested during hyperventilation. As mentioned above the loss of autoregulation in some of the patients involved all regions of the diseased hemisphere studied. For this reason we have also termed the phenomenon "global loss of autoregulation" [5]. Second, it would be interesting to know if the autoregulation was also lost in the contralateral hemisphere in such cases.

Paradoxical focal flow reactions during induced hypertension in patients with widespread loss of autoregulation. This was found in 3 of the NO-cases above mentioned. We interpret it as an "intracranial steal" effect which is also seen in patients with intracranial tumors during induced rise in arterial blood pressure [4]. Indeed the focal lesion in apoplexy, perhaps especially of the non-occlusive type (post-occlusive?) may well act as an acute "tumor", i. e. as a space occupying lesion producing the same pathophysiological derangement.

References

1. EASTON, J. D., and R. PALVÖLGYI: The dissociation of cerebral vasoconstrictor response to hypocapnia and hypertension. Scand. J. clin. Lab. Invest. Suppl. **102**, p. V:J (1968).
2. HÄGGENDAL, E.: Om autoreglering av hjärnans blodföde. Thesis. University of Göteborg, 1965.
3. LASSEN, N. A.: Brain extracellular pH: the main factor controlling Cerebral Blood Flow. Scand. J. clin. Lab. Invest., Editorial **22**, 247 (1968).
4. PAULSON, O. B.: Regional Blood Flow at rest and during functional tests in occlusive and non-occlusive cerebrovascular disease. Presented at the International CBF Symposium, Mainz, April 1969 (this volume, p. 111).
5. PALVÖLGYI, R.: Regional cerebral blood flow in patients with intracranial tumors, J. Neurosurg, August 1969, in press.

Discrepancies Between Autoregulation and CO₂ Reactivity of Cerebral Vessels

C. Fieschi, A. Agnoli, L. Bozzao, N. Battistini, and M. Prencipe

Department of Neurology and Psychiatry, University of Genoa

The problem of the "discrepancies" that we raised at the Bonn and Lund Symposia in 1968, can be expressed by saying that a rigorous separation between entirely "good" and "bad" vessels does not often hold in practice. When studying CBF in brain disease one should therefore qualify the concept of "vasoparalysis", in terms of:

a) the stimulus employed to test it; b) the threshold chosen to separate normal and abnormal responses, and the statistical basis for choosing such a threshold. Beside the errors of the method one should consider that the stimuli are seldom pure, since some CO_2 change usually accompanies changes in blood pressure and specially – vice-versa; furthermore, during a test situation a true steady state is often difficult to obtain.

Still, we are left with the important task of evaluating the regional cerebral vascular reactivity in disease states. By employing statistically "protected" criteria, and evaluating the entire desaturation curve of ^{85}Kr rather than the initial slope (see Agnoli et al., 1968; Fieschi et al., 1969), we found that in acute focal brain lesions the vasomotor responses are regionally altered, and often so in a discrepant manner. That is, we noted that about 50% of the regions had either impaired autoregulation with preserved response to CO_2, or vice-versa.

There are two alternative interpretations of such findings, 1. the discrepancies confute the unitarian theory of the cerebral vasomotor regulation, 2. the "pH" model still holds and these dissociations must be explained as an artefact.

Let us consider the last hypothesis. The parietal and occipital regions of the patient shown in Fig. 1 might include vessels that are normally regulated, whose response to CO_2 gives a sufficient vasodilatation to simulate a significant increase in the mean flow for the entire region. At the same time, other vessels may be unreacting, thus increasing the "mean" regional flow significantly when the blood pressure is raised.

In this case, the artefact depends on the fairly large size of the region explored by our probes (open face $2^1/_2$ cm), a region which includes vessels in different functional states. If this is so, then the smaller the region explored, the lower the frequency of such dissociated responses: however, as reported by Lassen this dissociation is frequently encountered even when using much smaller probes.

The opposite type of discrepancy, namely preservation of autoregulation with impaired response to CO_2, exemplified in the temporal region of the same patient and in the temporal region of case shown in Fig. 2, is more difficult to explain on the basis of artefacts, since the coexistence of vessels in different functional state should not lead to such dissociation. These vessels seem (for mechanical reasons?) uncapable to vasodilate, while they are still

32,5 | +13,3 | +18,8   29,6 | +0,8 | +7,2   27,8 | +8,3 | +1,7

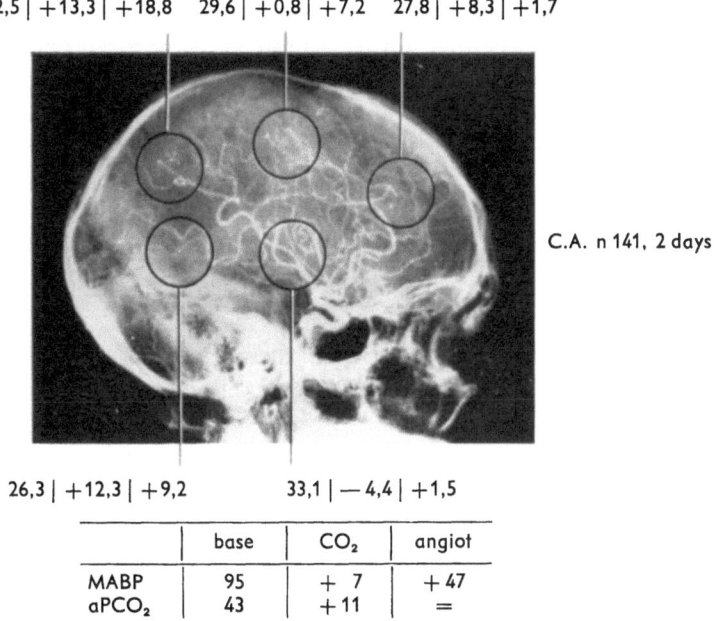

C.A. n 141, 2 days

26,3 | +12,3 | +9,2   33,1 | − 4,4 | +1,5

	base	CO_2	angiot
MABP	95	+ 7	+ 47
aPCO$_2$	43	+ 11	=

Fig. 1. Patient C. A., studied 2 days after onset of a right hemiplegia and aphasia due to hemispheric infarction. The numbers represent the mean regional blood flow (^{85}Kr intracarotid injection, compartmental analysis), and its absolute changes during CO_2 inhalation and during angiotensin infusion. Changes in aPCO$_2$ and MABP during tests are indicated at the bottom of the figure. The 5 regions behave unevenly: the parietal and occipital regions react to CO_2 but have lost autoregulation; the temporal region shows preserved autoregulation but reacts to CO_2 in a paradoxical manner; the rolandic region has lost both autoregulation and CO_2 reactivity; finally the frontal region reacts normally

30,6 | − 5,4 | − 4,2   21,5 | +4,8 | +4,0   23,6 | +7,1 | +4,9

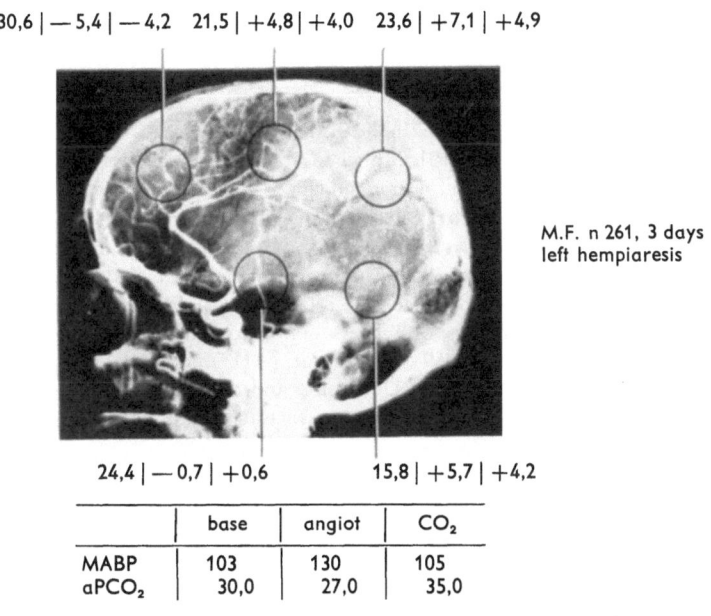

M.F. n 261, 3 days
left hempiaresis

24,4 | − 0,7 | +0,6   15,8 | +5,7 | +4,2

	base	angiot	CO_2
MABP	103	130	105
aPCO$_2$	30,0	27,0	35,0

Fig. 2. Patient M. F., studied 3 days after the onset of a left hemiparesis with occlusion of the right M.C.A. The numbers represent the mean regional blood flow and its absolute changes (in ml/100 g/min) during Angiotensin infusion and during CO_2 inhalation, Changes in MABP and aPCO$_2$ during tests are indicated at the bottom of the figure. Dissociated response (preserved autoregulation with lost response to CO_2) in the temporal region

able to respond to vasocontricting stimuli as suggested in the study by Brawley [4]. Another hypothesis however might be that autoregulation is only apparently preserved. The increased intracranial pressure due to the vasodilatation in the confining regions may indeed increase the extravascular resistances, enough to counteract the effects of the increased arterial pressure. In certain extreme cases it is possible to observe in such regions with severe vasoparalysis and brain edema a paradoxical decrease in flow during induced hypertension. Not all the available evidence is in favour of these hypotheses, and we should not discard the alternative explanation, namely that independent mechanisms may regulate the response of cerebral vessels to CO_2 and to perfusion pressure changes [5], and that these mechanisms may be effected at a different level in disease states.

These incongruous responses have certain therapeutic implications. This will be apparent in the papers presented by Battistini [2] and by Brock [3].

References

1. Agnoli, A., C. Fieschi, L. Bozzao, N. Battistini, and M. Prencipe: Autoregulation of cerebral blood flow: studies during drug induced hypertension in normal subjects and in patients with cerebrovascular diseases. Circulation **38**, 800 (1968).

2. Battistini, N., M. Casacchia, A. Bartolini, G. Bava, and C. Fieschi: Effects of hyperventilation on focal brain damage following middle cerebral artery occlusion. Presented at the International CBF Symposium, Mainz, April 1969 (this volume, p. 249).

3. Brawley, W. B., D. E. Standness, and A. Kelly: The physiologic response to the therapy in cerebral ischemia. Arch. Neurol. **17**, 180 (1967).

4. Brock, M., A. A. Hadjidimos, and K. Schürmann: Possible adverse effects of hyperventilation on rCBF during the acute phase of total proximal occlusion of a main cerebral artery. Presented at the International CBF Symposium, Mainz, April 1969 (this volume, p. 254)

5. Ekström-Jodal, B., and E. Häggendal: Cerebral blood flow and metabolism in patients with respiratory insufficiency, with special regard to induced acute changes of the blood gas situation. Presented at the International CBF Symposium, Mainz, April 1969 (this volume, p. 82).

6. Fieschi, C.: Regional cerebral blood flow in acute apoplexy, including pharmacodynamic studies. Scand. J. clin. Lab. Invest. Suppl. **102**, XVI:E (1968).

7. —, A. Agnoli, N. Battistini, L. Bozzao, M. Nardini, and M. Prencipe: Vasomotor responses of cerebral vessels in brain disease. In: E. Betz and R. Wüllenweber (Eds.), Pharmakologie der lokalen Gehirndurchblutung, p. 181. München: Verlag Banaschewski 1969.

8. — — — —, and M. Prencipe: Derangement of regional cerebral blood flow and of its regulatory mechanisms in acute cerebrovascular lesions. Neurol. **18**, 1166 (1968).

9. — —, M. Prencipe, N. Battistini, L. Bozzao, and M. Nardini: Impairment of the regional vasomotor response of cerebral vessels to hypercarbia in vascular diseases. Europ. Neurol. **2**, 13 (1969).

10. Lassen, N., and O. B. Paulson: Partial cerebral vasoparalysis in patients with apoplexy: dissociation between carbon dioxide responsiveness and autoregulation. Presented at the International CBF Symposium, Mainz, April 1969 (this volume, p. 117).

11. Symon, L., and R. Wüllenweber: Discussion and comments to section XIII: Intracerebral steal by vasodilators and inverse steal by vasoconstrictors. Scand. J. clin. Lab. Invest. Suppl. **102**, XIII:F (1968).

Microflow Patterns after Experimental Occlusion of Middle Cerebral Artery Demonstrated by Fluorescein Angiography and [133]Xenon Clearance

Y. L. Yamamoto, C. P. Hodge, K. M. Phillips, and W. Feindel

Cone Laboratory for Neurosurgical Research, Montreal
Neurological Institute, and McGill University, Montreal

During the past few years we have developed high resolution techniques for investigation of the structural and hemodynamic changes in cerebral blood flow on the exposed brain. The colour fluorescein photographic techniques of ciné films (24 frames per sec) and rapid sequence (3 per sec) single frames with a 1% Fluorescein solution injected into the carotid artery provide fine details of flow patterns in small cerebral vessels and in the cortical capillary bed on the surface of the brain [2].

In addition, the blood flow was measured directly from the exposed brain by Xenon-clearance using a multi-channel minisemiconductor detector system[1]. The detectors, made of lithium drift silicon with an active area of 4 mm- and 1 mm deep, have a signal response range within 1 cm radius in cerebral tissue with Xenon^{-133}.

With these experimental techniques, the details of which were described elsewhere [2, 3], we have examined the influence of arterial pCO_2 changes on the volume and pattern of microflow in an occluded area of the exposed brain distal to a branch of the middle cerebral artery which had been closed off temporarily by metal clip.

1. Influence of Arterial pCO₂ Changes on the Epicerebral Fluorescein Angiogram

No significant changes in the flow pattern in the normal or occluded areas were noted under normocapnic and hypocapnic conditions. But under hypercapnia, striking changes were observed with the direction of change in the occluded area being opposite to that in the normal cortex. In the normal cortex increase of arterial pCO_2 by inhalation of 5% CO_2 produced 1. more rapid circulation, 2. early filling of veins when the pCO_2 was over 50 mmHg and red veins when the pCO_2 was over 65 mmHg, 3. scattered and less dense filling by fluorescein in the capillary phase and shorter capillary circulation time. Conversely, in the occluded, the circulation time was paradoxically prolonged, but also with scattered and less dense filling by fluorescein. As noted in Table 1a, no significant change of fluorescein microflow patterns occurred between pCO_2 levels of 40 and 42. With the pCO_2 at 52, we

Supported by Grant MA-3174, Medical Research Council of Canada, and by the Cone Memorial Fund, McGill University and the Montreal Neurological Institute.

[1] Designed in collaboration with Simtec, Limited, Montreal.

observed early filling veins, shorter capillary circulation time, scattered and less dense filling of fluorescein in the normal cortex and a prolonged circulation time in the occluded area.

Table 1a. *Fluorescein circulation time Dog 69–2*
Effect of $PaCO_2$ *Changes following occlusion of middle cerebral artery*

	Time After Occlusion (min)	PCO_2	Normal Initial Vein Filling (sec)	Capillary Phase (sec)	Occluded Maximum Filling (sec)	Disappearance Time (sec)
Air Inhalation (Resp. Rate 8–9/min)	35	40	3.7	2.55–3.17 dense	5.48–7.81 dense	11
Hyperventilation (Resp. Rate 15–16/min)	62	24	3.7	2.78–3.74 dense	4.68–7.43 dense	10
5% CO_2 Inhalation Resp. Rate 9/min	137	52	2.5	2.57–2.91 scattered, less dense	8.94–12.82 scattered, less dense	23

2. Effect of Arterial pCO_2 Changes on $Xenon^{133}$ Clearance Studies

2 detectors of the 4-channel system were placed over the normal cortex and 2 over the occluded areas, these areas being easily defined during fluorescein angiography. With hyperventilation, the regional cerebral blood flow in the normal as well as in the occluded areas was reduced considerably in all cases (Table 1b). However, the reduction of blood perfusion rate in the normal cortex (probes 1 and 4) was much more intense than in the occluded area (probes 2 and 3). This could be related to loss of regulation of the blood vessels in the occluded area. With CO_2 inhalation, the blood perfusion rates were markedly increased in all 8 experiments, not only in the normal cortex but also in the occluded area, though the rate of increase in the occluded area was usually less.

Table 1b. *Xenon-133 clearance studies Dog 69–2*

	$PaCO_2$	Regional Blood Flow Probe 1	Probe 2	F/λ (infinite value) Probe 3	Probe 4
A. Before Occlusion	40	0.434	0.604	1.028	0.651
B. Hyperventilation	24	0.157 (−64%)[1]	0.276 (−54%)[1]	0.675 (−34%)[1]	0.114 (−82%)[1]
C. 5% CO_2 Inhalation	52	0.635 (+46%)[2] (+300%)[3]	0.693 (+15%)[2] (+150%)[3]	1.327 (+29%)[2] (+96%)[3]	0.706 (+8%)[2] (+520%)[3]

$$[1]\ \frac{A-B}{A} \qquad [2]\ \frac{C-A}{A} \qquad [3]\ \frac{C-B}{B}$$

Probe 1 and 4 Normal
Probe 1 and 3 Occluded

As noted in Table 1b, and Fig. 1, under hypocapnic condition the blood perfusion rate in the occluded area was reduced by 34% and 54% in the 2 probe areas. That in the normal cortex was reduced by 60% and 80% in the corresponding 2 sampling areas. When the arterial

pCO_2 was then increased to 52, the perfusion rate in the normal cortex was increased to over 300% and 500% as compared to the values obtained when the pCO_2 was 24. The rate of increase of blood perfusion, however, measured by the 2 probes in the occluded area was 96% and 150%. Nevertheless, the blood perfusion rate in the occluded area under hypercapnia showed a definite increase as compared to the rates under normocapnia.

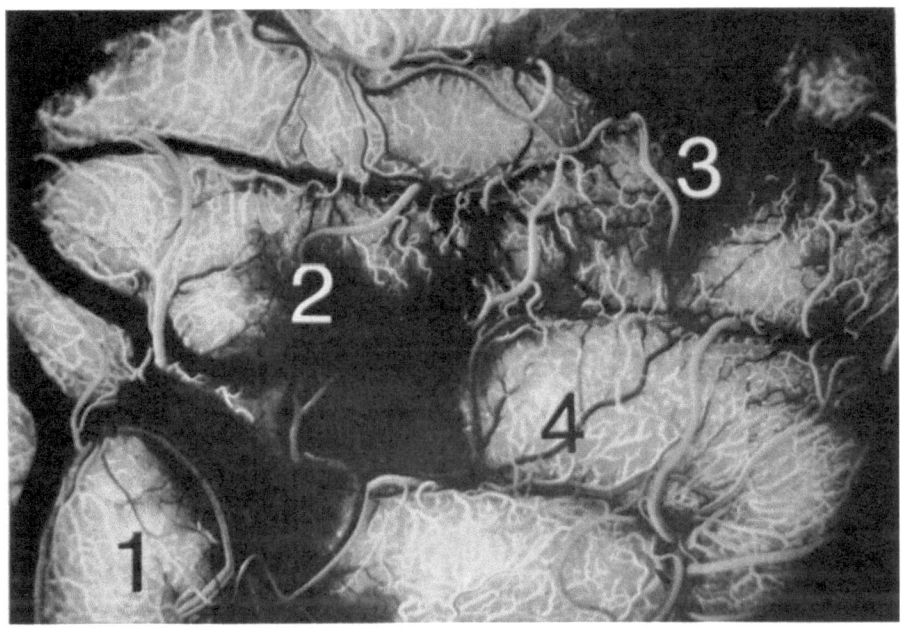

Fig. 1. Probes 1 and 4 are placed over the normal cortex, Probes 2 and 3 are placed over the occluded area. The details of Xenon-clearance studies are illustrated in Table 1 b

Summary

The main features in 8 experiments are summarized in Table 2. The most interesting finding was that hypercapnia gave prolonged circulation time in the occluded area by fluorescein angiography whereas a considerable increase in the blood perfusion rate in this same area as measured by Xenon-clearance.

Table 2. $PaCO_2$ *influence on the microregional CBF in 8 dogs*

| | | Normal | | Occluded | |
		Fluorescein	Xenon-133	Fluorescein	Xenon-133
$PaCO_2$	↓	no change	marked reduction	no change	moderate reduction
$PaCO_2$	↑	rapid circulation early filling veins. Red Vein scattered capillary phase	marked increase	slow prolonged circulation	moderate increase

Other workers have suggested that "intracerebral steal" took place in relation to ischemic areas during increased pCO_2 levels [1, 4]. Our present observations, made by Xenon-clearance technique and fluorescein angiography in the same experimental model, do indicate the presence of various degrees of vasoparalysis in the occluded area. But our data failed to demonstrate any evidence for "intracerebral steal" under hypercapnia. On the contrary, they show an increase of blood flow in both the normal and occluded regions during CO_2 inhalation and a definite reduction of blood flow during hyperventilation.

References

1. BRAWLEY, B. W.: The pathophysiology of intracerebral steal following carbon dioxide inhalation. An experimental study. Scand. J. clin. Lab. Invest. Suppl. **102**, p. XIII:B (1968).
2. FEINDEL, W., Y. L. YAMAMOTO, and C. P. HODGE: Intracarotid fluorescein angiography: a new method for examination of the epicerebral circulation in man. C.M.A.J. **96**, 1 (1967).
3. —, H. GARRETSON, Y. L. YAMAMOTO, C. HASLAM, and M. HEUFF: Analysis of the blood flow pattern in the pial and cortical circulation in man. Acta neurol. scand. Suppl. **14**, 187 (1965).
4. SYMON, L.: Experimental evidence for "intracerebral steal" following CO_2 inhalation. Scand. J. clin. Lab. Invest. Suppl. **102**, p. XIII:A (1968).

Cerebral Circulation Speed and EEG Frequency

L. A. Ectors

Neurological Clinic, Brussels

Cerebral blood flow is a function of brain metabolism [1, 2, 5, 8]. In the physiological state, blood flow is determined by brain activity [8, 9]. All the methods that increase EEG frequency increase blood flow, and vice versa [3, 4]. There is a correlation between EEG frequency and cerebral blood flow in the senile mental patient [11]; a correlation is often observed between EEG rhythm and blood flow in pathological conditions of the blood vessels of the brain [5].

We measured the speed of circulation in the brain by arteriography. Presumably the speed thus measured parallels blood flow and functional activity. The present study is confined to only a few cases in which artefact-free observations were made.

In 5 observations of thrombosis of the internal carotid on one side, and in 6 observations of atheroma of the internal carotid where the diameter of the vessel was diminished at its origin by more than 50%, a relationship was established between the magnitude of the circulatory lag and the mean frequency of the EEG (Fig. 1).

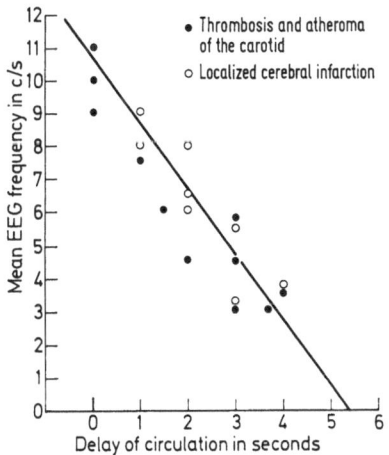

Fig. 1

In cases with unilateral internal carotid thrombosis, the circulatory speed was asymmetric with a lag on the affected side in 1 case, and a lag on both sides in 1 case; it was normal on both sides in 2 patients and showed an equal lag on both sides in 1 case. Apparently, in the 2 instances of normal flow, collateral circulation ensured satisfactory circulation to the affected hemisphere. The EEG rhythm was symmetric or asymmetric paralleling the circulatory lag.

In the case of atheroma of 1 internal carotid artery, there was a circulatory lag on both sides, greater on the affected side in 2 cases. Normal circulatory speed was seen in 1 case. In 1 patient there was a lag on the side opposite to the lesion, with normal speed on the affected side. There was no compensatory collateral circulation in these cases.

The foregoing observations do not include any case of cerebral infarct.

In 8 observations of cerebral infarct, without involvement of the carotid or vertebral arteries, a correlation was established between the local EEG frequency and the corresponding regional cerebral circulatory speed. The slowing down of circulation in these cases was not homogenous throughout the entire hemisphere but most often varied from one region to another. The correlation was identical with that established on the basis of the observation of carotid atheroma or thrombosis. The same correlation was found again in a study of a case of Alzheimer's disease.

In contrast, there was no correlation whatsoever between circulatory speed and EEG frequency when the circulation is not regulated by the functional metabolism of the neurons, as in cases of arteriovenous angioma and vascular paralysis or luxury perfusion.

References

1. Clasen, R., P. Cooke, S. Pandolfi, G. Carnecki, and G. Bryar: Hypertonic urea in experimental cerebral edema. Arch. Neurol. 12, 424 (1965).
2. Ectors, L., and J. Achslogh: Les suppléances vasculaires cérébrales et la vitesse circulatoire cérébrale. Acta clin. belg. 17, 189 (1962).
3. Gleichmann, U., D. H. Ingvar, N. A. Lassen, D. W. Lübbers, B. K. Siesjö, and G. Thews: Regional cortical metabolic rate of oxygen and carbon dioxide related to the EEG in anesthetized dog. Acta physiol. scand. 55, 82 (1962).
4. Ingvar, D. H., M. Baldy-Moulinier, I. Sulg, and S. Hörman: Regional cerebral blood flow related to EEG. Acta neurol. scand. Suppl. 14, 179 (1965).
5. —, The pathophysiology of occlusive cerebrovascular disorders. Acta neurol. scand. 43, Suppl. 31: 93 (1967).
6. —, and J. Risberg: Increase of regional cerebral blood flow during mental effort in normals and in patients with focal brain disorders. Exp. Brain Res. 3, 195 (1967).
7. —, and I. Sulg: Regional cerebral blood flow and EEG frequency content in man. 1967, Scand. J. Lab. clin. Invest., Suppl. 114, (1969).
8. Kety, S. S.: Blood flow and metabolism in the human brain in health and disease. In: Neurochemistry. E. Nurnberger (Ed.). p. 294. Springfield: Charles C. Thomas (1956).
9. Lassen, N. A.: Cerebral blood flow and oxygen consumption in man determined by the inert gas diffusion method. Copenhagen, 104 pp. (1958).
10. — The luxury-perfusion syndrome and its possible relation to acute metabolic localized acidosis within the brain. Lancet II, 1113 (1966).
11. Obrist, W. D.: Cerebral ischemia and the senescent electroencephalogram. In: Cerebral Ischemia, E. Simonson and Th. Mc. Gavack (Eds.), p. 71. Springfield: Charles C. Thomas (1964).

Correlations between Brain Gas Exchange and CSF Acid-Base Status in Patients with Cerebral Hemorrhage

A. E. Kaasik and R. Zupping

Department of Neurology and Neurosurgery, Tartu State University, Tartu

Earlier studies [1, 4] have shown that even a brief hypoxia leads to longlasting intracellular and extracellular lactacidosis in the brain. The CSF remains acidotic for prolonged periods even after a transient tissue hypoxia.

The dependence of CBF upon the brain extracellular pH has been directly established [3]. At the same time, there seems to be incomplete information about the relations between cerebral oxygenation and the CSF acid-base condition in patients with cerebral hemorrhage. Therefore, the aim of this work was to study the accumulation of lactate and the disappearance of bicarbonate in the CSF of the hemorrhage patients and to relate these changes to the concomitant alterations in arterial and cerebral venous blood gases and to the extent of brain damage.

Material and Methods

Investigations were carried out in 51 patients with intracerebral hemorrhage. 28 persons without cerebrovascular abnormalities constituted a control group for laboratory technique. The patients with hemorrhage were of various degrees of severity; during the acute stage of the disease 27 of them died.

The arterial (A) and internal jugular venous (VJ) blood samples were analysed for pH, pCO_2, bicarbonates and pO_2 by using the Astrup technique. The CSF pH was measured with a Radiometer electrode; the CSF pCO_2 was derived by means of a modified Siggaard-Andersen nomogram for pK = 6.13. The actual bicarbonate of the CSF was calculated with the help of the Henderson-Hasselbach equation. All samples were also analysed for lactate/pyruvate ratio. These analyses were performed with the colorimetric methods of Barker-Summerson and Friedemann-Haugen, respectively. The statistical process was performed by the use of the electronic computer Ural-4.

Results

The data given in Table 1 indicate that a rather noticeable respiratory alkalosis of both arterial and jugular venous blood was characteristic in the hemorrhage patients. The decrease of pCO_2 was especially remarkable in the lethal group. Both A pO_2 and VJ pO_2 were considerably decreased in the CH group. However, the cerebral venous hypoxaemia was remarkably more pronounced than the arterial. It was of interest that neither the arterial lactate, pyruvate and lactate/pyruvate ratio nor the corresponding cerebral venous values differed substantially from the corresponding control data.

Table 1. *Arterial, jugular venous and CSF acid-base parameters in brain hemorrhage (mean values ± SD of mean)*

	Arterial				Jugular venous				CSF					
	pH	pCO_2 mmHg	HCO_3^- mEq/l	pO_2 mmHg	pH	pCO_2 mmHg	HCO_3^- mEq/l	pO_2 mmHg	pH	pCO_2 mmHg	HCO_3^- mEq/l	La^- mEq/l	Py^- mEq/l	La^-/Py^-
Control patients	7.409 ±0.003	40.0 ± 0.6	24.5 ± 0.3	101.7 ± 1.9	7.360 ±0.005	48.4 ± 0.7	26.2 ± 0.5	43.0 ± 1.0	7.338 ±0.005	46.3 ± 0.6	23.5 ± 0.3	1.62[a] ±0.18	0.110[a] ±0.012	14.9[a] ± 1.3
Cerebral hemorrhage	7.464[b] ±0.004	32.4[b] ± 0.6	22.5[b] ± 0.2	83.0[b] ± 1.6	7.413[b] ±0.005	38.1[b] ± 0.7	23.4[b] ± 0.4	37.5[b] ± 0.9	7.302[b] ±0.007	41.6[b] ± 0.6	19.6[b] ± 0.4	5.49[b] ±0.32	0.174[b] ±0.014	32.7[b] ± 1.7
Survivors	7.458 ±0.006	34.8 ± 0.8	23.7 ± 0.4	88.3 ± 2.2	7.407 ±0.007	41.1 ± 1.0	24.9 ± 0.5	37.3 ± 1.4	7.322 ±0.006	43.7 ± 0.9	21.3 ± 0.5	4.83 ±0.66	0.138 ±0.018	34.1 ± 2.6
non survivors	7.472[c] ±0.006	29.3[b] ± 0.7	20.8[b] ± 0.5	78.6[b] ± 2.2	7.417 ±0.007	35.5[b] ± 0.9	22.0[b] ± 0.5	38.2 ± 1.4	7.283[b] ±0.013	39.7[b] ± 0.5	18.1[b] ± 0.5	5.85 ±0.44	0.194 ±0.036	31.9 ± 2.4

[a] Control values from PONTEN et al. (Scand. J. clin. Lab. Invest. 1968, Suppl. 102).
[b] p <0.01.
[c] p <0.05.

It was found that a considerable decrease of the CSF bicarbonate concentration was characteristic for the hemorrhage group. The concomitant decrease of the CSF pCO_2 was relatively less marked. Hence, the CSF acidosis was of metabolic origin and was caused by a marked increase of the CSF lactate concentration. The aforementioned changes were most remarkable in the cases with impaired consciousness and in patients who died. The CSF metabolic acidosis was in good correlation with the severity of coma. Although there was a considerable increase in the CSF pyruvate content, it remained relatively lower than the lactate values. The lactate/pyruvate ratio was, therefore, markedly increased.

Both arterial and venous hypoxaemia were most pronounced at the beginning of the disease; then there was a tendency to normalize, which was especially noticeable in cases with a good recovery. The same concerned the A and VJ pCO_2 values. In the surviving patients the CSF lactate usually normalized by the end of the first week. Although there was no considerable increase of CSF pCO_2 the CSF bicarbonate normalized on account of the lactate decrease.

Some correlation coefficients are worth mentioning. The CSF lactate increase correlated to the decrease of CSF bicarbonate ($r = -0.675$). Relatively smaller but still significant correlation was found between the increase of CSF lactate and decrease of CSF pH ($r = -0.373$). The correlation coefficient between the CSF lactate content and VJ pO_2 was -0.449, the correlation between the A pCO_2 and CSF pH was very small ($r = 0.291$).

Comments

The present study revealed that there is a remarkable lactacidosis in the CSF of patients with cerebral hemorrhage. This lactacidosis is caused by brain hypoxia which in its turn depends mostly on impaired CBF (the increase of CVR caused by the main disease and by the increase in intracranial pressure). Since the CBF is dependent on the brain extracellular pH, the results of this study indicate the biochemical basis for a comparatively longlasting relative hyperaemia [2] of the brain. The same mechanism is apparently responsible for the development of brain oedema. The results of this work show that the relatively more decreased V pO_2 (i. e. impaired brain oxygenation) was in most cases connected with the relatively more elevated CSF lactate concentration. However, in some deteriorating cases it was possible to find a very intensive CSF lactacidosis with concomitant increase in VJ pO_2 to supranormal values. The latter indicates relative hyperaemia of the brain.

CSF lactacidosis may give rise to the hyperventilation. However, the correlation coefficient between the A pCO_2 and CSF lactate values found in this study was rather small. Nevertheless, the data from another series, which consisted of brain injury cases, show a higher correlation ($r = -0.690$) between the mentioned values [5].

References

1. KAASIK, A. E., L. NILSSON, and B. K. SIESJÖ: Acid-Base and Lactate/Pyruvate Changes in Brain and CSF in Asphyxia and Stagnant Hypoxia. Scand. J. clin. Lab. Invest. Suppl. **102**, p. III:C (1968).
2. — Reduction of Cerebral Arteriovenous Oxygen Difference in Terminal Phase of Cerebral Haemorrhage. Scand. J. clin. Lab. Invest. Suppl. **102**, p. X:D (1968).
3. SIESJÖ, B., K. Å. KJÄLLQUIST, U. PONTEN, and N. ZWETNOW: Extracellular pH in the Brain and Cerebral Blood Flow. In: Progress in Brain Research. Vol. 30. W. LUYENDIJK (Ed.), Elsevier, p. 93, 1968.
4. THORN, W., H. SCHOLL, G. PFEIDERER u. B. MUELDENER: Stoffwechselvorgänge im Gehirn bei normaler und herabgesetzter Körpertemperatur unter ischämischer und anoxischer Belastung. J. Neurochem. **2**, 150 (1958).
5. ZUPPING, R., A. TIKK, and E. KROSS: Some Biochemical Mechanisms of Brain Damage in Cases with Brain Injury, 1969, in press.

9*

The Acid–Base–Equilibrium of the CSF and the Respiration in Cerebral Vascular Diseases

G. KIENLE

Neurological Department, Nordwest-Krankenhaus, Frankfurt/Main

The result of a cerebral ischemic insult is a disturbance in the autoregulation of the injured area. Vasodilator drugs may, in spite of an improvement of circulation in general, cause a deterioration of the circulation in such areas resulting in a "steal effect". LASSEN, PAULSON and others have suggested hyperventilation in cases of an acute ischemic insult in order to improve circulation in the injured area by means of a generalized constriction of cerebral vessels.

According to our experience, an ischaemic insult may cause hyperventilation and vasoconstriction physiologically. In several cases we have found a metabolic acidosis in the cerebrospinal fluid of such patients and an increased difference of the pCO_2 level between the cerebrospinal fluid and arterial blood (Fig. 1). This increase in difference is equivalent to a decrease in cerebral circulation, at least in some parts of the brain.

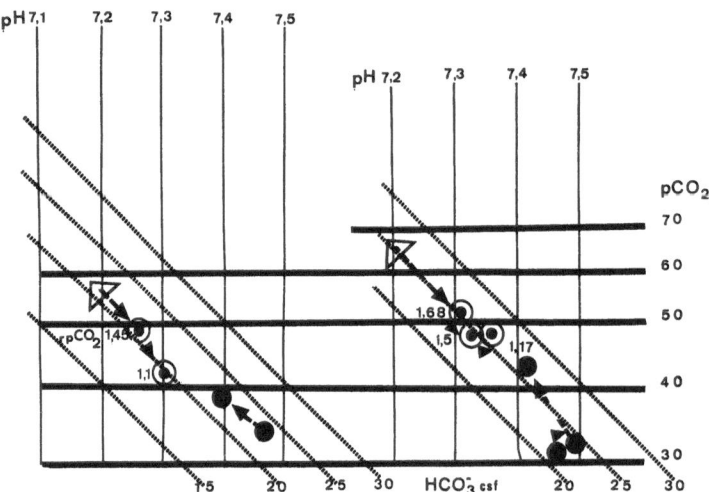

Fig. 1. Two patients with acute cerebrovascular insults. $\triangle$ CSF acute stage; $\odot$ CSF after treatment; $\bullet$ arterial values; $\rightarrow$ after treatment. Number $= r\,pCO_2 = pCO_2$ CSF/pCO_2 art.

Controls $\begin{cases} \text{CSF pH means values } (n = 70) \text{ 7.319} \\ r pCO_2 \text{ mean values } (n = 70) \text{ 1.242} \end{cases}$

In our opinion this change is based on the following mechanism: in the presence of a tissue hypoxia, a change of metabolism may occur resulting in accumulation of acid metabolites eventually causing vasodilatation. As soon as the acid metabolites, particularly lactic acid, have penetrated the cerebrospinal fluid a metabolic CSF acidosis occurs. As a consequence an increased breathing stimulus is triggered via the H-ion-sensitive area of the medulla oblongata. Furthermore in the infarction area a cerebral edema may occur which may throttle cerebral circulation. The decrease of the cerebral circulation will result in an increase of CO_2 concentration in the venous blood of the brain. In as much as the CSF pCO_2 follows the pCO_2 in the venous blood of the brain a CSF respiratory acidosis will result. Acid metabolites and the increase of the pCO_2 in the CSF will diminish the CSF pH and cause hyperventilation through the increased respiratory stimulus. The throttling of brain circulation through the respiratory stimuli will increase the pCO_2 difference between CSF and the arterial blood until the carbonic acid level has decreased to such an extent that, inspite of the increased difference between blood an CSF, the respiratory alkalosis in the CSF is equalized with regard to the metabolic acidosis, thus returning to an equilibrium. This mechanism will naturally not act in the case of an injury to the medulla oblongata when, in consequence of a central respiratory depression, an increase of the CSF H-ion concentration can no longer be met by an increased respiration.

Clinical Application of the Gamma Camera in the Evaluation of Patients with Cerebrovascular Disease

R. Janeway*, C. D. Maynard**, R. L. Witcofski**, R. Kemp***, and J. F. Toole*

Departments of Neurology*, Radiology**, and Anesthesiology*** of the Bowman Gray School of Medicine, Wake Forest University, Winston-Salem, North Carolina

We have performed more than 1500 "radioisotope arteriograms" by visualizing gamma emissions with a scintillation camera (Anger) during the first arterial circulation after the intravenous injection of a relatively non-diffusible radiopharmaceutical [2]. We believe that the technique is a rapid and efficient method for the identification of severe carotid artery stenosis and occlusion, middle cerebral artery occlusion, and arteriovenous malformation, and for post-operative evaluation of patients who have had carotid artery reconstructive surgery. It is an adjunct to rectilinear scanning that improves the probability of differentiating positive scans due to cerebral infarction, arteriovenous malformation and neoplasm. While the intravenous technique is useful for diagnosis, it cannot delineate quantitative variations in total or regional cerebral blood flow (rCBF). This report details our initial observations on the use of the scintillation camera and computer techniques to quantitate CBF. The techniques of radioisotope arteriography using Technetium-99m (^{99m}Tc), inhalation-recirculation, and arterial injection of Xenon-133 (^{133}Xe) will be compared.

Methods and Materials

Equipment. A scintillation camera (Anger)[1] equipped with a 1090 squarehole collimator, a 1600 channel multiparameter pulse height analyzer[2], and a magnetic tape transport and memory system[3] are interfaced for data collection and retrieval.

Data collection and retrieval. Permanent records are made by photographing the oscilloscope of the scintillation camera console with Polaroid or 35 mm film. For intravenous studies counts are collected and transferred to magnetic tape at 1 sec intervals for 24 sec. Data from inhalation-recirculation studies are accumulated for 30 sec intervals during the first 2 min of clearance and at 1 min intervals for the next 18 min. Data from arterial injections are accumulated at 1 sec intervals for 6 sec to establish the "input function", and then at the same intervals used for the inhalation-recirculation technique. Arterial samples for pO_2 and pCO_2 are obtained at the beginning and end of each arterial study.

Supported by USPHS Grant NB06655 and the Neurology Department Research and Development Fund.

[1] PhoGamma III, Nuclear Chicago Corp., des Plaines, Illinois, USA.

[2] RIDL 27503, Nuclear Chicago Corp.

[3] Model TM-7, Ampex Corp., Redwood City, California, USA.

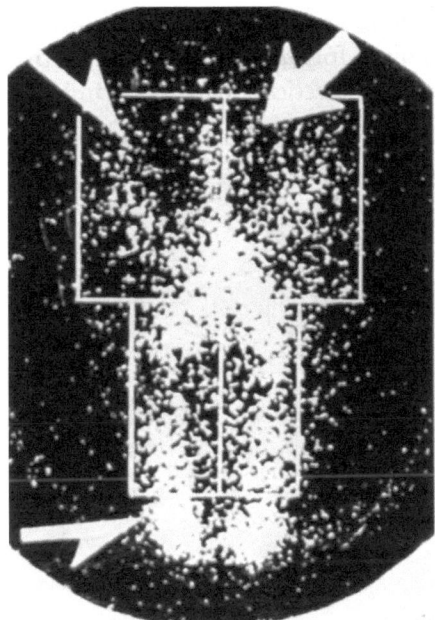

Fig. 1. Normal radioisotope arteriogram with "region of interest" overlay. Single frame exposed for 3 sec. Counts per second are compared for symmetry of distribution. Arrows indicate carotid, middle and anterior cerebral arteries

Technique

Radioisotope arteriography. The test is begun immediately after the rapid injection of less than 3 ml of saline solution containing 10–15 millicuries (mCi) of sterile ^{99m}Tc into an antecubital vein of a patient seated in a chair facing the crystal of the gamma camera. The arm is positoned at the level of the right atrium to facilitate rapid transit and to limit "smearing" as the bolus traverses the lungs and heart. A lead apron is placed across the chest and shoulders of the patient to limit scattering and background activity.

Inhalation-recirculation : The patient is viewed from the posterior projection, and breathes room air through a closed-circuit system containing a carbon dioxide absorbent. The distribution of a single breath of ^{133}Xe is visualized by the scintillation camera. The clearance of isotope is established by diverting the exhaled air through a one-way valve to an isolated collection system. When ventilation clearance is complete, 5 mCi of ^{133}Xe are injected into an antecubital vein and pulmonary perfusion is observed. The camera is then positioned to visualize clearance from the head.

Arterial injection. The patient may be viewed from any projection. One mCi of ^{133}Xe is rapidly injected through a catheter which is introduced via the femoral artery and positioned by fluoroscopic guidance.

Results

Radioisotope arteriography. In normal studies the carotid arteries and the middle and anterior cerebral arteries are well visualized and show even distribution. Carotid artery occlusion is easily visualized as is occlusion of the middle cerebral artery which can be recognized by sectorial defects in radioactivity within the anatomic distribution of the artery. The diagnostic accuracy of the technique can be extended by analyzing the count rate variation in selected homologous carotid and brain regions of interest during a 3 sec interval when the count rate is rapidly rising in the early arterial phase [1] (Fig. 1). The assignment of abnormality points for asymmetries greater than a 60%/40% distribution allows increasing accuracy in the detection of hemodynamically significant stenoses which

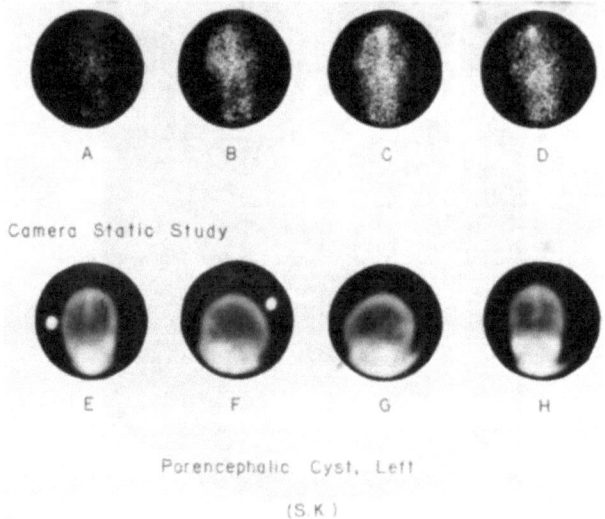

Fig. 2. Radioisotope arteriogram using ^{99m}Tc. Anterior projection. Region of markedly decreased activity superior to usual distribution of left middle cerebral artery. Diagnosis: porencephalic cyst (see Fig. 3)

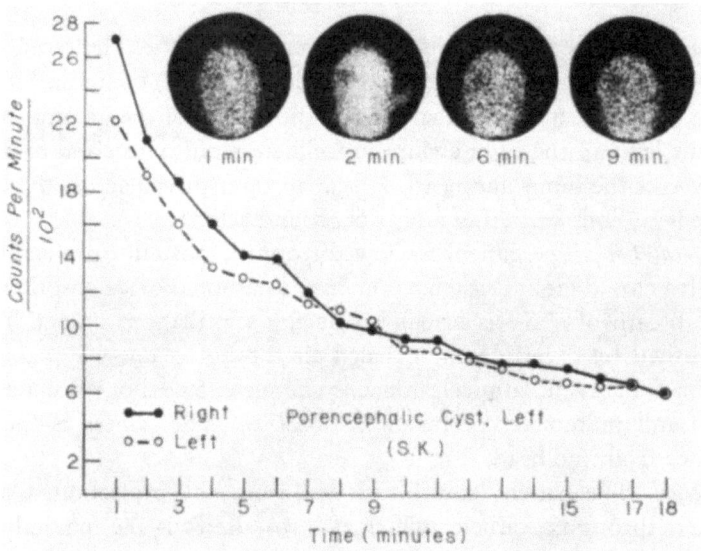

Fig. 3. Inhalation-recirculation technique using ^{133}Xe. Posterior projection. Cyst shows absence of perfusion. Clearance curves do not differ significantly because of extracerebral contamination.

are not amenable to visual interpretation. Arteriovenous malformations show a "washout phenomenon" during a sequential isotope arteriogram. This is characterized by an early increase in radioactivity which rapidly reaches a peak and fades to even distribution during later frames of the sequence. Smaller malformations can be recognized by the use of computer techniques [1].

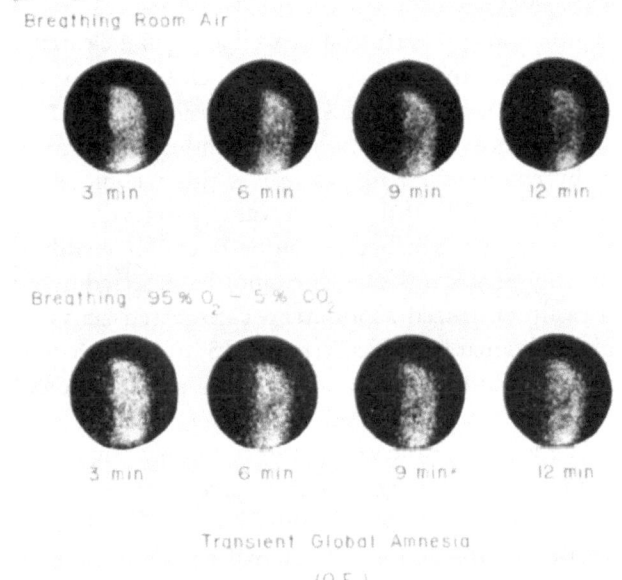

Fig. 4. Common carotid artery injection using ^{133}Xe. Anterior projection. External halo persists after hemispheric clearance is virtually complete. More prominent halo during CO_2 inhalation is photographic contrast artifact

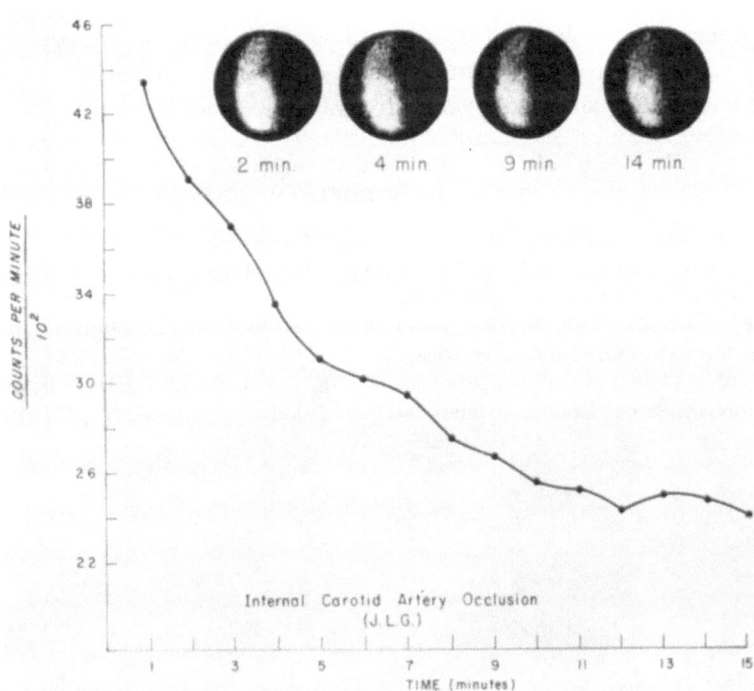

Fig. 5. External carotid artery injection using ^{133}Xe. Anterior projection. Patient had internal carotid artery occlusion. Persistence of activity shows striking similarity to Fig. 4. Areas of greatest activity are from face and are not involved in region of interest from which clearance is observed

Inhalation-recirculation. We have applied the inhalation-recirculation technique to the analysis of pulmonary ventilation and perfusion as well as cranial clearance to identify unrecognized pulmonary abnormalities that may affect cranial clearance curves. Pulmonary retention cysts and ventilation-perfusion discrepancies allow considerable recirculation of the isotope. Contamination of the clearance curve by pulmonary induced recirculation and the contribution of radioactivity by extracerebral tissues causes this technique to have relatively little value in the determination of CBF. It does have value, however, in the atraumatic visualization of porencephalic cysts into which the isotope does not readily diffuse. Since the cyst is surrounded by normally perfused tissues, it cannot be detected by external detection techniques alone, but is easily visualized as an area of decreased activity with both a non-diffusible and diffusible radiopharmaceutical (Figs. 2, 3). Clearance curves may show only insignificant difference in total count rates and yet the area of decreased perfusion may be obvious.

Arterial injection. We are using the arterial injection technique to determine the degree of extracerebral contamination attendant upon common carotid or aortic arch injection of ^{133}Xe to provide a standard of comparison for inhalation studies. Only patients who have had angiographic visualization of the entire aortocranial circulation are evaluated by this technique. Selective injections of the common and internal carotid arteries have shown a surprising lack of contralateral hemisphere perfusion, and in patients with normal pulmonary function, no visually detectable recirculation contamination occurs. Common carotid injection shows an extracerebral halo of radioactivity which persists after hemispheric clearance is complete. This renders accurate rCBF determinations impossible [3] (Fig. 4). In a patient with angiographically verified internal carotid artery occlusion who had no detectable extracranial-intracranial anastomoses, selective external carotid injection produced a clearance curve which was virtually undifferentiable from internal carotid artery injections (Fig. 5). A similar phenomenon was observed after ascending aortic injection in a patient with internal carotid artery occlusion on one side and external carotid artery occlusion on the other. We have not attempted to calculate flow values from the clearance curves.

References

1. Addario, D.: Computer Analysis of the "Radioisotope Arteriogram". Dissertation in partial fulfillment of the degree of Doctor of Medicine. Bowman Gray School of Medicine of Wake Forest University, Winston-Salem, North Carolina, April 1969.
2. Janeway, R., C. D. Maynard, and R. L. Witcofski: Radioisotope arteriography: a new diagnostic technique for evaluation of cerebrovascular disorders. Circulation **38**, 107 (1968).
3. Oldendorf, W. H., and Y. Iisaka: Interference of scalp and skull with external measurements of brain isotope content. Part 1: Isotope content of scalp and skull. J. nucl. Med. **10**, 177 (1969).

Iconoscopic Evaluation of Cerebral Vascular Disease

J. C. KENNADY

UCLA School of Medicine, Harbor General Hospital, Los Angeles, California

Introduction

A major problem in patients with cerebral vascular disease has been correlating the neurological status with results from electroencephalography, carotid angiography, brain scanning and cerebral blood flow studies. The EEG is limited in regional localization when the involved area is small or if the adjacent collateral blood supply is sufficient. Blood vessels smaller than 0.5 mm in diameter cannot be visualized by carotid angiography because of resolution limitations. Unless there is anatomical alteration of the vascular bed, especially at the capillary level, the brain scan using 197 Hg or ^{99m}Tc pertechnetate will not show regions of increased radioactivity. Cerebral blood flow studies have been more precise in regional localization, especially when multiple collimated detectors and a tape recording system are used. However, inordinate time is required to retrieve and calculate these results. In contrast, iconoscopic study of the cerebral hemisphere [1, 2] has provided a rapid (1 min) procedure which has been used in 100 patients with clinical evidence of intracranial pathology. The radioisotope studies were performed following carotid angiography, using the same indwelling internal carotid artery catheter. 30 of these patients had cerebral vascular disease.

Method and Equipment

Scintiangiophotography with the iconoscope is a method to continuously visualize the lateral hemisphere and record rapid changing events on videotape at a rate of 60 frames per sec. Following a fast bolus injection of 1 ml containing 3 millicuries 99m technetium labeled albumin microaggregates (^{99m}Tc AA), 1–8 micron size, the regional distribution and velocity index of the non-diffusing intravascular test agent can be seen immediately on a television monitor. These data can be retrieved from the permanent video tape record as polaroid pictures of single (0.03 sec) or integrated frames, also quantified as histograms of large or small designated regions of the brain.

The iconoscope quantifier is initially calibrated using a 92 microcurie 57 Cobalt standard source placed flush to the 3 mm hole collimator in a 1.5 cm^2 area. This field is purposely small in order to improve the spacial resolution of the video record. Phantom studies show that the iconoscope resolution of a static field is 3 mm and can delineate individual coils of tubing (1.5 mm diameter) spaced 3–5 mm apart through which a 99m technetium pertechnetate solution is flowing. The iconoscope sensitivity can be determined by using a

Supported by contract AT-Gen-12 between the U. S. Atomic Energy Commission and the University of California at Los Angeles.

pinhole collimator with a single 3 cm diameter hole and a voltmeter attached to the quantifier output. By adding an increasing number [15] of 57 Cobalt pellets (10 μc) to the collimator field and plotting the quantifier output voltage versus the microcuries of 57 Cobalt present, the results show a *linear* incremental dose response.

Clinical Results and Interpretation

a) Controls. In 5 patients, ^{99m}TcAA appears uniformly distributed throughout the hemisphere within 3 sec. Regional radiohistograms indicate the sylvian area has a higher peak radioactivity level than the midtemporal region which in turn has a slightly higher radioactivity peak than the frontal and parietal areas. However, the transit (velocity index) of the test agent through the temporal region is slower (0.5–1.5 sec) compared with the frontal and parietal areas in 3 of these patients. The lateral sinus is seen within 3.5 sec with a peak radioactivity level in 4.5–5 sec. No significant ^{98m}Tc AA remains in the hemisphere after 8–10 sec.

b) Acute Vascular Disease. In this group are 5 patients with neurological symptoms ranging from 12 h to 7 days and angiographic evidence of anterior or middle cerebral artery trunk or branch occlusions. The occluded areas have a decreased ^{99m}Tc AA concentration in contrast to adjacent brain regions in the video polaroid pictures. Radiohistograms of these areas show a 0.5–1.5 sec delay in the appearance of the microaggregates, lower peak radioactivity and rapid disappearance of the test agent compared with other regions of the hemisphere.

c) Chronic Vascular Disease. 25 patients presented a history of cerebral vascular disease of several months to 4 years duration. Only those patients with definite anterior or middle cerebral artery trunk or probable distal branch involvement by angiography are considered here. The iconoscope studies show clearly delineated regions of increased ^{99m}Tc AA concentration in the areas indicated by the neurological deficit. Radiohistograms of these regions show an earlier appearance, higher peak radioactivity and delay in the disappearance (2–2.5 sec) of ^{99m}Tc AA than observed elsewhere in the hemisphere. Patient H. M., age 37 years, exemplifies this group of patients. He complained of left sided headaches for several months, then gradually became lethargic, confused and dysarthric. No peripheral motor or sensory abnormalities were detected on examination. The EEG indicated a destructive focus in the left anterior and mid-temporal regions. Left carotid angiography showed retrograde filling of the left posterior temporal artery. Results of the iconoscope studies are illustrated in Fig. 1. The 3.5–4.0 sec interval video polaroid picture shows the increased ^{99m}Tc AA concentration in the anterior and mid-temporal regions (arrow). The numbers indicate the areas where the radiohistograms were obtained. The frontal (1), posterior (2) and parietal (3) regions do not show high peak radioactivity levels or the delayed disappearance of ^{99m}Tc AA seen in the anterior temporal area [3]. The rapid appearance and only the initial portion of the high temporal radioactivity peak can be ascribed to the ^{99m}Tc AA bolus in the deep underlying carotid syphon. These results would indicate an increase in the microvasculature, possible loss of autoregulation and a delay in capillary-venous drainage from the temporal region.

Conclusions

Iconoscopic cerebral studies provide a relatively comfortable, rapid (1 min) assessment of the radiopharmaceuticals distribution and velocity index through the hemisphere. The appearance peak radioactivity and the disappearance of ^{99m}Tc AA in any given area can be

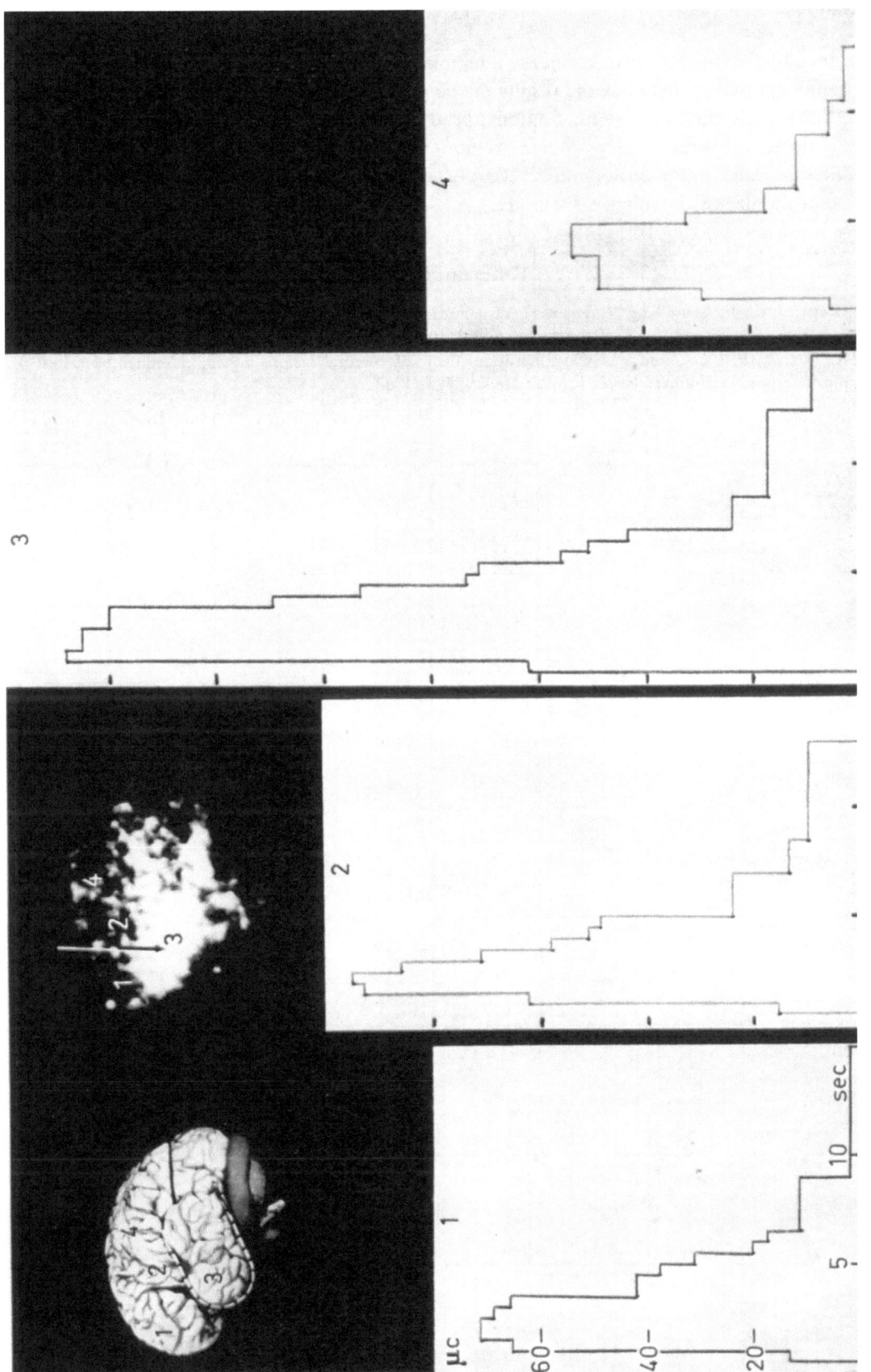

Fig. 1. Patient H.M., age 37 years. Occlusion of left temporal artery. Left intracarotid injection of 3 mc ^{99m}Tc AA. Temporal lobe shows increased radioactivity (arrow). Numbers indicate regions where corresponding radiohistograms were obtained. Radiohistograms show μc per 1.5 cm^2 area vs. 0.5 sec sampling times. Phantom brain orients the 3.5–4 sec interval videopolaroid picture. (see text)

determined by the iconoscope quantifier as a regional radiohistogram. Admittedly the number of patients reported is small but the results of the 2 respective groups with cerebral vascular disease have been quite consistent. Regions distal to an acutely occluded vessel are seen as areas of decreased radioactivity whereas regions involved by chronic vascular pathology have increased and more prolonged ^{99m}Tc AA concentration and retention, indicating an increase in the size of the microvasculature.

References

1. KENNADY, J. C., F. CHIN, and R. POTTER: Visualization of cerebral blood flow with the scintillation camera in man. Scand. J. clin. Lab. Invest. Suppl. **102**, p. XI–K (1968).
2. —, R. POTTER, F. CHIN, and L. SWANSON: Assessment of cerebral lesions by rapid sequential scinti-photography. Preliminary Note. J. nucl. Med. **9**, 423 (1968).

Correlations between CBF and Clinical Results in Cerebrovascular Patients. A Statistical Analysis

P. D'Amico, I. Sanguinetti, and M. Minazzi

S. Gerardo dei Tintori Hospital, Monza

The problem of the clinical meaning of CBF values found in focal ischemic cerebrovascular lesions has not yet been solved, so that it has been judged interesting to analyse statistically the correlations between the CBF data and the clinical and instrumental ones.

Sixty-four CBF determinations were carried out in 37 patients, 10 of which were affected by internal carotid artery thrombosis. CBF of the controlateral hemisphere was studied in this last group, while in the others the study was bilateral. Xenon[133] was employed as tracer: it was introduced into the internal carotid in an average amount of 0.4 mCi. The recording with 2 scintillation detectors (2-inches crystals) was performed simultaneously over both parieto-rolandico-temporal regions. The calculation was that proposed by Lassen and Ingvar [4]. The ratio hO/hC between the greatest heights of the 2 curves simultaneously recorded was also computed, hO/hC being the ratio of the injected to the non-injected side.

For statistical analysis, the variables were divided according to the different patient categories (Table 1).

Table 1. *Groups of clinical and instrumental data*

a) extension of the ischemic zone, inferred from clinical and scintigraphic data:

    1. small – e. g. a terminal branch
    2. mean – e. g. one of the main arterial fields
    3. large – the whole field of one of the main intracranial arteries;

b) evolution:

    1. recovery within 2 days
    2. recovery within 2 weeks
    3. recovery after 2 weeks
    4. serious damage or death

c) EEG:

    1. normal
    2. slight-mean abnormal, focal
    3. very abnormal, focal
    4. slight-mean abnormal, diffuse
    5. very abnormal, diffuse

d) EEG, carotid compression:

    1. not significant
    2. delta, homolateral
    3. delta, diffuse

e) carotid arteriography:

    1. internal carotid a. thrombosis, with absent or scanty collateral circulation
    2. internal carotid a. thrombosis, with good coll. circulation
    3. mild arteriosclerosis
    4. serious arteriosclerosis
    5. occlusion of a main branch of the internal carotid a.

Groups of CBF data

$\bar{f}$	1.	$\geqq 40$ cm³/100 g/min	f_1	1.	$\geqq 64$ cm³/100 g/min
	2.	30–40		2.	54–64
	3.	20–30		3.	44–54
	4.	10–20		4.	34–44
				5.	24–34
f_2	1.	$\geqq 13$ cm³/100 g/min	hO/hC	1.	2.5
	2.	11–13		2.	2–2.5
	3.	9–11		3.	1.5–2
	4.	7– 9		4.	1–1.5
	5.	5– 7			

Variance analysis with the method of multiple regressions was then carried out with the 7040 IBM computer of the University Computing Center, Milan.

Results and Discussion

Several correlations appeared, of which only the most significant are reported.

In the injured cerebral hemisphere $\bar{f}$ is strictly related to the extension of the ischemic zone ($P < 0.001$; $r = 0.81$). A lesser degree of correlation ($P < 0.01$; $r = 0.52$) binds $\bar{f}$ to the clinical evaluation. A weak relation is found between $\bar{f}$ and the arteriographical data. f_1 flow estimate of fast component behaves like $\bar{f}$.

In the hemisphere controlateral to the lesion the values of $\bar{f}$ are also correlated with the extension of the ischemic zone ($P < 0.001$; $r = 0.65$) as well as with the clinical evolution ($P < 0.001$; $r = 0.65$). The correlations of f_1 appear less significant. The ratio hO/hC is correlated with the clinical evolution, which gets worse with the decrease of the ratio.

On the basis of our data, we would conclude that decreasing values of $\bar{f}$ and f_1 correspond to proportional greater extensions of ischemic zones. Similarly, more severe cases are correlated with low values of $\bar{f}$ and f_1 found in both cerebral hemispheres. However, the low coefficients of correlation underline that other elements also affect the clinical evolution.

The decrease of the ratio hO/hC in the CBF of the controlateral hemisphere means a passage of tracer to the damaged hemisphere, i. e. a collateral circulation from one carotid system to the other [2]. The fact that this has a weak correlation with the worsening of the clinical course, in opposition to angiographical studies [6], may suggest both a poor collateral circulation from other channels (external carotid and vertebrobasilar arteries) and a noxius action on the circulatory dynamics of the contralateral hemisphere brought about by an "interhemispheric steal".

Very weak correlations were found, however, between the data of CBF and EEG and angiographical ones, confirming what had already been observed by Loeb and Fieschi [5] for EEG and by Ingvar et al. [3] and Cronqvist et al. [1] for angiography.

In conclusion, statistical analysis has validated significant correlations mostly between CBF and various clinical parameters in cases of focal ischemic cerebrovascular lesions.

References

1. Cronqvist, S., D. H. Ingvar, and N. A. Lassen: Quantitative measurements of regional cerebral blood flow related to neuroradiological findings. Acta Radiol. (diagn.) 5, 760 (1966).
2. D'Amico, P., and M. Minazzi: Il flusso sanguigno cerebrale studiato con Xe¹³³ in soggetti normali ed in pazienti affetti da trombosi della carotide interna. Riv. Pat. nerv. ment. 87, 376 (1966).

3. INGVAR, D. H., E. HÄGGENDAL, N. J. NILSSON, P. SOURANDER, I. WICKBOM, and N. A. LASSEN: Cerebral circulation and metabolism in a comatose patient, studied with a new method. Arch. Neurol. **11**, 13 (1964).
4. LASSEN, N. A., and D. H. INGVAR: Regional cerebral blood flow measurement in man. Arch. Neurol. **9**, 615 (1963).
5. LOEB, C., and C. FIESCHI: Electroencephalograms and regional cerebral blood flow in cases of brain infarction. Electroenceph. Clin. Neurophysiol. Suppl. **25**, 111 (1967).
6. SANGUINETTI, I., A. PIATTI, and D. ZERBI: Correlazioni clinico-angiografiche nella trombosi della carotida interna. Arch. psicol. Neurol. **25**, 600 (1964).

rCBF Studies in Patients With Arteriovenous Malformations of the Brain

D. Oeconomos, B. Kosmaoglou, and A. Prossalentis

Neurological Center, Polyclinic of Athens

The use of rCBF measurements in cases with arteriovenous malformations of the brain has a double aim for the neurosurgeon:

1. To quantitate the tissue perfusion in the region of the malformation, i. e. to demonstrate whether this flow is adequate or not. Information of this type may be of importance for the decision to operate or not.

2. To assess the results of the operation by a) proving the disappearance of the shunt, and b) by measuring the rCBF postoperatively in and around the site of the malformation.

This is a report of 5 patients who were operated upon for arteriovenous malformations and who were submitted to rCBF studies pre- and postoperatively. One patient with a deeply situated angioma (not operated) on the left side will also be considered. In addition, observations in 3 patients with intracerebral hematomas due to bleeding from small sized angiomas will also be discussed.

For the rCBF measurements we have used a 6 detector equipment with narrow collimators with a diameter of 15 mm. The detectors were placed contralaterally, and homolaterally to the malformation over its periphery, as well as remotely from it, in accordance with the findings on a preceding serial brain scan with ^{99m}Tc pertechnetate, and in arteriograms. In two patients repeated rCBF studies were carried out in order to cover the diseased hemisphere as widely as possible.

Results

As described by Häggendal et al. [1], isotope clearance curves over an arteriovenous shunt show a large initial peak. Such peaks usually have a broad base which may last about 3 sec and they are superimposed upon the slower clearance curve from the tissue underlying the shunt. Peaked curves of this type create problems for the calculation of rCBF according to the height over area method.

It is our experience that peaked curves are more easily recognized if plotted on semilog paper in which case it is also easy to plot the starting point of the tissue curve underlying the peak. Occasionally it has been possible to identify very fast tissue flows in such curves, which are followed by clearance curves representing tissue flows of medium or slow type.

Superfast tissue flows of the above mentioned type were observed in highly vascularized tumours, and, also, in young adults and children following CO_2 inhalation. They may represent a fast perfusion in angiomatous tissue or in normal brain tissue in which a pronounced vasodilatation has been provoked by carbon dioxide.

In the patients with arteriovenous malformations it was demonstrated postoperatively that the shunt peaks disappeared completely and – as pointed out by other workers – this constitutes a proof that the malformation has been excised in toto.

The deeply situated angiomas will now be considered. In one of these cases the angiographic analysis disclosed an angioma in the posterior part of the left thalamus, but no shunt or tissue peaks were recorded in the clearance curves. In these studies the lateral border of the angioma was situated about 5.3 cm from the rCBF detector. Apparently this large distance makes it impossible to record a shunt peak. Absence of an expected peak was also noted in another case in which the rCBF was measured over a partially excised arteriovenous malfor-

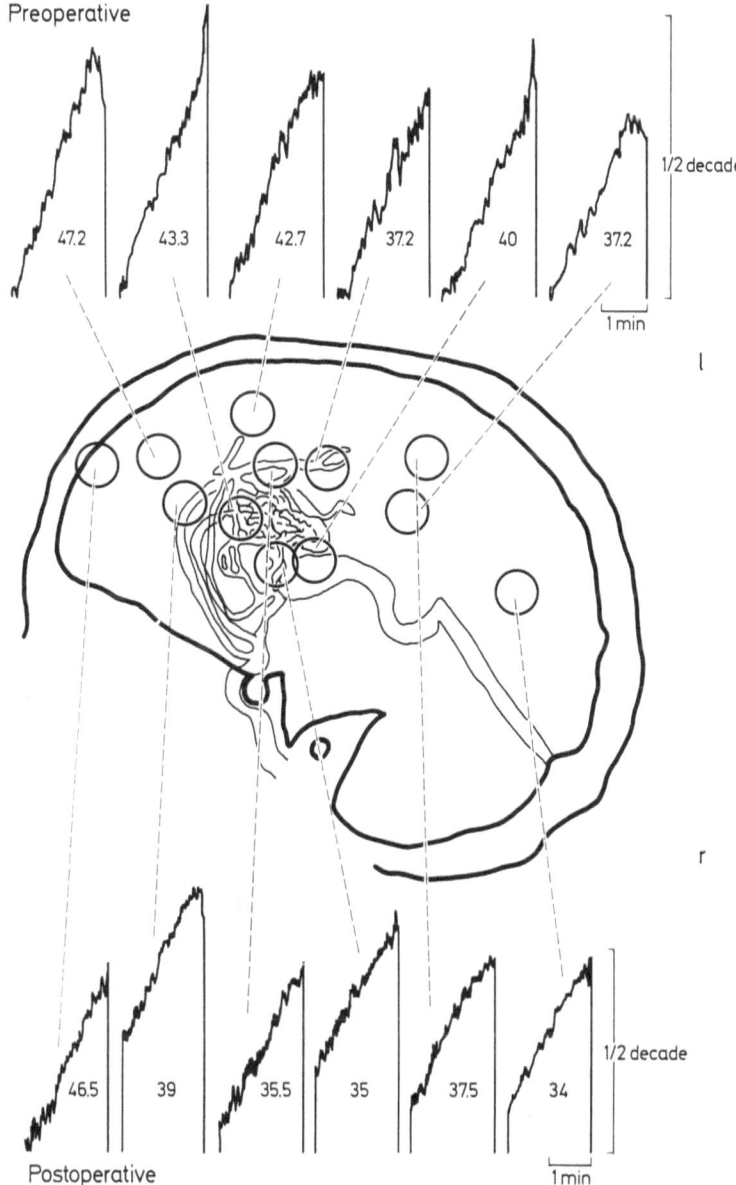

Fig. 1. Pre- and postoperative rCBF measurements in a case of deeply seated arteriovenous malformation. Total extirpation was achieved. Before surgery a shunt peak was disclosed over the lesion proper, as well as over the draining huge internal cerebral vein

mation, in spite of the fact that one of the detectors faced a small part of the remaining angioma at a depth of about 4.0 cm (Fig. 1).

In the 3 cases with intracerebral hematomas from small angiomas no peaks were recorded over the hematomas.

From the observations reported it is concluded that shunt peaks may be absent over arteriovenous malformations or angiomas if these are situated in deeper parts of the hemisphere.

Pre- and Postoperative rCBF Measurements

Preoperatively the mean rCBF in the five cases operated upon was 48.4 $\mp$ 4.0 ml/ 100 g/min. No significant changes of the flow were found "behind" or beside the malformation, provided there was not an underlying intracerebral hematoma.

On carbon dioxide inhalation it was not possible to demonstrate any steal phenomena over the lesion itself or over parts of the brain distant to it. There was therefore no evidence which could favor the hypothesis that transient regional circulatory events of the "steal" or "inverse steal" type could be responsible for intermittent epileptic symptoms on patients of this type. Such steal effects are under similar conditions easily detected in clearance curves from patients with brain tumors.

Postoperatively, a slight general decrease of the rCBF, with good preservation of autoregulation, was found in 2 patients studied 20 and 32 days, respectively, following the operation. In another patient, studied 60 days following operation, flow rates almost identical to the preoperative ones, were recorded except from over the lesion itself where a decrease of about 60 per cent was noted.

The rCBF studies carried out, have thus failed to demonstrate any particular change in the brain perfusion following excision of the lesion, provided the surgical procedure did not give rise to any major tissue damage. The ablation of the malformation did not appear to change the hemodynamics of the diseased or the healthy brain tissue. This is in agreement with the common clinical experience that even big arteriovenous malformations can be excised without consequent functional disturbances except those caused by the surgical approach itself and the ensuing manipulation of the brain.

Concluding Remarks

The present small series of rCBF studies in patients with arteriovenous malformations indicates that there are in such cases no significant changes of the rCBF which can be attributable to the malformation, nor can any rCBF changes be recorded following excision of the malformation. However, the shunt peaks disappear in the postoperative rCBF curves.

Small or deeply situated shunts may be missed by the rCBF technique. It is an advantage if the rCBF study is made repeatedly so that the detectors can cover the whole surface of the hemisphere.

It might also be pointed out that the presence of peaked clearance curves in expanding lesions of the brain may demonstrate the presence of an angioma, but such curves do not exclude an avascular glioblastoma.

Reference

1. Häggendal, E., D. H. Ingvar, N. A. Lassen, N. J. Nilsson, G. Norlén, A. Wickbom, and N. Zwetnow: Pre- and postoperative measurements of regional cerebral blood flow in three cases of intracranial atreriovenous aneurysma. J. Neurosurg. 22, 1 (1965).

ATP Influences on Cerebral Hemodynamics and Metabolism in Patients With Ischaemic Cerebral Disturbances

D. Hadjiev, E. Witscheff, and Z. Marinova

Research Institute of Neurology and Psychiatry, Sofia

There is experimental evidence that, in situations of cerebral ischaemia and anoxia, the synthesis of ATP is disturbed and its concentration in the brain tissues decreases rapidly and significantly. At the 15th min. of incubation of ischaemic slices of brain tissue the concentration of ATP within grey matter decreases 28 times and that within white matter about 12 times [2, 5].

The experiments of AIZAWA et al. [1] with the nitrous oxide method showed that, following intravenous administration of ATP and hypertonic solution of glucose, cerebral oxygen consumption and cerebral blood flow increase. By using the same method GOTTSTEIN [4] observed opposite effects of ATP on cerebral hemodynamics, cerebral blood flow remaining unaltered.

These data induced us to investigate the effects of ATP on cerebral blood flow and glucose metabolism in patients with cerebral ischaemic lesions in order to assess the possible therapeutic value of this substance.

Material and Methods

The studies were carried out in 19 patients with cerebral ischaemic disturbances due to atherosclerosis and hypertension. Table 1 contains the basic clinical patterns of the patients. CBF of the affected hemisphere was studied prior to and during the 10 min following intravenous application of 20 mg of ATP in 10 ml saline. The radioisotope venous dilution method of SOLTI et al. was used for measuring CBF. The venous pressure in the superior bulb of the internal jugular vein was measured and the perfusion pressure of the brain was calculated.

Table 1

Number of patients	19
Males	9
Females	10
Age in years (average)	60,3 (41–87)
Etiology of cerebral blood flow disturbance:	
hypertension and atherosclerosis	10
atherosclerosis	9
Type of cerebral blood flow disturbance:	
cerebral infarction	12
transient ischaemic attacks	7

After application of ATP blood was withdrawn from the bulb of the jugular vein and from the femoral artery in the 10th and 15th min. Arterial and venous blood samples were analysed for oxygen, carbon dioxide (microgasometric method of NATELSON) and glucose.

Results

The changes in CBF and CVR after application of ATP are presented in Table 2. In some cases CBF increased and CVR decreased and in others (with normal CBF value) the opposite effect was observed. Perfusion pressure in the brain changes equivocally following i.v. application of ATP. We found no significant correlation between derivations of CBF and perfusion pressure.

Table 2

| | CBF increased and CVR decreased | | | |
| | before | | after 10 min | |
	A	B	C	D
$\bar{x} \pm s\bar{x}$	255.6 ± 30.6	28.8 ± 3.9	370.0 ± 39.5	19.4 ± 2.1
s	68.2	3.9	86.8	4.77
R	$155.0 - 346.3$	$21.5 - 43.3$	$292.4 - 515.7$	$14.8 - 25.5$

A vs C $p < 0.025$
B vs D $0,01 > p > 0.05$

| | CBF decreased and CVR increased | | | |
| | before | | after 10 min | |
	E	F	G	H
$\bar{x} \pm s\bar{x}$	347.6 ± 43.0	22.1 ± 4.1	257.0 ± 39.2	27.7 ± 8.1
s	104.8	10.1	97.7	19.8
R	$164.3 - 460.2$	$14.0 - 42.0$	$151.2 - 422.0$	$16.8 - 45.6$

E vs G $p > 0.10$
F vs H $p > 0.10$

The effects of ATP on the brain metabolism are presented in Table 3. The changes between the 10th and 15th min are presented in Fig. 1. These data permit the conclusion that, parallel to enhancement of anaerobic glycolysis at the 15th min, an activation of oxydative cerebral metabolism takes place. We found no significant correlation between the changes of RQ and the glucose-oxygen ratio on the one hand, and CBF, CVR and perfusion pressure in the brain on the other, excepting the correlation between perfusion pressure and respiratory quotient ($r = +0.677$).

These data suggest that the changes in RQ and the glucose-oxygen ratio in the brain following i.v. applications of ATP are directly related to brain metabolism. The lack of parallelism between brain hemodynamics and metabolism showed that their connection in cerebral ischaemia is disturbed.

EEG investigations have shown disturbances of the normal close correlation between bioelectrical activity of the brain and regional CBF [6] in such cases.

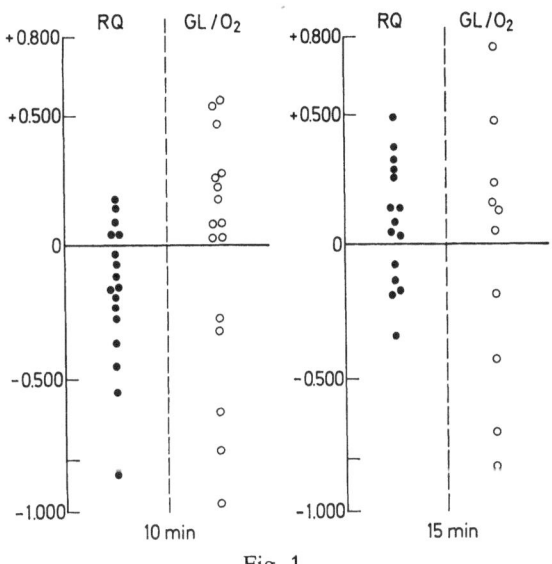

Fig. 1

Table 3

| | Respiratory quotient of the brain | | | Glucose-oxygen ratio | | |
| | before | after 10 min | after 15 min | before | after 10 min | after 15 min |
	A	B	C	D	E	F
$x \pm s\bar{x}$	0.542 ± 0.047	0.393 ± 0.045	0.598 ± 0.102	0.611 ± 0.102	0.761 ± 0.106	0.614 ± 0.102
s	0.206	0.198	0.313	0.448	0.448	0.380

A vs B	$0.02 > p > 0.01$	D vs E	$p > 0.10$
A vs C	$p > 0.10$	D vs F	$p > 0.10$
B vs C	$0.02 > p > 0.01$	E vs F	$p > 0.10$

The present results indicate that isolated improvement of cerebral hemodynamics in ischaemic lesions may have no effect on the disturbances of brain metabolism and function. Therapy with cardiotonic, vasoactive and metabolic active drugs could be more promising. Furthermore, ATP has seldom been used as a cerebral vasodilator [3, 4] and may find its place in the therapy of metabolic disorders in cerebral ischaemia.

References

1. AIZAWA, T., J. TAZAKI, and F. GOTHO: Cerebral circulation in cerebrovascular disease. World Neurol. **2**, 635 (1961).
2. COHEN, M. M.: The effect of anoxia on the chemistry and morphology of cerebral cortex slices in vitro. J. Neurochem. **9**, 337 (1962).
3. FERLIGA, G.: Considerazioni per una terapia eclettica della insufficienza circolatoria cerebrale. Giorn. Geront. **15**, 621 (1966).
4. GOTTSTEIN, U.: Der Hirnkreislauf unter dem Einfluß vasoaktiver Substanzen. Heidelberg: Hüthig-Verlag 1962.
5. KIRSCH, W. M., and J. W. LEITNER: Glycolytic metabolites and co-factors in human cerebral cortex and white matter during complete ischemia. Brain Res. **4**, 358 (1967).
6. SULG, I. A., and D. H. INGVAR: Regional cerebral blood flow and EEG in occlusions of the middle cerebral artery. Scand. J. clin. Lab. Invest. Suppl. **102**, p. XVI–D (1968).

A New Therapeutical Method for Cerebral Circulatory Disturbances

V. G. Chişu

Neurosurgical Service, State Hospital No. 1, Timisoara

This paper reports the results obtained on 25 patients hospitalized for cerebrovascular disease treated exclusively by intracarotid injection of K-Mg-aspartate (Table 1). Intracarotid

Table 1

	No. of CASES	SEX	AGE	BLOOD PRESSURE IN Mg.	BEGINNING OF SYMPTOMATOLOGY (WITH LOSS OF CONSCIOUSNESS / WITHOUT LOSS OF CONSCIOUSNESS)	CLINICAL CONDITION (HEMIPLEGIA / HEMIPARESIS FACIAL BRA. / HEMIPARESIS CRURALIS / APHASIA)	PATHOGENESIS (SPASMUS / STENOSIS / OCCLUSION)	ILL BLOOD VESSEL (CAROTIS INTERNA / SIPHON CAROTIS / CEREBRALIS ANTERIOR / CEREBRALIS MEDIA)	PERIOD UP TO THE TREATMENT DAYS	TREATMENT WITH K-Mg – ASPARTAT (INTRA CAROTIS INJECTION / ENDO VENOUS PERFUSION)
MUCH IMPROVED	1	M	35	100/55					10	I. II. III.
	2	M	47	110/60					40	I. II. III.
	3	M	54	180/90					2	I. II. III. IV.
	4	M	35	170/90					12	I. II. III.
	5	M	61	160/100					12	I. II. III.
	6	W	56	155/85					9	I. II. III. IV.
	7	W	58	220/110					21	I. II. III.
	8	W	17	125/80					30	I. II. III.	1-2
	9	W	18	120/70					21	I. II. III. IV. V. VI.
	10	M	37	150/80					60	I. II. III. IV. V. VI.
	11	M	31	130/70					16	I. II.	1-2-3-4
IMPROVED	12	M	53	180/90					49	I. II. III.
	13	W	47	165/80					2 YEARS	I. II. III.
	14	M	64	175/80					30	I. II. III. IV.
	15	W	73	190/100					7	I. II. III.
	16	W	63	160/70					10	I. II. III.
	17	M	58	170/90					16	I. II. III. IV.
	18	W	56	110/50					30	I.	1-2-3
	19	W	45	155/90					45	I. II. III. IV.
UNCHANGED	20	M	41	140/80					7	I.
	21	M	60	200/110					21	I.
	22	W	47	145/80					90	I. II. III.
	23	W	60	200/110					4	I. II.
	24	M	45	155/70					7 MONTHS	I. II. III. IV.
	25	W	57	110/85					4 MONTHS	I.	1-2-3

injections combined with endo-venous perfusions with the same substance were used in 3 cases. All cases were clinically and angiographically studied before and at the end of the treatment. In some special cases EEG and TEG examinations were performed.

The etiology of the disease was angiographically verified and localized. 10–15 min following angiography, 10 ml of K-Mg-aspartate[1] at body temperature were slowly injected into the common carotid artery. The patients immediately felt a strong sensation of warmth in the corresponding hemiface and hemicranium. Vegetative changes, as facial blush, frontal sweating, bradycardia and masticatory movements, sometimes even movements of deglutition (without having the character of a seizure) were observed 30 sec. to 2 min. Usually the therapeutic procedure was repeated 4 days later. Clinical amelioration could be observed after the second injection in some cases. In average, 3–4 intracarotid injections were performed per patient. Only in 2 cases 6 injections were made.

Complications did not occur during this treatment. Three patients refused a repeated injection after the first or second administration, invoking the strong sensation of warmth.

For the purpose of evaluation of the results obtained with this new therapeutical method, our patients were classed in 3 distinct groups:

1. Much improved: 11 cases which left the hospital with almost complete disappearance of the deficiencies. In 3 cases carotid angiography showed partial repermeabilisation of the involved vessel.

2. Improved: 8 cases which left the hospital with partial disappearance of the deficiencies.

3. Unchanged: 6 cases, clinically and angiographically unaltered.

Conclusions

1. The treatment with intracarotid K-Mg-aspartate may be of value in the treatment of various forms of cerebrovascular disease.

2. This treatment is more promising if the interval from the onset of symptomatology to the administration of the first injection is shorter.

---

[1] This product has kindly been put at our disposal by Trommsdorff, Chem. Fabrik, Aachen.

The author is indebted to the members of the Neurologic Clinic, Faculty of Medicine, Timisora (Prof. Dr. AL. SOFLETEA) as well as to the group of neurosurgeons and to the radiologist of the hospital.

Comments to Chapter III

The first 4 papers were discussed as a group, and various explanations for the loss of auto-regulation and CO_2 responsiveness in regions of acute cerebral infarction were considered. It was agreed that while both types of vasomotor responsiveness were usually lost together, the regional loss might be dissociated, e. g., maintenance of CO_2 responsiveness but loss of autoregulation.

The following points were raised in connection with this topic.

1. The loss of autoregulation may be associated with edema and with increased tissue pressure resulting in an increased cerebrovascular resistance which is overcome by the increase in perfusion pressure when the blood pressure is raised while no change in cerebrovascular resistance is produced by CO_2 inhalation (FIESCHI).

2. Changes in regional reponsiveness to CO_2 inhalation due to adaptation to prolonged regional increases in $PaCO_2$ (see, in this volume, EKSTRÖM-JODAL and HÄGGENDAL, Cerebral blood flow and metabolism in patients with chronic respiratory insufficiency with special regard to induced acute changes of the blood gas situation).

3. In areas of severe cerebral ischemia, the PO_2 may be so reduced that hyperventilation causes sufficient further regional anoxia to cause paradoxical vasodilatation due to the Bohr effect (HÄGGENDAL and MEYER). Furthermore, local cerebral anoxia causes local accumulation of lactic acid which also may influence regional pH and hence alter CO_2 responsiveness.

4. It seemed agreed by all that, while autoregulation is a pressure dependent phenomenon, separate from the cerebral vasomotor effects due to changes in regional PCO_2 and PO_2, the 3 mechanisms work together and mutually influence each other. For example, the "rheostat" for autoregulation may be influenced by abnormally high or low regional PCO_2 and PO_2 levels.

In the discussion SKINHØJ reported a case of cerebral vasoconstriction during the "aura" of migraine and (in another case) increased blood flow during the "headache" phase of migraine confirming the observations described by O'BRIEN.

Discussion of the loss of regional cerebral autoregulation caused by transient ischemic attacks followed. It is probable that regional CBF temporarily decreases during transient ischemic attacks and regional loss of autoregulation in the ischemic area may persist for some time despite a return to normal of blood flow values.

The group of FEINDEL reported on experimental studies of occlusion of the middle cerebral artery in which CO_2 inhalation increased regional blood flow to the ischemic area without evidence of intracerebral steal. After various groups reported apparently conflicting data relating to this point, it seemed generally agreed that: I. induced intracerebral steal in an area of cerebral ischemia by CO_2 inhalation was a transient phenomenon of 2–3 days duration if it occurred, II. that it only occurs if severe experimental ischemia and infarction is induced, usually by proximal occlusion of the middle cerebral artery which permits little collateral blood flow, III. that in less severe cerebral ischemia, CO_2 inhalation increases collateral blood flow rather than inducing a steal phenomenon.

Doctors YAMAMOTO and KETY discussed the use of positron emitter isotopes and coincidence counting for measuring regional cerebral blood flow as a promising area for future research, permitting a resolution of regional blood flow not presently available.

J. S. MEYER

Cerebral Blood Flow Studies During Carotid Surgery

G. Boysen, H. J. Ladegaard-Pedersen, and H. C. Engell

Surgical Laboratory of Circulation Research, Rigshospitalet, Copenhagen

In operation for stenosis of the carotid artery, the artery must either be clamped during endarterectomy, or a temporary shunt inserted while the endarterectomy is performed. To insert a shunt, however, may technically complicate the operation and may even be hazardous by damaging the intima. For these reasons it has not been used routinely. We assume that the neurological complications sometimes associated with operation for stenosis of the carotid artery are due mainly to cerebral vascular insufficiency during clamping of the artery.

In patients treated for ruptured intracranial aneurysms by ligation of the carotid artery JENNETT [3] found that a reduction in CBF of 25% resulted in hemiplegia. In normal, conscious persons FINNERTY et al. [2] found that neurological disturbances, provoked by acutely induced hypotension, developed at a critical lower limit of CBF, which for elderly persons was found to be 29 ml/100 g . min. The present investigation was undertaken in order to examine whether the critical lower limit of CBF could be related to the neurological complications sometimes seen after operation for carotid stenosis.

Material and Method

In 9 patients operated upon for stenosis of the internal carotid artery preoperative CBF measurements were performed before, during and after the endarterectomy.

We have used the ^{133}Xe clearance method, the injections being made directly into the internal carotid artery after surgical exposure. The measurement during clamping was done in the following way: after injection, the clearance curve was followed visually, and when the maximal height was reached, the artery was clamped. By this procedure trapping of radioactivity in the internal carotid artery was avoided. The clearance curve was recorded by a single scintillation detector placed over the temporal region. The curves were written out both linearly and logarithmically.

Calculation was done according to the formula

$$CBF_{10} = \frac{H - H_{10}}{A_{10\,(corr)}} \cdot \bar{\lambda} \cdot 100 \, ml/100 \, g \cdot min$$

where H is the initial height of the clearance curve, H_{10} the height at 10 min, $A_{10(corr)}$ is the area below the curve corrected for recirculation and rest activity, if any. $\bar{\lambda}$ is the mean blood-tissue partition coefficient of Xenon.

Another principle of calculation given by PAULSON et al. [4] is based on the fact that the semilogarithmic curve is practically rectilinear during the first $1^1/_2$–2 min. This has been found to be the case also during clamping of the carotid artery. Using the slope of the rectilinear curve, d_0, expressed as the fraction of a decade which the curve decreases per minute, CBF was estimated as

$$\text{Initial slope index (ISI)} = d_0 \cdot 200.$$

For d_0 values below 0.30 there is a fairly good correlation between ISI and CBF_{10}. Above this value the ISI method gives higher values than CBF_{10} (to be published later). Using this method an estimation of CBF can be made after only 2 min of recording.

The patients were anesthetized with halothane and ventilated via a non-rebreathing system. Blood pressure was monitored via a catheter in the radial artery and arterial pCO_2 was measured just before each CBF measurement.

Results

As seen in Fig. 1 neurological complications occurred in 3 patients. Of these the 2 cases with the lowest CBF values showed aggravation of already existing hemiparesis. The third patient had a normal preoperative neurological examination, but, on awakening from anesthesia, he had a contralateral hemiparesis, which gradually remitted. One patient had a questionable exacerbation of an existing paresis, but made a quick and complete recovery.

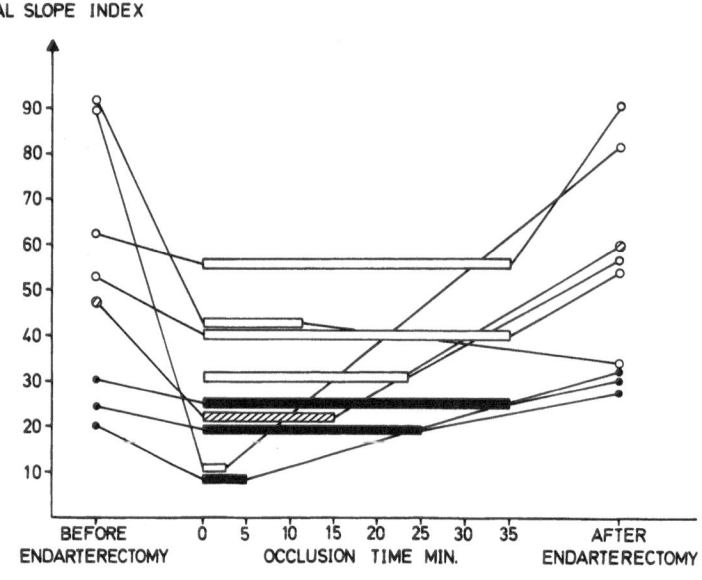

Fig. 1. Graphical representation of ISI values in 9 patients before, during and after carotid endarterectomy. The length of the clamping period is indicated in the middle section. The open circles indicate the cases without, and the filled circles the cases with neurological complications. The hatched figure represents a case with questionable exacerbations of an existing paresis. The 2 cases with the shortest occlusion times had temporary shunts inserted. The cases with ISI values above 60 and the one with the lowest ISI value were hypoventilated and had an arterial pCO_2 about 55 mmHg. In the other cases arterial pCO_2 was kept at a normal or a little below normal level through all measurements

A reduction of CBF was always found during clamping. The reduction of ISI (8 patients) ranged from 11–89%, mean reduction: 39%. The reduction of CBF_{10} (6 patients) ranged from 8–58%, mean reduction: 29%. There was no correlation between the relative or the absolute flow reduction and development of neurological deficits. The neurological complications occurred in cases with the combination of low flow value during clamping and a relatively long period of reduced flow.

As seen from Fig. 1 CBF was almost the same before and after endarterectomy. In the patient with a decrease in CBF after endarterectomy a fall in arterial pCO_2 from 54–42 mmHg had taken place. In the other patients the arterial pCO_2 varied maximally 6 mmHg between the 2 measurements.

Discussion and Conclusion

Reduction of CBF below a certain lower limit for a sufficiently long time inevitably will be followed by symptoms of cerebral vascular insufficiency. The shorter the period in which the flow is reduced, the greater the reduction of CBF that will be tolerated. In Fig. 2, where the ISI values during clamping are plotted against occlusion time, the suggested position of the critical lower limit is indicated by the dotted line. The curve flattens at an ISI value of about 30. This is in accordance with the findings of FINNERTY et al., and we assume that the artery could be clamped infinitely above this level. Below this level there will be a correlation between the size of any further reduction of flow and the length of time for which it is tolerated.

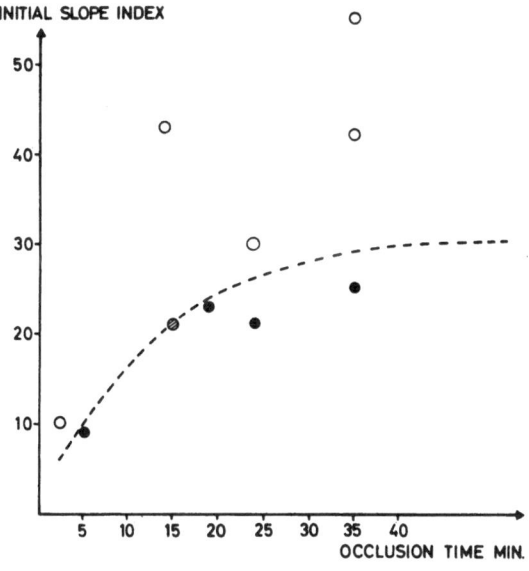

Fig. 2. ISI values during clamping plotted against the period of occlusion of the carotid artery. Open circles represent the cases without and the filled circles the cases with neurological complications. The dotted line indicates the assumed critical lower limit of cerebral blood flow

In our material 2 patients with initial ISI values below 30 and with preoperatively persisting neurological deficits showed an aggravation of their neurological symptoms after the operation, during which the artery was clamped for 5 and 25 min respectively. In 1 patient without persisting neurological symptoms, a reduction of ISI from 30–25 for 35 min was followed by a transient hemiparesis, while a reduction from 90–10 for $2^{1}/_{2}$ min was well tolerated in another patient.

The present material suggests the existence of a critical time/flow curve. However, the definite shape of this curve will need a greater number of observations.

In carotid surgery, CBF measurement during a short test clamping before endarterectomy may give a reliable estimate not only of CBF during clamping but also of the time available for surgery.

References

1. Boysen, G., H. J. Ladegaard-Pedersen, N. Valentin, and H.-C. Engell: Preoperative cerebral blood flow measurements. Cerebral ^{133}Xenon clearance during reconstruction of the carotid and subclavian arteries. Scand. J. clin. Lab. Invest., 1969, in press.
2. Finnerty, F. A., L. Witkin, and J. F. Fazekas: Cerebral hemodynamics during cerebral ischemia induced by acute hypotension. J. clin. Invest. 33, 1227 (1954).
3. Jennett, W. B., M. A. Harper, and F. C. Gillespie: Measurement of regional cerebral blood-flow during carotid ligation. Lancet, II, 1162 (1966).
4. Paulson, O. B., S. Cronquist, J. Risberg, and F. I. Jeppesen: Regional cerebral blood flow. A comparison of 8-detector and 16-detector instrumentation. J. nucl. Med., 10, 164 (1969).

The Effect of Hyperbaric Oxygen During Carotid Surgery

W. B. Jennett, I. Mc.A. Ledingham, A. M. Harper, G. D. Smellie, and J. D. Miller

Department of Neurosurgery, Institute of Neurosurgical Sciences, Glasgow

Hyperbaric oxygen, by increasing the dissolved oxygen content of the blood, might be expected to improve oxygenation in areas of the brain suffering threatened or incomplete ischaemia but which are still being perfused sufficiently for blood to reach them, although in inadequate amounts. It might also increase the period of occlusion of major neck vessels which the brain will tolerate, such as clamping of the carotid artery during surgical procedures. However, previous studies in man [3] and in dogs [1] showed a reduction of 24% and 21% in CBF at 2 ATA; it seemed possible that this vasoconstriction might largely cancel out any beneficial effect if it were also to operate in pathological situations.

The opportunity was therefore taken to measure rCBF in the course of operations on the carotid artery in a walk-in pressure chamber. These included carotid endarterectomies, carotid ligations and excision of carotid body tumours (14 patients in all). The patients were anesthetised with trilene vaporised in air or oxygen and controlled ventilation established and adjusted to obtain PCO_2 values within the range 35–45 mmHg; arterial and jugular bulb samples of blood were obtained for estimations of PO_2 and PCO_2. Collimated scintillation counters were used to determine rCBF from the frontal and temporal regions after injection of ^{133}Xe into the exposed carotid arteries, using the inert gas clearance method; injections were either into the internal carotid directly or into the common carotid during clamping of the external carotid.

Pairs of measurements of rCBF were made at 2 ATA in order to compare the blood flow during air breathing and oxygen breathing; 10 min stabilization was always allowed after changing the inspired gas. On breathing air the inspired oxygen fraction was the equivalent of 40% at 1 ATA; this gave a PO_2 from 150–200 mmHg, rising on 100% O_2 to about 1.000 mmHg. In 20 pairs of observations there was no significant change (Table 1). The change in jugular venous oxygen tension in 14 pairs of observations was 46.9 ± 1.76 (S.E.) to 77.6 ± 4.59. In these patients therefore, no change in flow was recorded, although the error of the method would not be expected to detect a change of ±10%. Observations were also made on the rCBF before and after endarterectomy and again no change was found; the postoperative estimation was made immediately on completion of the procedure with the patient still under anaesthesia. Comparison of the jugular venous oxygen before and after endarterectomy in 11 pairs showed 56.5 ± 4.18 (S.E.) and 60.8 ± 1.67 respectively; in only one instance was there a marked increase in jugular venous oxygen, from 36–45 mmHg on air and from 54–98 mmHg on oxygen.

If there is no perfusion hyperbaric oxygen can clearly bestow no benefit. This was strikingly indicated in a patient in whom the opposite carotid artery was known to be completely occluded, although the vertebrals were patent. Occlusion of the common carotid below the stenosed segment in the internal, immediately after the injection of xenon, caused complete arrest of clearance from that hemisphere and within 6 sec the EEG was almost flat bilaterally. On releasing the clamp after 2 min a normal flow was established and the EEG recovered [2].

Table 1. *Changes in rCBF (ml/100 g/min) during carotid surgery*

	Air at 2 ATA	Oxygen at 2 ATA	Endarterectomy Before	After
SD	48.9 ± 19.5 (20 pairs of observations)	47.85 ± 22.5	54.2 ± 20.5 (12 observations)	58.9 ± 22.3
	Controlled ventilation, monitored $PaCO_2$ Trilene anaesthesia rCBF by clearance after intra-carotid ^{133}Xe			

It is always difficult to assess the benefits conferred by one therapeutic component in as complex a situation as carotid endarterectomy, where the general state of the patient and the local lesion differ so widely and surgical expertise may vary even within a single series. It can only be recorded that we have not encountered a single incident of post-operative neurological deficit in this series of 14 patients, which included some in their sixties and seventies, and bad risk cases such as that just described and another, with complete occlusion of both the opposite carotid artery and the basilar artery.

References

1. JACOBSON, I., A. M. HARPER, and D. G. McDOWALL: The effects of oxygen under pressure on cerebral blood flow and cerebral venous oxygen tension. Lancet II, 549 (1963).
2. JENNETT, W. B.: Proc. roy. Soc. Med. 61, 606 (1968).
3. LAMBERSTEN, C. J., R. H. KOUGH, D. Y. COOPER, A. L. EMMAL, H. H. LOESCHKE, and C. F. SCHMIDT: Oxygen toxicity. Effects in man of oxygen inhalation at 1 and 3.5 atmospheres upon blood gas transport, cerebral circulation and cerebral metabolism. J. appl. Physiol. 5, 471 (1953).

Pre- and Postoperative Cerebral Blood Flow in Patients with Carotid Artery Stenoses

S. E. BREGENTZ, E. HÄGGENDAL, and N. J. NILSSON

Departments of Surgery and Clinical Physiology, University of Göteborg

In order to analyse the immediate effect of the operative removal of stenosis of the internal carotid artery we have determined cerebral blood flow and oxygen consumption in the patients a few days before and about a month after the operation. They have also been examined psychiatrically and by EEG besides the usual clinical and neurological examination.

CBF was determined by KETY's nitrous oxide method with bilateral blood sampling from the jugular bulbs, the mean of the values from the 2 sides being given. Double determinations of arterial carbon dioxide pressure as well as of oxygen saturation and content in arterial and cerebral venous blood were also made during the procedure. Since the study aims at an analysis of the effect of the stenosis as such, we have excluded patients in whom other probable causes for flow alterations exist, i. e. persons with greater change in carbon dioxide tension between measurements than 5 mmHg or with lower hemoglobin oxygen capacity at any examinations than 15.5 vol. %. At the moment, therefore, the material consists of 7 patients, all males, between 57 and 71 years old. As the study is still going on, the results given here are preliminary.

Results

On the basis of the other examinations mentioned above the patients have been tentatively divided into 2 groups, "improved" and "not improved", a classification based upon general clinical impression as well as on neurological, electroencephalographical and psychiatric examinations, definite improvement in at least 2 of the respects being required for the patient to be placed in the "improved" group. In no case has the state of the patient deteriorated after the operation.

The CBF values (Table 1), averaging 41.7 ml/100 g/min at a mean arterial carbon dioxide tension of 40.0 mmHg before the operation, are lower than normal, even considering the age of the patients. The same can be said of the oxygen consumption of 2.76 ml/100 g/min. After the operation the mean CBF value is 42.7 ml/100 g/min, obviously not significantly different from the preoperative value. This applies also to the very small differences in oxygen consumption and in hemoglobin oxygen saturation of the cerebral venous blood (SvO_2).

When the 2 groups of improved and unimproved are regarded separately, it is evident that both the flow increase of 3.3 ml/100 g min in the improved group and the decrease of 2.0 ml/100 g/min in the unimproved group is not significant. The averaged absolute flow value for the unimproved group before the operation is 12.2 ml/100 g/min greater than

162 S. E. Bregentz, E. Häggendal, and N. J. Nilsson

that of the improved patients. This is an interesting observation, the significance of which, however, must still be regarded as doubtful in view of the smallness of the material.

Table 1

	Age	CBF ml/100 g/min		CMRO$_2$ ml/100 g/min		S$_{VO_2}$ %		Pa$_{CO_2}$ mmHg	
		pre-op.	post-op.	pre-op.	post-op.	pre-op.	post-op.	pre-op.	post-op.
Improved									
R.K.	66	37	33	2.80	2.15	58.0	59.0	38	42
E.J.	57	31	45	2.50	3.25	53.5	57.5	44	46
L.P.	58	34	41	2.15	2.35	58.5	62.0	36	40
A.L.	68	44	40	2.86	2.55	53.3	61.6	37	38
Mean	62	36.5	39.8	2.58	2.58	55.8	60.0	38.8	41.5
Not improved									
L.L.	56	63	58	3.51	3.60	58.5	58.5	44	49
G.B.	55	43	40	2.54	2.36	62.7	60.6	41	37
H.L.	71	40	42	2.94	3.71	54.6	52.5	40	38
Mean	61	48.7	46.7	3.00	3.22	58.6	57.2	41.7	41.3
Mean of whole material	62	41.7	42.7	2.76	2.85	57.0	58.8	40.0	41.4

Flow measurements in patients with carotid artery stenosis have a special interest because there is an obvious and removable obstruction to the flow of blood combined with signs of cerebral damage, which sometimes develop rather rapidly. Therefore in this condition one might expect to find a cerebral tissue actually suffering from an insufficiency of blood supply, which might be at least partially reversible.

From this point of view the most striking finding from the present material is the lack of flow augmentation after the operation. Even this small material would seem to justify the conclusion that an operative elimination of a carotid artery stenosis does not per se augment even a reduced cerebral blood flow.

The results therefore speak against a common presence in these cases of cerebral tissue suffering from chronic malnutrition. This is in concordance with the fact that the signs usually appear in attacks, the state of the patient having become stabilized at the time of examination. However, the change in the arterial wall is usually regarded as progressing more evenly. Thus probably the signs of the disease are not caused solely by the flow-impeding effect of the stenosis but need a second contributing factor, the influence of which is comparatively large. Such a factor may be periodically appearing showers of small emboli from the stenosing atheroma or transient falls in arterial blood pressure.

It is still an open question if there exists a possibility of further analysis. The material might consist of 2 groups, the stenosis contributing more to the state of the patient in one group than in the other with a corresponding postoperative flow increase as a consequence. Since, however, the present material does not warrant such a conclusion, we intend to continue the investigation with special consideration of this problem.

Cerebral Blood Flow Measurements in the Evaluation of Carotid Surgery

H. O. Christensen Lou and F. von Wowern

Department of Neuromedicine, Gentofte Hospital, Hellerup

Some attempts have been made to measure rCBF before and after operation for stenosis of the internal carotid artery. Nilsson [5] reported preliminary results, possibly indicating a correlation between clinical improvement and increased CBF, as measured by the original Kety technique. Another approach was made by O'Brien et al. [6], who were unable to demonstrate any correlation between CBF values and the clinical results of operation. The latter authors used the ^{133}Xe inhalation technique measuring semiquantitatively the flow in each carotid territory (internal + external).

We have measured rCBF immediately before and after internal carotid endarterectomy, under control of arterial blood pressure in 2 patients. The 1% halothane anaesthesia was combined with inhalation of 4–5 % CO_2 in an open circuit system as described by Christensen et al. [1] in order to obtain cerebral vasodilation. Arterial pCO_2 measurements during the entire procedure gave values of 50 ± 1 mmHg. Possibly higher pCO_2 values might have been preferable, but this was considered to be hazardous to the patient. rCBF-determination were carried out according to Lassen et al. [3]. 1 mCi of ^{133}Xe was injected and its clearance recorded by a single scintillation detector. During injection the external carotid artery was clamped. Two patients were studied. Selection was based on the criteria of Marshall [4]:

1. transient episodes of neurological dysfunction of one cerebral hemisphere, leaving at most only slight sequels;
2. severe stenosis of the corresponding extracranial internal carotid artery;
3. no other major arteriosclerotic lesion of extra or intracranial cerebral vessels.

Table 1. *rCBF measurements during 1 % Halothane Anaesthesia and 5 % CO_2 inhalation*

	Before internal carotid endarterectomy	After endarterectomy
Patient CB		
Flow calculated from fast component ("grey matter")	63 ml/100 g/min	91 ml/100 g/min
Flow calculated from slow component ("white matter")	7.4 ml/100 g/min	20 ml/100 g/min
Patient SO		
"grey matter"	49 ml/100 g/min	57 ml/100 g/min
"white matter"	9.4 ml/100 g/min	13 ml/100 g/min

Both patients SO and CB had suffered from several transient episodes during 6 months prior to operation. The values of ophthalmodynamometry and ultrasonic carotid echography

11*

[2] were significantly decreased on the side corresponding to the symptoms. Carotid angiography revealed in both cases a severe stenosis reducing the lumen of the affected vessels to about 1 mm².

Result

rCBF measurements revealed an increase in flow after carotid endarterectomy in both cases (Table 1).

Control angiography, 6 months following operation, showed almost normal conditions. The patients did not experience further episodes following operation, and clinical examination revealed normal conditions. Ophthalmodynanometry and ultrasonic carotid echography were also normal.

References

1. Christensen, M. S., K. Høedt-Rasmussen, and N. A. Lassen: The cerebral blood flow during halothane anesthesia. Acta neurol. scand. Suppl. 14, 152 (1965).
2. Kristensen, J. K., and J. Kvist: Ultrasonic pulse detection applied to carotid and vertebral arteries. Scand. J. thorac. cardiovasc. Surg. 1, 178 (1967).
3. Lassen, N. A., K. Høedt-Rasmussen, S. C. Sørensen, E. Skinhøj, S. Cronqvist, B. Bodforss, and D. H. Ingvar: Regional cerebral blood flow in man determined by ⁸⁵Kr. Neurology 13, 719 (1963).
4. Marshall, J.: The mangement of cerebrovascular disease. London: Churchill 1968.
5. Nilsson, N. J.: in the discussion of S. S. Kety, Fundamental Aspects of the Human Cerebral Circulation and their Implications for Cerebrovascular Disease. In: Thule International Symposia. Stroke, Nordiska bokhandel förlag, Stockholm, 1967.
6. O'Brien, M. D., N. Veall, R. J. Luch, and W. T. Irvine: Cerebral cortex perfusion-rates in extracranial cerebrovascular disease and the effects of operation. Lancet II, 382 (1967).

The Influence of Carotid Stenosis on Cortex Perfusion

M. D. O'BRIEN and N. VEALL

Newcastle General Hospital, Newcastle, and Guy's Hospital Medical School, London

Regional cerebral cortex perfusion (CPR) rates were measured in 34 patients who were being considered for extracranial vascular surgery in the Professorial Surgical Unit at St. Mary's Hospital, London. 10 patients eventually had an operation and the measurements were then repeated postoperatively. The cerebral blood flow data were not used in the selection of patients for surgery. Simultaneous bilateral measurements of regional cerebral cortex perfusion rates were estimated by the inhalation method. This technique is particularly suitable for the study of the effect of carotid lesions on cortex perfusion rates and has some advantages over other methods of blood flow measurement currently available. First it gives simultaneous bilateral measurements, and in the present series this has shown to give valuable additional information, secondly all sources of blood supply to the hemisphere are labelled with tracer, third no carotid puncture or internal carotid artery catheterization is necessary. This is particularly useful for carrying out postoperative measurements after the acute effect of the operation has passed and where there is no indication for repeat angiography. In addition the response to the inhalation of 5% CO_2 was measured in 10 patients, in 4 of these the measurements were repeated postoperatively. The lesions were divided into 4 grades according to the single-plane angiographic appearances; Grade I being atheromatous irregularity only, Grade II a moderate stenosis but not significant in terms of local vessel flow, Grade III stenoses were thought to be significant in terms of local blood flow and Grade IV were complete occlusions.

Results

There was no constant relationship between a "significant" arterial lesion and a reduced flow in the ipsilateral hemisphere compared to normal values for the method, as might be expected if these lesions directly affected cortex perfusion rates. Similarly it might be expected that where there was an asymmetry in flow between hemispheres, the lower perfusion rate would be on the side of the lesion, but this too was not found, in fact, the cortex perfusion rate was actually higher on the side of a carotid occlusion compared to the other side in 2 patients. Also, perfusion rates in the clinically affected hemisphere was not consistently lower than in the clinically normal hemisphere. Eight of the 10 patients selected for surgery had a Grade III stenosis, but removal of these did not increase the CPR. Since this original series several more patients have also been measured pre- and postoperatively and in one of these there was a significant increase in cortex perfusion, it is of interest that this patient had a complete occlusion on one side and a severe stenosis on the other.

Ten patients also had CPR measurements following stimulation with 5% CO_2; 4 of these eventually had an operation and it was possible to repeat the measurement of response to CO_2 in 3 patients postoperatively.

The results of these experiments show that patients with extracranial cerebrovascular disease have a normal percentage response to CO_2 when the blood flow figure is taken as a mean between both hemispheres; this applies even when the initial CPR measurements are low. However, if the 2 hemispheres are considered separately there is a considerable within patient variation. The results in the 3 patients whose response to CO_2 was measured pre- and postoperatively show that this between hemisphere variation may be profoundly affected by operation.

Conclusions

The following conclusions have been drawn from these data:

1. that most atheromatous lesions in the extracranial cerebral vessels, including Grade III stenoses and complete occlusions, have no constant effect on blood flow in the cerebral cortical territory of the affected vessel. This is probably because of adequate collateral supplies above the stenosis or occlusion;

2. that even if the lesion is significant in terms of local vessel flow, its slow development minimises its effect on cerebral blood flow because of the development of alternative sources of blood supply;

3. that removal of these lesions does not increase cerebral blood flow unless the total cranial input is impaired and this means severe stenoses of more than one vessel. Lowe (1962) has pointed out that for the extracranial cerebrovascular resistance to equal the intracranial cerebrovascular resistance there has to be stenosis to about 10% of all four major vessels, or complete occlusion of 3 of the 4 major vessels;

4. that asymmetry between hemispheres is a good indication of disease at or above the Circle of Willis;

5. that it is the state of the small vessels which determines flow. This follows from the fact that it is not the state of the main arteries in the neck which determines cortex perfusion rate. Small vessels in this context are of course those distal to effective collateral circulation; it is lesions in these small vessels which probably determine the site of an infarct, not due to emboli, which might occur with a generalized fall in perfusion pressure;

6. it is clear that the changes in response to CO_2 cannot be satisfactorily interpreted by themselves, but considerable additional information from angiography is necessary to explain the changes. It seems that removal of a significant stenosis does not necessarily increase the blood supply to the brain unless the total available supply is in jeopardy, though it does seem to increase the peak flows obtainable under maximal vasodilation; this discrepancy may be due to the reduced metabolic demands of these patients. The response to CO_2 depends on the degree of stenosis, that is whether the stenosis is significant at normal flow rates, becomes significant with increased flow rates or is not significant at all; this applies not only to the obvious carotid lesions but also to smaller stenoses more distally. Nor can the development of asymmetry following 5% CO_2 inhalation be necessarily ascribed to a lesion in the major vessels, this may well in fact be demonstrating an inadequacy at the Circle of Willis or disease above this;

7. that simultaneous bilateral measurements of CPR using a method which labels all sources of blood supply to the area seen by the counter is necessary to evaluate the blood flow changes which occur as a result of vascular stenoses and the effect of CO_2 inhalation.

Acknowledgements. We are grateful to Professor W. T. IRVINE and Mr. R. J. LUCK for their co-operation and permission to study their patients. This investigation was carried out in the Medical Research Council's Radioisotopes Laboratory, Department of Medicine, Guy's Hospital Medical School, London.

References

1. O'BRIEN, M. D., N. VEALL, R. J. LUCK, and W. T. IRVINE: Cerebral Cortex Perfusion Rates in Extracranial Cerebrovascular Disease and the Effects of Operation. Lancet II, 392–395 (1967).
2. LOWE, R. D.: Adaption of the Circle of Willis to occlusions of the carotid or vertebral artery; its implications in caroticovertebral stenoses. Lancet I, 395 (1962).

Comments to Chapter IV

Carotid Surgery

The first 4 papers of this section were concerned with studies of rCBF in relation to surgery of the internal carotid artery. BOYSEN et al. reported that clamping of the artery always reduced CBF and if this fell to a level of 20–30 ml/100 g/min, neurological complications occurred. JENNETT in discussion reported that in his experience a reduction to 25 per cent of the pre-clamp level, represented a critical level.

The other reports showed that the rCBF (or CBF) was not significantly increased after endarterectomy in the majority of cases, and the suggestion was made that this indicated that the stenosis produced clinical deficit by providing a source of emboli rather than by reducing flow. There was clearly a danger of an oversimplified approach to the problem. Initially it had been thought that stenoses produced their effect by interfering with flow (the haemodynamic crisis hypothesis); later the importance of stenosis as a source of emboli began to be appreciated (the embolic hypothesis). The 2 hypotheses are not mutually exclusive, however, and both mechanisms probably occur. The reports available to-date do not permit the sub-division of cases into groups according to the severity of the stenosis and the condition of the other arteries to cast further light on this problem.

The report on carotid surgery under hyperbaric oxygen suggests this may confer some protection, though in view of the limited availability of chambers the place of this, other than as an experimental tool, is doubtful of clinical value.

The overall impression of the section was that rCBF studies are now beginning to yield information in the clinical situation. To date this information is different from what might have been expected. Extra-cranial stenoses do not appear to produce the anticipated reduction in cerebral blood flow. Since it is clear that in some cases flow is reduced, it remains to define more clearly the circumstances in which this may occur. The use of provocative tests such as alterations in the level of CO_2 and blood pressure will have to be applied more extensively. More work is needed, particularly in relation to the type of stenosis and the condition of the other arteries. If the clinical material is carefully classified according to these criteria, rCBF studies may be expected to further our knowledge and help in the assessment of patients for surgery.

J. MARSHALL

Regional Cerebral Blood Flow in Cases of Brain Tumor

M. Brock, A. A. Hadjidimos, K. Schürmann, M. Ellger, and F. Fischer

Neurosurgical Department, University of Mainz

The present series consists of 21 patients with brain tumors: 12 malignant gliomas, 4 oligo-dendro gliomas, 3 meningeomas and 2 metastases, studied by the ^{133}Xe-gamma-clearance rCBF method. In 16 patients at least 3 determinations of rCBF were performed: a) in the resting state, R; b) during hyperventilation, HV, and c) during hypertension, HT. Of the remaining 5 patients, 1 was studied only during R, while the other 4 were submitted either to HV or to HT additionally to the R study. The average age of this series of patients was 47 years, the youngest being 19 and the oldest 64. rCBF studies had to be performed under general anesthesia ($N_2O + O_2$, Engström respirator) in 5 instances in which the patients were un-cooperative. MABP was continuously recorded through the internal carotid catheter. End-expiratory CO_2 was also monitored (and recorded) by means of an infrared analyser in selected cases. Arterial (and in some cases also cerebral venous) pH, pCO_2 and pO_2 were determined at least once for each Xenon injection.

Results

Average rCBF and focal changes: Average hemisphere blood flow in the whole series was 31.5 ml/100 g/min. All patients except 1 (without focal changes) had average rCBF values below 42 ml/100 g/min, the lowest value being 14.5 ml/100g/min.

In 19 cases focal rCBF disturbances could be detected in agreement with the tumor location as verified directly or by other diagnostic procedures. In 13 cases a relative hyperemia was detected at the areas corresponding to the tumor site. In 6 cases there were ischemic rCBF focal changes corresponding to the tumor location.

Cases with ischemic focus: As stated, in 6 cases rCBF values at the tumor site were lower than at the surrounding areas. In 3 of these cases there was a perifocal hyperemia, i. e., the tumor site was surrounded by areas with higher rCBF than the rest of the hemisphere.

Reactivity to HV varied very much in these 6 cases, ranging from an excessive rCBF reduction at the tumor site (case 39/68) or over its periphery (case 45/68) to a "paradoxical" focal CBF increase (case 22/69). In all 6 cases but 1 (22/69) HV caused a reduction of inter-regional rCBF differences (inter-channel SD) with homogeneisation of mean hemispheral CBF at a lower level than during rest (this phenomenon will be discussed further on).

A more or less pronounced impairment of autoregulation (AR) was present in all 6 cases – either at the tumor site or at its near or remote periphery – the number of cases being still too small to allow systematization.

Cases with hyperemic focus: In 13 cases there was a hyperemic focus over the tumor location.

In contrast to the ischemic group, there was a clear increase in interregional CBF differences (inter-channel SD) in most cases (6 out of 10) with a hyperemic focus corresponding to the tumor location. The mechanism governing both types of changes seems

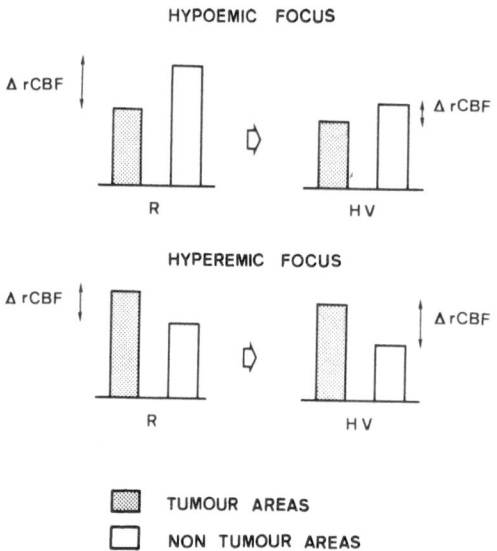

Fig. 1. rCBF changes caused by hyperventilation in cases of brain tumor. The non tumor areas react more to HV than the tumor areas. The consequence of this is a decrease of interregional CBF differences in cases with hypœmic focus

Fig. 2. Measurement of multichannel rCBF during the operation of brain tumors

to be the same, namely a deficient $ApCO_2$ reactivity over the tumor areas. When there is a ischemic focus, its decreased $ApCO_2$ reactivity will contribute towards homogeneisation of flow under HV. On the other hand, a hyperemic area will become more distinct if the surrounding tissue further decreases its rCBF during hyperventilation (Fig. 1).

In 4 cases we observed an increase in rCBF at the tumor location or at its periphery during HV. Furthermore, in 3 cases there was a very bad recovery from HV effects, CBF still being very much below control values after 30 min, at a time when $ApCO_2$ had already returned to normal levels.

Autoregulation was preserved in only 1 case. It was more or less impaired in the remaining 11 cases in which HT had been induced. Impairment of AR was observed at the "tumor areas" in 5 cases, while the autoregulation was impaired in the immediate tumor vicinity in 7 cases, and in the remote areas in 5 instances.

It is worthwhile mentioning briefly the possibility of performing rCBF studies during surgical interventions (Fig. 2), and during the postoperative evolution. Although our results with such studies are still too few to warrant conclusions, it seems that the per- and postoperative follow up of rCBF changes might be of help in the assessment of immediate prognosis after operation of brain tumors.

Summary and Comments

1. In our experience it is an exception that a tumor of the cerebral hemispheres, whatever its nature, does not cause focal rCBF changes either at the resting state or during functional tests.

2. According to the rCBF values, brain tumors can be classed into 3 groups: those causing no focal changes, those causing an ischemic focus, and those causing focal hyperemia at the tumor site. The first type of tumours (no focal changes) constitutes an exception, while the last (focal hyperemia) is the most frequent.

3. In most instances, in cases with focal rCBF changes, there is a decrease in $ApCO_2$ reactivity in the areas corresponding to the location of the tumor. As discussed in the text (Fig. 1) in cases of an ischemic focus this decreased reactivity will cause a homogeneisation of hemisphere CBF during IIV. The opposite will happen in cases of a hyperemic focus.

4. Even in cases with satisfactory CBF decrease in response to HV there may be a lag in returning to control CBF values after $ApCO_2$ has returned to the resting levels.

5. Autoregulation to hypertension is impaired in practically all cases of both focal groups, mainly around the "tumor areas".

rCBF Studies in Intracranial Tumors

D. Oeconomos, B. Kosmaoglou, and A. Prossalentis

Neurological Center, Polyclinic of Athens

When the neurosurgeon looks at the possible advantages of flowmetric studies he is mostly thinking of approaching the problem of the histology of the lesion with the help of rCBF information.

Lately many papers have been devoted to this subject and variable results have been reported [1, 2, 3, 4, 5].

The present investigation was made in order to clarify the practical interest of flowmetric studies for the neurosurgeon.

Material and Method

30 gliomas (27 astrocytomas type II–IV, 3 astrocytomas type I), 11 meningiomas (8 benign meningiomas, 3 meningosarcomas), and 15 metastatic tumors, were reviewed. 4 more cases of expanding lesions of different histology (1 tuberculoma, 1 case of sarcoidosis, 2 brain abscesses), were also included in this series.

Since we are using a 6-detectors equipment, the positioning of the narrowly collimated probes (15 mm wide crystals) was made according to the data obtained from serial scans with ^{99}Tc-pertechnate. The distribution of the detectors over the hemisphere was made in such a way that data from the tumor proper, from paratumoral tissue and from non-tumor tissue could be simultaneously collected. Arteriographic data were also correlated with the scans in establishing the geometry of the lesion. A few cases have also been studied after the ablation of the tumor. Clearance curves were submitted to stochastic analysis and the slope index calculation.

Gliomas

The mean hemisphere flow in glioma cases was found to be 37 (S.D. 10) ml/100 g/min. This general decrease of flow rate was disclosed also in the other tumor groups. In this group the mean $rCBF_{10}$ over the tumor was 37.7 ml/100 g/min, in the peritumoral area 33.5 ml/100 g/min, and over the non-tumor tissue 35.8 ml/100 g/min. These figures have no absolute value since the partition coefficient for gliomas in unknown. We used the partition coefficient of normal brain tissue. Nevertheless it seems that there is a more marked decrease of flow over the non-tumor tissue than over the tumor itself.

In fact, there are individual variations of the tumor flow as compared to the non-tumor tissue flow. In some cases the former is higher while in others the reverse is true.

Gliomas

Over the tumor itself the flow is more often increased in vascularized tumors as compared to the avascular ones. The clearance curves may show a very fast initial component in the form of the so called tissue peak [6]. Those peaks are disclosed mostly, but not exclusively, over highly vascularized lesions. For instance, in our material, tissue peaks were present in 6 cases of hypervascularized grade II–IV astrocytomas and in one avascular grade III glioma. The clearance curves with such peaks had either high tumor flows of 67 ml/100 g/min and 57 ml/100 g/min as was the case in 2 patients, or a decreased tumor flow ranging from 35–43 ml/100 g/min in 5 cases. It appears, therefore, that high flow values, tissue peaks and vascularity of the lesion are related but not strictly interdependent findings.

The reactivity to CO_2 inhalation showed variable responses over the tumor. Very few were the tumors with a marked increase of perfusion rates after CO_2 administration. Most of them showed a poor reactivity or no change of the flow values. Intratumoral steal has been observed as well as tumor to non tumor tissue steal.

Low Grade Astrocytomas

We had but 3 cases of low grade gliomas. In 2 of them slight abnormalities to functional tests were found while in 1 the flow and functional responses were normal.

Meningiomas

The mean hemisphere blood flow was 47.6 $\pm$ 15.3 ml/100 g/min. The mean rCBF was respectively 52.2 ml/100 g/min over the tumor, 51.2 ml/100 g/min over the peritumoral tissue and 41.2 ml/100 g/min over the remote non-tumor tissue. All of the cases showed the usual angiographic pattern of meningiomas following selective injection of the opaque material into the internal carotid artery. 3 cases were also fed by the external carotid artery according to the operative findings.

2 of the meningioma cases had angiographic evidence of intratumoral shunts. Both of them were benign meningiomas and had high tumor values ranging from 51–79 ml/100 g/min in the one case and 48–113 ml/100 g/min in the other.

The 3 meningosarcomas had tumor flow rates of 34, 36 and 62 ml/100 g/min respectively.

Intratumoral steal was observed 3 times, tumor to peritumoral tissue steal once, tumor and peritumoral tissue combined to normal tissue steal once, normal tissue to tumor once. Fluctuations of flow – with the CO_2 test – were found in 2 cases over the tumor, in 2 cases over the normal tissue and in 3 instances over the peritumoral tissue. 2 cases of deep seated meningiomas, 1 olfactory and 1 intraventricular, showed fluctuations of flow following the CO_2 test on the area overlying the lesion and the clearly paratumoral tissue.

We mention the above observations in order to stress the fact that not only the brain parenchyma but also the tumor tissue proper may show abnormal and paradoxical reactivity to CO_2. Such an example is given by an almost round, 5–6 cm thick, meningioma showing variable responses to CO_2. While 1 area of the tumor, at its thickest part, has a 120% rise in flow, the next to it gives but a 30% increase. At the same time, 1 detector looking at the periphery of the lesion shows a fall by 30% in perfusion rate; 2 other detectors looking at the paratumoral tissue show, respectively, a decrease by 19% and an increase by 9%. Finally, the normal tissue has an increase of flow of 20%.

Metastatic Lesions

The mean hemispheric flow was found to be 35.7 ± 6.1 ml/100 g/min.

In 3 cases of medium and large sized tumors the flow was higher over the tumor than over the non-tumor tissue. This difference was moderate and slight in the other 2 cases.

5 cases had a tumor flow lower than the non-tumor tissue. In 7 cases there was non significant difference between those 2 flow rates. Only 7 cases showed a positive CO_2 reactivity, 5 showed no changes with CO_2-inhalation or even a paradoxial slight fall of perfusion rates over the tumor and non-tumor areas. In only 1 case we had the impression that a steal of tumor to normal tissue was present. These findings indicate that although 12 out of 15 cases were small lesions, less than 3–4 cm in diameter, the flow was generally more depressed than in the glioma and meningioma groups.

The other types of expanding processes included in this review did not disclose any specific and different characteristics from those already described in the other groups.

Comments

Although we are using a small number of detectors and, consequently, missing a great deal of information, the review of our series gave evidence that the gliomas produce definite diffuse changes in flow rates of the diseased hemisphere and definite focal abnormalities in CO_2-reactivity and in the other functional tests. These changes are not specific for the glioma group. They are also disclosed in the other tumor groups. The more pronounced depressions in flow rate of non-tumor tissue were observed in metastatic cases, although in all of them but 2 the lesion was small to medium-sized. Each of the 2 last mentioned cases had 2 small-sized metastatic foci.

This more pronounced diffuse hemispheric depression in metastatic lesions may be related to the well-known tendency of metastases to cause brain edema. It should be mentioned again that in patients harboring a meningioma the non-tumor tissue shows a smaller decrease in flow than in the case of gliomas. This finding may be explained on the basis that gliomas, which are intrinsic brain tumors, give a more diffuse brain edema, to which eventually a more pronounced neurogenic flow depression in areas remote from the lesion is added. Finally, abnormal responses to functional tests are found over the tumor proper. This observation underlines the necessity of routine use of the functional tests in order to disclose focal flow abnormalities.

The number of low grade astrocytomas in this series is too small to permit definite conclusions.

Although the CBF studies in tumor cases do not lead to practical neurosurgical conclusions, they nevertheless give the opportunity of detecting abnormal reactivity to functional tests which, in their turn, may be of value in establishing the diagnosis in neurosurgical cases.

References

1. Cronqvist, S., and N. Lundberg: Regional Cerebral Blood Flow in intracranial tumors with special regard to cases with intracranial hypertension. Scand. J. clin. Lab. Invest. Suppl. 102, p. XV–A (1968).

2. —, and D. H. INGVAR: Regional Blood Flow in patients with brain tumors: Discussion and comments. Scand. J. clin. Lab. Invest. Suppl. **102**, p. XV–E (1968).

3. —, and F. AGEE: rCBF in intracranial tumors. Acta Radiol. **7**, 393 (1968).

4. ESPAGNO, J., and Y. LAZORTHES: Cerebral Blood Flow in brain tumors. Scand. J. clin. Lab. Invest. Suppl. **102**, p. XV-C, (1968).

5. PALVÖLGYI, R.: Regional Cerebral Blood Flow in tumor patients. Scand. J. clin. Lab. Invest. Suppl. **102**, p. XV-B (1968).

6. PAULSON, O. B., S. CRONQVIST, J. RISBERG, and F. JEPPESEN: Regional cerebral blood flow. A comparison of 8-detectors and 16-detectors equipment. J. nucl. Med., **10**, 164 (1969).

Paradoxical rCBF Reactions in Intracranial Tumors

R. Palvölgyi

Radiological Clinic, Faculty of Medicine, University of Budapest,
and Department of Clinical Physiology, Bispebjerg Hospital, Copenhagen

The "intercerebral steal" and "inverse steal" represent in reality different forms of the same functional damage of the cerebral vessels, which may be revealed by changes of the $aPCO_2$ tension.

The damage of the vessels manifests itself in different ways. In some cases we find in the pathological region a qualitatively normal but quantitatively diminished reaction; in other cases we cannot find any reaction at all, eventually the loss of function manifests itself many times in opposite (paradoxical) reactions. The form of manifestation is determined by the intracranial hemodynamic situation and by the sensitivity of the equipment applied.

This opinion appears to be supported by our observations in tumor patients.

In Copenhagen we investigated the regional cerebral blood flow of 26 tumor patients using the intra-arterial Xenon-133 clearance method. In 10 cases, showing diffuse loss of autoregulation, paradoxical reactions were provoked by artificially induced blood pressure changes. In 8 cases the hypertensive state provoked in some regions a paradoxical decrease of blood flow. In 2 cases the hypotension was followed by enormous increase of the rCBF in several regions.

On the base of these facts it seems to be very likely that the changes of the $ApCO_2$ tension and the blood pressure, which are harmless for a normal brain, can cause similar paradoxical reactions and therefore may be harmful for the patients suffering from local cerebral lesion. Hypercapnia or hypertension can cause local tissue damage as a result of the paradoxical decrease of the flow. On the other hand, possibly, a local bleeding may be caused by the paradoxically increased flow.

We can rightly suppose that the use of angiography or gamma camera techniques would be much more efficient for the representation of cerebral tumours in the state of hyperventilation in consequence of the inverse steal effect.

References

1. Palvölgyi, R.: Die Messung der regionalen Hirndurchblutung mittels intraarterieller Xenon-133 Injektion beim Menschen. Arch. Psychiat. Nervenkr. **212**, 8 (1968).
2. — Regional cerebral blood flow in tumour patients. Scand. J. clin. Lab. Invest. Suppl. **102**, p. XV-B (1968).

Cerebrovascular Dilatation and Compression in Intracranial Hypertension

T. W. Langfitt, J. D. Weinstein, N. F. Kassell, and H. M. Shapiro

Department of Neurosurgery, Hospital of the University of Pennsylvania, Philadelphia

Increased intracranial pressure (ICP) causes dilatation of pial vessels [5]. Cerebral blood flow (CBF) may be maintained at high levels of ICP produced by infusion of the CSF spaces [7]. Ultimately, CBF begins to decline as ICP continues to rise, and when ICP equals the systemic arterial pressure (SAP), CBF ceases [3]. The cerebral vasodilatation that maintains CBF during rising ICP appears to be a form of autoregulation. The subsequent decline in CBF is due to vascular compression, and an unresolved issue is where the compression occurs. In patients in whom ICP equals SAP, opaque substances injected into the carotid or vertebral arteries does not enter the intracranial space [4]. This suggests collapse of these vessels as they penetrate the dura, but the results can be explained equally by collapse of the distal cerebrovascular tree; that is, opaque substances will not enter the intracranial space irrespective of the site of occlusion if CBF has ceased. Previous studies of cerebral venous pressure have suggested that both cerebral veins and the sagittal sinus are compressed by increased ICP [1, 6].

This report is a summary of 2 series of experiments in monkeys. In the first series the surface of both cerebral hemispheres was observed and photographed through a transparent water-tight calvarium during elevation of ICP by expansion of a subdural balloon or lumbar subarachnoid infusion of saline. SAP and systemic venous pressure (SVP), anterior sagittal sinus pressure (SSP), and ICP were recorded continuously. Internal carotid artery blood flow (ICBF) was metered in several animals. Only acute changes in ICP have been studied sofar.

A rapid rise in ICP produced by subarachnoid infusion produced dilatation of cortical arteries and veins accompanied by maintainance of ICBF at near normal values for a brief period. As ICBF began to fall with continued elevation of ICP, vessels on the surface of gyri were compressed against the overlying calvarium, but vessels in sulci remained dilated. Expansion of the subdural balloon caused compression of vessels adjacent to the balloon that gradually spread, often in a circumferential manner, to involve the cortex of the compressed hemisphere. At the same time vessels remote from the balloon were dilating. Again, vessels on the surface of gyri were always collapsed before those in sulci. The changes in SSP with both balloon expansion and subarachnoid infusion were complex and were interpreted to indicate vascular compression both proximal and distal to the recording catheter in the anterior sagittal sinus.

Pressure waves occurred frequently during expansion of the balloon and were accompanied by progressive brain swelling and cerebral vasomotor paralysis [2]. When ICP equalled MAP, with the balloon still expanded, all cortical vessels were collapsed. Rapid evacuation of the balloon was accompanied by a rush of red blood through both arteries and veins, usually with cortical hemorrhage. ICP fell to zero during the evacuation, but as the cerebral vessels filled with blood, ICP rebounded within a few seconds to the MAP, and carotid vessels were again compressed. ICBF showed a precise inverse relationship to ICP under these circumstances. ICBF was zero prior to evacuation of the balloon, increased to very high values during evacuation of the balloon, and returned to zero as the ICP rebounded to the level of the SAP.

In the 2nd series of experiments brain swelling was produced by gradual expansion of a subdural balloon, as described above. In the end stage of decompensation the balloon was evacuated and removed and ICP rebounded to the SAP. The heads were frozen in liquid nitrogen and sectioned in coronal planes. Most cortical vessels were collapsed. The sagittal and straight sinuses were reduced to narrow slits. Large subcortical vessels also were affected. In 1 animal the anterior cerebral artery was collapsed between swollen cingulate gyri, in its course over the corpus callosum.

These observations demonstrate again that rising ICP causes cerebrovascular dilatation that in turn appears to be responsible for maintenance of CBF. However, the cause of the intracranial hypertension is as important as the level of ICP in determining CBF, and vessels remote from an expanding mass may be dilated at a time when those adjacent to the mass are collapsed. The first vessels compressed in the transparent calvarium experiments were those on the surface of gyri, arteries and veins alike. Cortical veins in sulci remained dilated, perhaps in part due to constriction at their junction with the sinuses. In our experiments these junctions could not be visualized, because the bone over the sagittal sinus was not removed. The SSP recordings, however, did suggest venous constriction proximal to the sinus at a time when the cortical veins were dilated. In the presence of severe brain swelling, sufficient to raise ICP to the SAP, there is diffuse collapse of the cerebral vessels, including the sinuses and large arteries. Cerebral vasomotor paralysis is defined as loss of the vaso-constrictor tone of cerebral vessels causing release of the arterial pressure head into the capacitance system. When vasomotor paralysis is complete ICP equals MAP. The brain is massively swollen due to increased intravascular blood volume and edema, and as demonstrated here, hemorrhage also is common.

References

1. Greenfield, J. C., and G. T. Tindall: Effect of acute increase in intracranial pressure on blood flow in the internal carotid artery of man. J. clin. Invest. 44, 1343 (1965).
2. Langfitt, T. W., J. D. Weinstein, and N. F. Kassell: Cerebral vasomotor paralysis produced by intra-cranial hypertension. Neurology 15, 622 (1965).
3. —, N. F. Kassell, and J. D. Weinstein: Cerebral blood flow with intracranial hypertension. Neurology 15, 761 (1965).
4. — — Non-filling of cerebral vessels during angiography: Correlation with intracranial pressure. Acta Neurochir. 14, 96 (1966).
5. Wolff, H. G., and H. S. Forbes: The cerebral circulation. V. Observations of the pial circulation during changes in intracranial pressure. Arch. Neurol. Psychiat. 20, 1035 (1928).
6. Wright, R. D.: Experimental observations on increased intracranial pressure. The Australian and N. Z. J. Surg. 7–8, 215 (1938).
7. Zwetnow, N.: Cerebral blood flow autoregulation to blood pressure and intracranial pressure variations. Scand. J. clin. Lab. Invest. 1968, Suppl. 102, p. V-A (1968).

Studies on the CBF and Problems of the Determination of the Cerebral Metabolism in Patients with Localized Cerebral Lesions

H. D. HERRMANN, H. PALLESKE, and J. DITTMANN

Department of Neurosurgery, University of Homburg/Saar

Measuring in patients with brain tumors the transit of a non-diffusible isotope for each hemisphere separately we found preoperatively in the phase of increased intracranial pressure a reduction of the blood flow especially in the tumor hemisphere. Postoperatively after sufficient pressure relief we found a transient reactive hyperemia. This luxury perfusion however seems to depend on the intracranial pressure. The dilated vessels can easily be compressed during the postoperative swelling phase and thus the flow over the operated hemisphere can even be reduced below normal values. The different regulatory mechanisms therefore can cause a different blood flow at the site of the lesion and the uninjured brain.

We have furthermore studied the metabolism of brain tumors in vitro and found that gliomas have a different metabolism from normal brain. The quotient of respiration is lower and the quotient of aerobic glycolysis is higher, the aerobic lactate production therefore is significantly increased compared with normal brain and even with meningiomas.

If these metabolic differences could also be found quantitatively in vivo it could be of value in the diagnosis of brain tumors. In patients with unilateral cerebral lesions we therefore determined the CBF simultaneously as a mean value of both hemispheres and of each hemisphere separately and at the same time pCO_2, pO_2, pH, vol. % O_2, glucose, lactate and pyruvate in the arterial blood and in the cerebral venous blood. For the cerebral venous blood the jugular bulb either on the side of the lesion or on both sides was punctured. For the calculation of the metabolism the A-V differences of O_2, glucose, lactate and pyruvate have to be multiplied with the CBF.

In tumor cases with greater flow differences between both hemispheres the A-V difference seems to correlate better to the mean flow than to the flow of the side of the lesion, compared to the normal O_2 and glucose consumption. Several cases however seem to correspond better to the higher flow of the healthy hemisphere. Exact quantitative results which could be of pathognomonic value could not be obtained in these cases with non-homogenous CBF.

In the determination of the metabolism in cases with non-homogenous flow several problems arise:

Within the jugular bulb on the side of the slow flow a part of the blood from the healthy hemisphere with faster flow and normal metabolism appears. This amount of admixture from the opposite side is unknown. Therefore the metabolism appears too low if the A-V difference is correlated to the slow flow. Furthermore, often malignant tumors have A-V shunts. With none of the present methods of CBF-measurement a shunt flow can be measured exactly. Therefore the diminution of the A-V difference due to the shunt blood cannot be estimated.

Besides these difficulties from inhomogenous flow we have to consider a source of error due to inflow of blood to the jugular bulb from the basilar artery if we measure the hemisphere flow and determine the A-V difference from the jugular bulb and from admixture of extracerebral blood to the sinuses.

This physiological source of error can be demonstrated. In 2 patients with circulatory arrest which was proven by angiography and isotope measurement we could not find an A-V difference for O_2 but we found a significant difference for glucose and lactate.

In face of these mentioned difficulties we want to try to establish some criteria which seem to us to be essential for a method of determination of the cerebral metabolism in man with unilateral technique.

If an A-V difference of a substance shall be used for calculation of the metabolism all of the blood which is drained into the vein where the blood sample is taken from has to be measured and must have irrigated a defined and known brain area. Within this brain area the flow must be homogenous. No A-V shunts may exist unless the shunted portion is exactly known. No extracerebral admixture of blood to the vein may exist.

With circumscribed space occupying lesions and cerebral venous blood samples obtained from the jugular bulb these criteria are not fulfilled as we could show. An exact determination of the metabolism in these cases therefore is not possible!

Since no other vein is accessible for taking the blood sample this direct way of measuring the metabolism must be rejected. Our efforts therefore should rather be directed to finding an indirect way of determining the regional metabolism by extracranial monitoring if possible by registrating, within the brain, the isotope uptake or clearance of substances which are in a known dependance from the cerebral metabolism or which are metabolized themselves.

Until such a new way is found, we have to be satisfied with in vitro measurements with all their problems and objections.

Total CO_2 in Brain Biopsies Taken from Patients with Intracranial Tumors and Cerebrovascular Disease after Freezing of the Tissue in Situ

J. OLESEN

Departments of Neurosurgery and Clinical Physiology, Bispebjerg Hospital, Copenhagen

The focal- or perifocal hyperemia ("luxury perfusion syndrome" [4]) often seen in patients with apoplexy and brain tumor, and the demonstration of a vasoparalytic condition in these areas [2, 3, 5, 7] has stimulated the interest in brain and CSF acid/base conditions, because tissue acidosis is thought to be the basis of the vasoparalysis. In the present study, the total carbon dioxide in human brain tissue has been determined in these diseases.

A simple apparatus which makes it possible to take frozen biopsies in vivo was constructed. It uses liquid nitrogen as the source of cold and suction to fix the frozen brain tissue to the apparatus while removing it surgically. The apparatus can be sterilized and it prohibits contamination with liquid nitrogen which cannot be considered sterile.

Biopsies were taken from 10 patients, 3 with subcortically localized glioblastomas, 1 with astrocytoma, 1 with meningeoma, 2 with metastatic carcinoma, 1 with old cerebral infarction and 2 with intracerebral hematoma (Table 1). Immediately after the opening of the dura a biopsy was taken from the perifocal area. Biopsies from subcortical tumor-tissue were removed surgically and immediately dropped into liquid nitrogen.

The total CO_2 was determined titrimetrically [8]. The results are shown in Table 1 together with relevant clinical data. The normal total CO_2 content is about 14 μmole per. g tissue in rats [9] and due to the fairly low solubility of CO_2 in brain tissue, gross variation of total CO_2 is indicative of similar variation of tissue bicarbonate. Subnormal values of this parameter indicates accumulation of acid metabolites (lactic acid primarily). On this basis we interpret distinctly subnormal total CO_2 values as indicative of brain tissue acidosis.

The preliminary data here reported show that tumor cases with very high intracranial pressure (papillary edema) have brain tissue acidosis. On the contrary patients without clinical signs of increased pressure had nearly normal total CO_2 content, i. e. no tissue acidosis. The tumor-tissue itself does not show acidosis at normal intracranial pressure.

Two cases of intracerebral hematoma did both show a low total CO_2 content. Especially case no. 11, who was operated 2 days after the onset of symptoms is clearcut. The angiogram showed a very early filling marginal vein, and the biopsy was taken over a pink cortical vein. The CO_2 content was 9.67 μmole per. g which is $2/3$ of the normal value. rCBF was not measured, but perifocal hyperemia is characteristically seen in this category of patients [1, 6]. The results obtained so far are in agreement with the concept of a central role of brain tissue acidosis in the pathogenesis of focal- or perifocal "luxury perfusion".

Table 1. *Total CO_2 in brain biopsies*

Case No.	Histological diagnosis	Signs of increased intracranial pressure and other data of special interest	Biopsy taken from	Total CO_2* μmole/g
1	Meningeoma	No papilledema	Normally looking peritumoral cortex after some manipulation	14.5
3	Glioblastoma	No papilledema	Cortical tissue over the tumor	14.0
6	Glioblastoma	No papilledema	Normally looking cortex near tumor	12.8
			Tumor itself, removed surgically and frozen immediately after	15.7
4	Metastatic carcinoma	No papilledema	Normally looking cortex	14.0
			Soft tissue over cyst in the tumor	12.6
8	Metastatic carcinoma	No papilledema	Edematous cortex over tumor	14.5
			The same area but removed surgically and frozen immediately after	14.0
2	Glioblastoma	Slight papilledema	Very edematous cortex over tumor	9.1
			From other area of very edematous cortex over tumor	10.4
			Tumor itself removed surgically and frozen immediately after	12.2
5	Oligodendroglioma	Slight papilledema	Cortex over tumor	10.7
7	Several months old cerebral infarct	No papilledema	Yellow necrotic area	14.7
9	Intracerebral hematoma, 1–2 months old	No papilledema	Cortex over hematoma	11.2
10	Intracerebral hematoma, about 2 days old	Papilledema Big early filling vein	Cortex over hematoma	9.8

* Arterial pCO_2 not measured at time of biopsy. In no cases was assisted ventilation with hyperventilation used. Arterial pCO_2 can consequently be estimated to have been in the range of 30–45 mmHg.

References

1. AGNOLI, A.: Relationships between regional hemodynamic findings in acute cerebrovascular lesions, and clinicopathological aspects. Contribution to Fourth International Symposium of the Research Group on Cerebral Circulation, Salzburg, Austria, Sept. 1968.
2. FIESCHI, C., A. AGNOLI, N. BATTISTINI, L. BOZZAO, and M. PRENCIPE: Derangement of the regional cerebral blood flow and of its regulatory mechanisms in acute cerebro-vascular lesions. Neurology 18, 1166 (1968).
3. HØEDT-RASMUSSEN, K., E. SKINHØJ, O. PAULSON, J. EWALD, J. K. BJERRUM, A. FAHRENKRUG, and N. A. LASSEN: Regional cerebral blood flow in acute apoplexy. The "luxury perfusion syndrome" of brain tissue. Arch. Neurol. 17, 271 (1967).
4. LASSEN, N. A.: The luxury perfusion syndrome and its possible relation to acute metabolic acidosis localized within the brain. Lancet II, 1113 (1966).

5. Palvölgyi, R .:Regional cerebral blood flow in patients with intracranial tumors. J. Neurosurg, Aug. 1969, in press.

6. Paulson, O. B., S. Cronqvist, J. Risberg, and F. I. Jeppesen: Regional cerebral blood flow: A comparison of 8-detector and 16-detector instrumentation. J. nucl. Med., **10**, 164 (1969).

7. — Regional cerebral blood flow at rest and during functional test in occlusive and non-occlusive cerebrovascular disease. Presented at the International CBF Symposium, Mainz, April 1969 (this volume, p. 111).

8. Pontén, U., and B. K. Siesjö: A method for the determination of the total carbon dioxide content of frozen tissues. Acta physiol. scand. **60**, 297 (1964).

9. — — Acid labile carbon dioxide of rat brain after freezing the tissue in situ. Acta physiol. scand. **60**, 309 (1964).

The Effect of Intracranial Pressure on Perifocal Hyperemia

R. Christ, R. Heipertz, M. Brock, and A. A. Hadjidimos

Department of Neurosurgery, University of Mainz

Introduction

It is known that a *reversible* cortical trauma caused by local brain compression in the cat is accompanied by a pronounced decrease of rCBF at the compressed area and by a transient perifocal hyperemia [1].

When an *irreversible* lesion with a diameter of 7 mm is induced by means of circumscribed freezing of the suprasylvian gyrus to $-180\ °C$ with liquid O_2 in craniotomized cats there is a marked and progressive decrease of rCBF at the lesion and a pronounced hyperemia around the frozen area [2] (Fig. 1).

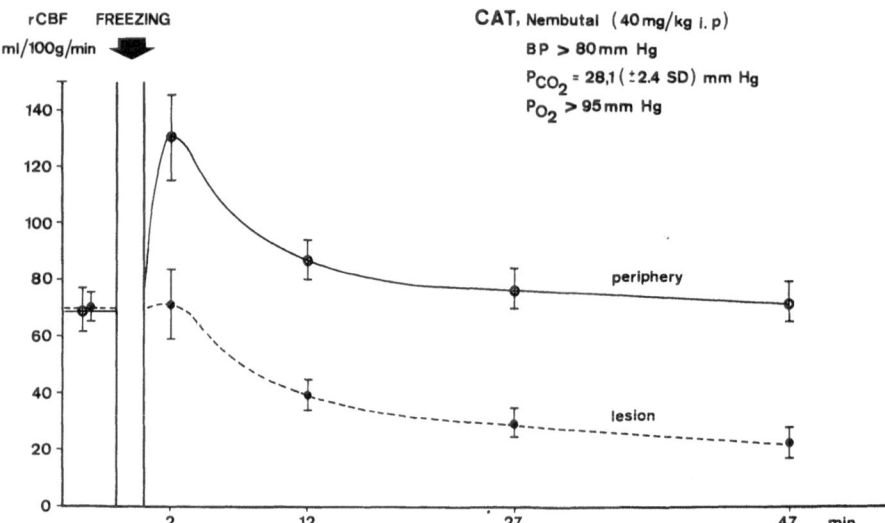

Fig. 1. Changes in cortical rCBF (^{85}Kr-beta-clearance) following a freezing lesion at the suprasylvian gyrus in the cat. While rCBF progressively decreases at the lesion proper, there is a clearcut hyperemia at its periphery

Intravital microscopic observation of cortical vessels showed the development of widened and congested blood vessels, as well as a complete stasis of corpuscular elements within the lesion. Histologically there is a sharply delineated cortical lesion with congestion and extravasation of erythrocytes. PAS positive material was found between the fibres of the white matter but was confined to the site of the lesion. This picture is in contrast with the mild changes (vascular engorgement and loosening of tissue texture) observed after cortical compression [1].

Material and Methods

Sixteen Nembutal anesthetized and artificially ventilated cats (mean body weight 2650 $\pm$ 200 g, mean $ApCO_2$ 30.2 $\pm$ 1.9 mmHg, ApO_2 80–120 mmHg) were employed in the present study. Hemispheric CBF was measured by the ^{133}Xe-clearance-technique. After establishing control values a circumscribed cortical freezing lesion identical to those described by Heipertz [2] was induced through the intact calvarium. CBF was alternatively measured over either hemisphere. Intracranial pressure (ICP) was monitored by means of a subarachnoid catheter in 8 animals. Alternative ^{133}Xe injections were made at intervals of 10 min. (The size of the cat's brain only allows hemispheric CBF comparisons.)

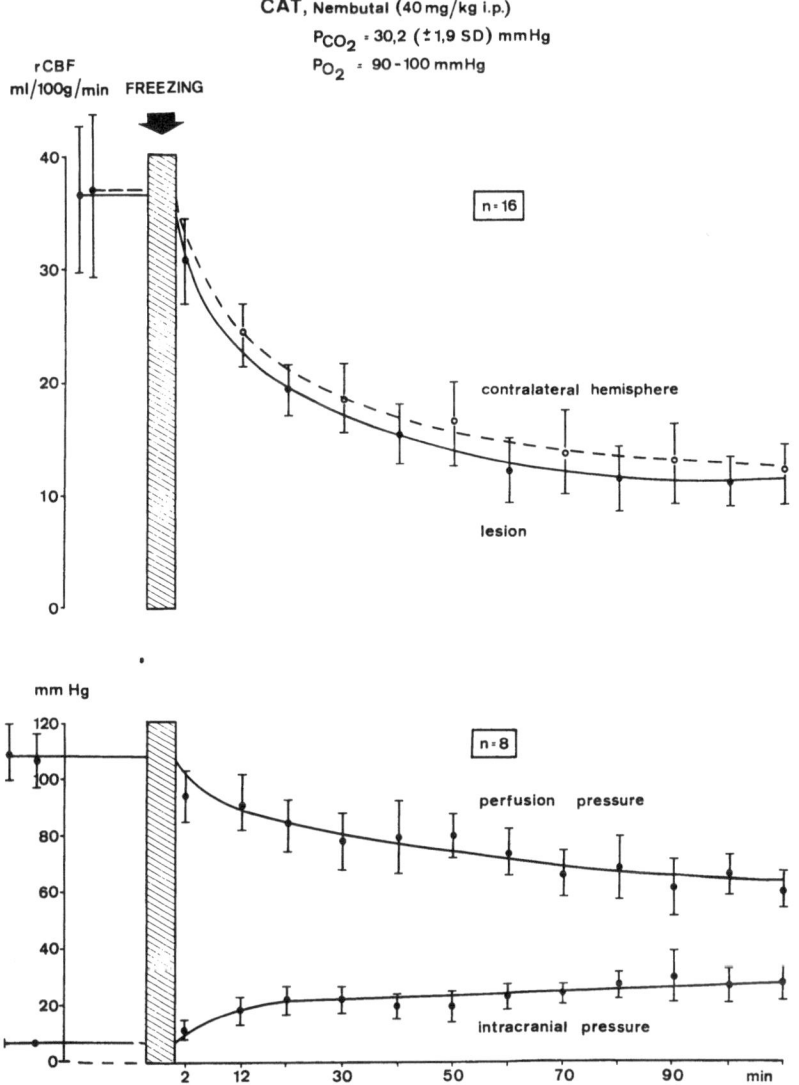

Fig. 2. When a freezing lesion is induced through the intact skull, the marked and abrupt increase in intracranial pressure, and the consequent decrease in perfusion (lower graph), do not allow perifocal hyperemia to take place; rCBF (^{133}Xe-gamma-clearance) falls both at the injured hemisphere and at the contralateral side (upper graph). A perifocal hyperemia does take place, however, if the same procedure is preceded by a decompressive craniotomy

Results

The anatomical aspect of the lesions corresponded to the findings of Heipertz [2].

After freezing the cortex of one side through the intact calvarium hemisphere CBF changed in a parallel way on both sides, there being no significant differences. About 30 min after the lesion CBF had decreased to half the control values. At this time ICP had risen to about 310 mm H_2O. Some 80 min after the lesion CBF became stabilized at about $1/3$ of the control values, while ICP was about 410 mm H_2O (Fig. 2).

Discussion

Two aspects of our results seem to deserve special attention.

First, we observed a more abrupt CBF decrease than authors who have induced a *gradual* ICP increase by inflation of a subdural balloon [3]. It is probable that under such circumstances – in progressive rise in ICP – there is a possibility of adaptation for the brain, making CBF changes less drastic.

The second aspect – and the main result of this series – is that no perifocal hyperemia as described by Heipertz [2] can develop in the presence of increased ICP. A marked hyperemia on the damaged side can however be observed by the same method in craniotomized cats before the increasing edema causes the brain to protrude through the craniotomy.

References

1. Brock, M.: Regional cerebral blood flow (rCBF) changes following local brain compression in the cat. Scand. J. clin. Lab. Invest. Suppl. **102**, XIV-A (1968).
2. Heipertz, R.: The effects of local cortical freezing on rCBF in the cat. Scand. J. clin. Lab. Invest. Suppl. **102**, XIV-D (1968).
3. Langfitt, T. W., N. F. Kassell, and J. D. Weinstein: Cerebral blood flow with intracranial hypertension. Neurology **15**, 761 (1965).

Transit Time and Microflow Measured by Fluorescein Angiography and Radioisotopes in Brain Tumor with Red Veins

W. Feindel, Y. L. Yamamoto, and C. P. Hodge

Cone Laboratory for Neurosurgical Research, McGill University,
and The Montreal Neurological Institute, Montreal

At the Glasgow conference, 2 years ago, we reviewed our evidence for the multi-compartmental nature of blood flow through tumors and nearby cortex, as studied by fluorescein angiography, ^{191}Hg transit times and ^{133}Xe clearance curves from the exposed brain during craniotomy [1].

We ascribed a wide spectrum of flow rates in various parts of the same tumor and nearby brain including

1. high by-pass flow through arterio-venous shunts producing "red veins" and "arterial" peaks on the venous part of the transit time curves – The "cerebral steal" [2],
2. slow filling and clearance of the tumor bed with reduced blood flow,
3. almost zero flow in areas of avascular parts of the tumor [3].

These differences were more striking and more complex than indicated by standard X-ray arteriography or by external radioisotopic flow studies, since both techniques have limitations of resolution which fail to demonstrate such fine local changes.

Patient G. T., with a glioma of the central area, showed an early-filling rolandic vein which was red at operation. Transit time studies with 4 mini-probes showed shunt peaks at the lower and upper points of this vein, at 1.6 and 2.4 sec, the middle cerebral artery peak being 1.0 sec and the normal venous peak 3.5 sec. ^{133}Xe flow curves showed a rate of flow through the tumor of more than 30% that of 2 normal areas of cortex. Another feature which we wish to point out is the phenomenon of laminar flow: this is normally present in the epicerebral veins as shown by fluorescein angiography. Each tributary vein, when joining a collecting vein, carries with it, after its embouchement, its own clearly defined stream [3]. These may be traced as fine ribbons, each remaining discrete from the other, until they reach the sagittal sinus. Of great interest is our observation that in red veins, flow seems not to be laminar, but turbulent in nature. In patient G. T., such mixing occurs as several smaller shunting red veins over the tumor enter the central rolandic vein. But a jet-like flow into the same vein higher up, some distance away from the tumor also caused mixing of flow. Thus, the red vein showed no evidence of segregated laminar streams, although these were clearly defined in nearby venous tributaries.

A more dramatic example of mixing flow is afforded by patient N.V. with an angioma of the 2nd temporal convolution, 3 cm in diameter. Mixing here was most complex, with

Supported by the Medical Research Council of Canada, grant MA-3174.

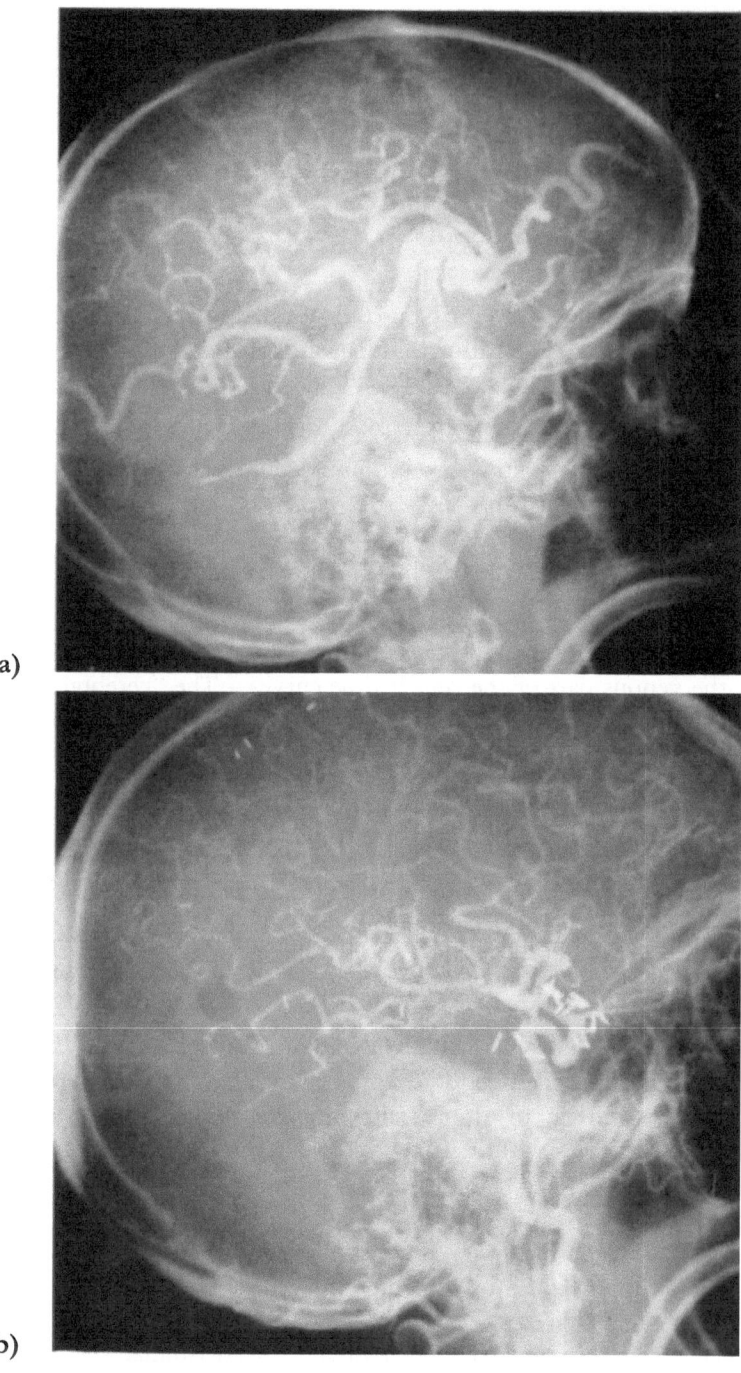

Fig. 1. a) Before and b) after excision of angioma in case N. V.

spiral streaming, both mural and axial, feeding into the venous system of the surface of the entire hemisphere (Fig. 1a and b) and isolation and closure of the 2 large feeding arteries supplying most of the flow through the angioma caused the flow in these epicerebral veins to revert to a more distinct laminar pattern. At the same time the blood flow measured by ^{13}Xe clearance showed an increase over the normal cortex but also over the residual portion of the angioma which had not yet been excised. The main middle cerebral artery during this manoeuver was patent.

Summary

1. Brain tumors, as studied on the exposed brain during operation by fluorescein angiography and radioisotopic blood flow curves, show multicompartmental flow rates which vary from almost zero to high shunt by-pass flow.

2. Red early-filling veins show a disturbance of normal laminar flow as indicated by fluorescein streaming patterns.

3. The mixing or turbulent flow reverted to laminar flow after closure of the shunts.

References

1. FEINDEL, W., Y. L. YAMAMOTO, and C. P. HODGE: Preferential shunting and multi-compartmental blood flow in the human brain. In: Blood Flow in Tissues and Organs, Ed. BAIN and HARPER. p. 232. Edinburgh: Livingstone 1968
2. —, and P. PEROT: Red cerebral veins. J. Neurosurg. **22**, 315 (1965).
3. —, Y. L. YAMAMOTO, and C. P. HODGE: Intracarotid fluorescein angiography: a new method for examination of the epicerebral circulation in man. Canad. med. Ass. J. **96**, 1 (1967).

Correlation between rCBF, Angiography, EEG and Scanning in Brain Tumors

A. A. HADJIDIMOS, M. BROCK, J. P. HAAS, H. DIETZ, R. WOLF, M. ELLGER, F. FISCHER, and K. SCHÜRMANN

Departments of Neurosurgery, Radiology and Anesthesiology, University of Mainz

The findings with techniques of different sensitivity, and which investigate different aspects of brain structure and/or function are difficult to correlate. However, such a comparison should help in elucidating the physiopathological mechanisms of brain lesions and the post-operative clinical evolution and prognosis in cases of brain tumors.

Material and Methods

The present study is based on 21 brain tumor patients studied pre-operatively; 3 cases were also followed post-operatively. Results from the following investigations have been correlated:

1. Regional Cerebral Blood Flow (rCBF) measurements, by means of the intra-carotid ^{133}Xe injection technique [1, 5] under local or general anaesthesia, in the resting state as well as during functional tests [1, 2, 5].

2. Serial angiography, performed just after or at least 2 days prior to the rCBF study.

3. Electroencephalogram (EEG) with 8 or 16 channel equipment.

4. Brain scanning, following intracarotid injection of ^{131}I-marked albumin macro-aggregates (MAA) [4]. A volume of 0.3–0.5 ml, corresponding to 100–150 mCi of radioactive suspension was usually administered, the mean caliber of the particles being 10–15 microns. It is known that such particles do not permeate normal brain capillaries.

5. Brain scanning, following intra-venous administration of ^{99}Tc-Fe^{++}-complex.

Reliable data in all the above examinations were obtained in 16 cases, 2 of which post-operatively.

Results

The efficiency of each investigation method in detecting focal abnormalities is demonstrated in Table 1, which shows that such abnormalities were detected by one or more among the methods used in 78—92% of the cases. Uncertain or negative results were encountered in cases of tumor recurrencies, metastases and deeply located tumors.

From the 14 pre-operative studies with all 5 techniques 7 showed a hyperemic rCBF focus, 6 a ischemic one, and 1 did not show any focal abnormality in rCBF, but a generalized hyperemia. In the latter case, a deeply seated malignant astrocytoma, the EEG and the clinical status were also diffusely disturbed. On the other hand, the angiogram and both the Tc and the MAA scans showed a local abnormality.

Table 1. *24 studies in 21 patients. Pre-operative (21) and post-operative (3) abnormalities*

Investigation	Focal abnormality	Dubious focal abnormalities	General or absent abnormalities
Clinical sympt. ($n = 24$)	21 (87.5 %)	3 (12.5 %)	0 (0 %)
E.E.G. ($n = 23$)	20 (87 %)	2 (8.7 %)	1 (4.3 %)
^{99}Te-Fe^{++}-Compl. scan ($n = 18$)	14 (78 %)	4 (22 %)	0 (0 %)
Carot. Angiogr. ($n = 24$)	20 (83 %)	4 (17 %)	0 (0 %)
^{131}I-MAA scan ($n = 20$)	18 (90 %)	2 (10 %)	0 (0 %)
rCBF ($n = 24$)	22 (92 %)	1 (4 %)	1 (4 %)

Among cases with hyperemic rCBF foci, a case of deeply located brain *metastasis* (bronchial carcinoma) showed a clearcut rCBF focus with "paradoxical" (hyperemic) reaction to hyperventilation, correlating well with the clinical abnormalities, while all other examinations were dubious. The *Meningiomas* (2 cases), known as well vascularized tumors, showed focal hyperemia in the rCBF, focal damage of blood brain barrier in the Tc-scan and focal lack of particle arrival in the MAA-scan, as well as a focal avascular area on the angiogram. Further, perifocal angiographic blush and localized EEG delta waves were also present. Hyperemia, in such a case, seems to be mainly peri-focal [7]. In the *Gliomas* (4 cases) focal hyperemia corresponded to a clearcut positivity of all other examinations. In 3 cases the MAA-scan revealed a strong focal particle accumulation suggesting a preferential shunting of particles towards the tumor (see case report below).

Concerning ischemic rCBF foci, *primary gliomas*, mainly those deeply located, confirmed the fact [3] that a ischemic tumor can appear well vascularized on the angiogram. The glioma-recurrencies gave uncertain results.

The practical value offered by the present investigation will be briefly illustrated by the report of a case.

Case report: A 35 years old male, who had had an epileptic status prior to admission, developed right sided pyramidal tract symptoms, papilledema on the right side, speech disturbances and confusion. *Preoperatively,* the EEG showed a paroxysmal theta and delta wave focus in the left temporo-parietal region with occipital irradiation and diffuse general slowing. The Tc-scan revealed a well circumscribed activity accumulation temporo-parieto-occipitally on the left. Left-sided carotid angiography showed a highly vascularized brain tumor at the same location as seen in the Tc-scan. An extremely dilated middle cerebral artery fed the tumor. Further, a very poor filling of the anterior cerebral artery branches was noted. rCBF values, in the resting state were significantly decreased in the whole hemisphere (mean=37 ml/100 g/min). A marked hyperemic focus (mean=65 ml/100 g/min) was found in the temporo-parieto-occipital area. The ^{131}I-MAA-scan revealed a well limited focal activity *accumulation* in a circumscribed area corresponding to the angiographic location of the tumor. No radioactive material was detected in the remaining hemisphere, thus suggesting a shunting of particles to the tumor through the dilated middle cerebral artery. Although, in this case, the EEG and the Tc-scan gave evidence of a tumor in the left temporo-parieto-occipital region, the angiogram, the rCBF values and data from the MAA-scan further enriched our diagnostic knowledge by showing that there was a shunting mechanism towards the tumor area. Furthermore, during hyperventilation a series of rCBF channels corresponding to the angiographic projection of the dilated middle cerebral artery showed abnormally decreased reaction (partial vasoparalysis?) while their close periphery became strongly ischemic. The clearance curves recorded over the projection of the dilated media, presented typical "vascular shunt-peaks", indicating the "stealing route" of blood to the tumor. Only limited, remote, disturbances of autoregulation were present at this stage.

At operation, a highly vascularized oliogodendroglioma was totally removed. A marked peri-tumoral edema was found during craniotomy. The patient was *reinvestigated* on the 15th post-operative day, after clinical recovery. There was a slight paresis of the right leg, slight speech disturbances and almost no mental confusion. In the EEG the alpha rhythm was improved bilaterally and the pre-operative temporo-parietal focus on the left was now essentially absent. Flattening was observed on the left parieto-accipital region. The Tc-scan, revealed no focus. Carotid angiography showed a normal filling of the anterior cerebral artery while the caliber of the middle cerebral artery had returned to normal. Average rCBF at rest, was still significantly decreased (mean=33 ml/100 g/min), mainly due to a ischemic focus corresponding to the site of the extirpated tumor (mean=25 ml/100 g/min). While circumscribed hyporeactivity to hyperventilation was found in the central areas, in contrast to the pre-operative study, a general loss of autoregulation was now present. The MAA-scan showed a homogenous distribution of particles over the *whole* left hemisphere except in a well limited parieto-occipital region corresponding to the extirpated tumor, where a decreased uptake was seen.

A comparison of the "non-tumor" areas (Fig. 1) in the pre- and post-operative rCBF studies during rest, indicates that rCBF had increased about 27% 15 days after the operation. This finding correlates well with the data obtained from the post-operative clinical examination, from EEG, angiography and scanning.

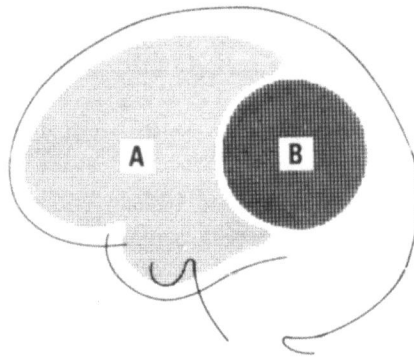

Rest	Topography	Pre-operative $\overline{M}$ rCBF$_{std}$	Post-operative $\overline{M}$ rCBF$_{std}$	% Difference
$\overline{M}$r CBF (std) =	(A + B) Hemisphere	37	33	− 8%
$\dfrac{\overline{M} \text{ rCBF}}{1 + 0.025 \, (ApCO_2\text{-}40)}$	(A) Periphery	28	36	+27%
in ml/100 g/min	(B) Focus	65	25	−62%

$\overline{M}$ = mean   Fig. 1

In another case (Hadjidimos et al., in preparation), examined pre- and post-operatively, the opposite situation has been observed, namely a slow post-operative recovery which corresponded to a post-operative decrease (21%) in rCBF and to specific alterations in all other investigations.

In conclusion, it is appropriate to stress that our results show that the correlation of all above reported investigations, if systematically performed pre- and post-operatively, may help in establishing postoperative immediate prognosis, local diagnosis and in understanding local circulatory changes induced by brain tumors.

References

1. BROCK, M., A. A. HADJIDIMOS, K. SCHÜRMANN, M. ELLGER, and F. FISCHER: Regional cerebral blood flow in brain tumors. Presented at the International CBF Symposium, Mainz, April 1969 (this volume, p. 169).

2. CRONQVIST, S., D. H. INGVAR, and N. A. LASSEN: Quantitative measurements of regional cerebral blood flow related to neuroradiological findings. Acta Radiol. 5, 760 (1966).

3. —, and F. AGEE: Regional blood flow in intracranial tumors. Acta Radiol. 7, 393 (1968).

4. DIETZ, H., J. P. HAAS u. R. WOLF: Ergebnisse der Hirntumor-Diagnostik mit ^{131}I-Albumin-Makro-aggregaten. Der Radiologe 8, 397 (1968).

5. HADJIDIMOS, A. A., D. S. OECONOMOS, N. A. LASSEN et D. H. INGVAR: Débit sanguin cérébral regional et ses aspects cliniques. Rev. Neurol. 119, 211 (1968).

6. LASSEN, N. A., and R. PALVÖLGYI: Cerebral steal during hypercapnia and the inverse reaction during hypocapnia observed by the ^{133}Xenon technique in man. Scand. J. clin. Lab. Invest. Suppl. 102, p. XIII-D (1968).

7. PALVÖLGYI, R.: Regional Cerebral Blood Flow in Tumor Patients. Scand. J. clin. Lab. Invest., Suppl. 102, p. XV-B (1968).

Iconoscope Assessment of Meningiomas, Gliomas and Metastatic Brain Tumors

J. C. KENNADY

UCLA School of Medicine, Harbor General Hospital, Los Angeles, California

There are three primary problems related to the diagnosis of brain tumors. First is the ability to detect a small tumor (less than 1.5 cm in diameter). Secondly, accurate preoperative determination of the type of neoplasm and lastly, evaluation of the tumor for possible neuro-surgical excision. Localization of the primary artery supply, the magnitude of the vasculature within and surrounding the neoplasm and its effect on the entire hemisphere hemodynamics are of prime importance. In order to supplement serial carotid angiography and gain further assessment of the brain tumors in 19 patients, iconoscopic studies were done using the same indwelling carotid catheter [1]. The neoplasms were verified at operation or necropsy.

Method and Equipment

The iconoscopic instrumentation (Ter-Pogossian type of image intensifier gamma-detection system), and iconoscope quantifier calibrations have been described earlier in this volume. A rapid bolus injection of 1 ml containing 3 millicuries Technetium-99m labelled albumin microaggregates (^{99m}Tc AA), 1–8 micron size, and Technetium-99m pertechnetate (^{99m}Tc 04) are given via the carotid catheter. These data are permanently recorded on video tape and replayed as polaroid pictures of single (0.03 sec) or integrated frames to show the regional distribution of the test agent in the injected hemisphere. Also these data can be processed by the quantifier (0.2 and 0.5 sec sampling time) as radiohistograms to demonstrate the appearance, peak radioactivity level and disappearance of the radiopharmaceutical in specific areas.

Clinical Results and Interpretations

a) Meningiomas

These are well encapsulated tumors that can derive blood from both the external and internal carotid arteries. In 5 patients with frontal [3] and parietal [2] tumors, the iconoscope studies show a slower filling and smaller tumor area on external carotid compared with internal carotid injection of ^{99m}Tc AA and ^{99m}Tc 04. The delineation quality of the hemispheric regions in sequential videopolaroid pictures appears inversely proportional to the size of the meningioma. Radiohistograms of the tumor region indicate a rapid appearance,

Supported by contract AT-Gen-12 between the U.S. Atomic Energy Commission and the University of California at Los Angeles.

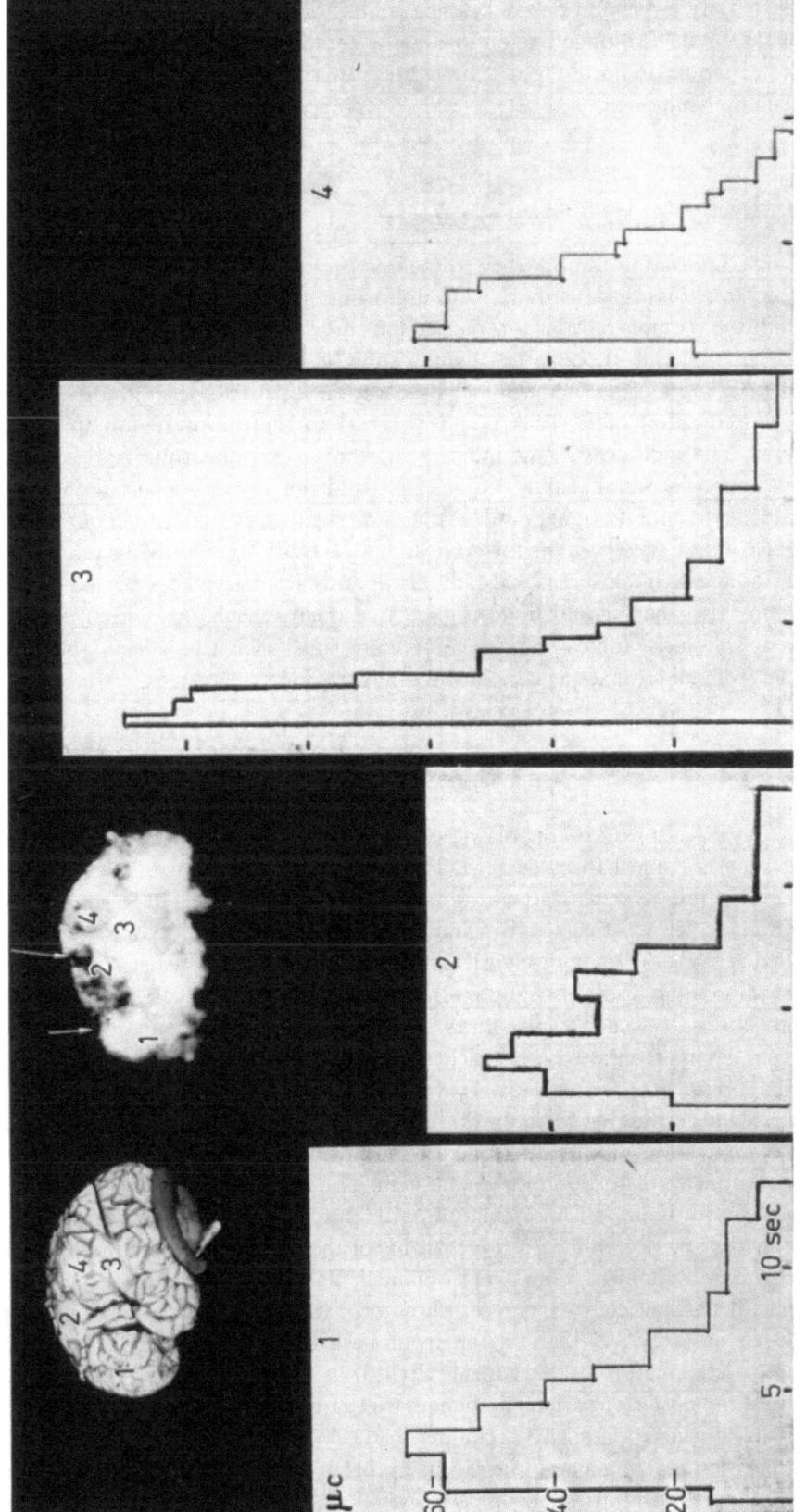

Fig. 1. Patient J.H., age 54 years. Metastatic lesion in left frontoparietal area from lung. Phantom brain orients the 3–4 sec interval video polaroid picture. Left intracarotid injection of 3 mc ^{99m}Tc AA. Arrows show margins of lesion and numbers indicate regions where corresponding radiohistograms were obtained. Radiohistograms show μc per 1.5 cm^2 area vs. 0.5 sec sampling time. (See text)

high peak radioactivity levels and delayed disappearance of both test agents compared with other areas of the brain. Prolonged retention in the tumor region is particularly apparent with ^{99m}Tc 04 and would indicate rapid extravasation of the test agent. This has not been observed in the other tumor groups.

b) Gliomas

These invasive, infiltrating tumors vary from slow (grade I) to rapid growing, malignant (grade IV) neoplasms. Inasmuch as there is no uniformity of the heterogeneous cell growth the final classification requires fastidious serial section study of the whole tumor. In 2 patients with astrocytomas, grades I–II, both test agents show less propensity to enter the tumor region than in other areas of the hemisphere. By contrast, 2 patients with grade IV gliomas show greater concentration of ^{99m}Tc AA and ^{99m}Tc 04 in the tumor region in the video polaroid pictures. The radiohistograms indicate the tumor and non tumor areas are comparable in appearance and disappearance times only and not in peak levels with both test agents. In 1 patient (temporal glioma grade II) with markedly elevated intracranial pressure, the tumor region continues to show the lowest peak radioactivity level; only the disappearance times of the 2 test agents in all 6 regional radiohistograms are delayed for 25 sec.

The results in this group would indicate increased microvasculature within or adjacent to the malignant gliomas. Increased intracranial pressure primarily affects the capillary venous drainage from all areas, especially in the tumor region.

c) Metastatic Tumors

In the hematogenous spread of neoplastic cells, the metastases begin in the first cortical layer and later show a conical form with a wide base at the pial surface. They are usually well circumscribed, poorly vascularized within and could easily be separated from the adjacent brain at operation. In all 6 patients with cerebral metastasis, the iconoscope studies show clearly delineated regions of low radiodensity in comparison with the other areas of cortex. In general, corresponding radiohistograms show a $^{1}/_{4}$ to $^{1}/_{2}$ the peak values recorded in other regions but with similar transit times. Patient, J. H., age 54 years, exemplifies this group. She began having frontal type headaches, then weakness on her right side especially in the lower extremity. Progressive lethargy followed with an elevated temperature. The EEG demonstrated a destructive focus in the left frontal region. Left carotid angiography showed a slight depression of the anterior cerebral artery above the genu of the corpus callosum and questionable distortion of a frontal vein. The iconoscopic studies are illustrated in Fig. 1. The phantom brain is for orientation with the 3–4 sec interval polaroid picture. The arrows show the anterior and posterior margins of the tumor and the numbers indicate the regions where the radiohistograms were obtained. Radiohistograms of the frontal (1), temporoparietal (3) and parietal (4) regions show an earlier appearance and higher peak radioactivity levels of ^{99m}Tc AA than in the frontoparietal tumor area (2). The distinctly higher temporoparietal region peak radioactivity (3) is probably caused by the transit ^{99m}Tc AA bolus in the posterior parietal and angular arteries on the pial surface. Left frontal craniotomy disclosed a necrotic adenocarcinoma in the location and of a size indicated by the video polaroid pictures. In general, the results in this group indicate a significant decrease in the microvasculature within these metastases.

Conclusion

Iconoscopic studies following intracarotid injection of ^{99m}Tc AA and ^{99m}Tc 04 in patients with meningiomas, show as video polaroid pictures and regional radiohistograms, more rapid appearance, higher peak radioactivity and delayed disappearance in the tumor region compared with other areas of brain. In low grade gliomas (I–II) the tumor regions show low radioactivity whereas in malignant gliomas (grade IV) they are more radioactive than other areas. This indicates changes in the microvasculature. Clearly delineated regions of low radioactivity are seen in patients with brain metastases signifying a decrease in the microvasculature within these tumors.

The accuracy of the iconoscope studies in showing the size, location and vascularity of these tumors was confirmed in each patient at operation.

Reference

1. KENNADY, J. C., R. POTTER, F. CHIN, and L. SWANSON: Assessment of cerebral lesions by rapid sequential scintiphotography. Preliminary Note. J. nuc. Med. 9, 423 (1968).

Comments to Chapter V

Tumors and Intracranial Pressure

Patterns of cerebral blood flow in brain tumor patients were described by several investigators. The reactivity of the cerebral vessels in and around the tumor and the influence of changes in cerebral tissue tension and intracranial pressure on rCBF were the principal items of discussion.

rCBF was found to vary greatly within and adjacent to tumors. BROCK et al. and OECONOMOS et al. described increased or decreased flow within the tumors and either a hyperemic or ischemic focus in the peritumoral tissue. rCBF in the remainder of the brain was normal, or, at times, diffusely decreased, the latter presumably due to increased intracranial pressure. Autoregulation was impaired in the peritumoral tissue in all of BROCK's patients, and LASSEN described a patient with a meningioma who had loss of autoregulation in the entire ipsilateral hemisphere. The reason for this wide-spread loss of vascular reactivity is obscure. Tissue acidosis was discussed but seems unlikely in a patient whose clinical manifestations of the tumor are slight. Paradoxical responses to CO_2 and arterial hypertension occurred frequently in the patients of BROCK, OECONOMOS, and PALVÖLGYI. In some regions, usually adjacent to the tumor, inhalation of CO_2 and hypertension caused a decrease in rCBF. PALVÖLGYI noted 2 cases in whom hypotension was accompanied by an increase in rCBF in some regions. The paradoxical responses were generally attributed to a local increase in tissue tension or diffuse intracranial hypertension, probably due to edema. LANGFITT discussed some of the effects of increased cerebral tissue pressure and intracranial pressure on cerebral blood flow. Dilatation of resistance vessels (CO_2, loss of autoregulation) promotes cerebral edema by increasing the intravascular pressure in the capillaries and veins. The edema in turn increases local tissue tension tending to compress the capacitance vessels. rCBF is determined by the resistance in series of the vascular bed. If the increase in resistance in the capillaries and veins, produced by edema, exceeds the decrease in resistance at the arterioles, rCBF will fall when autoregulation is lost and the blood pressure is elevated. This sequence of events is still largely hypothetical but is one explanation for the paradoxical responses observed in the brain tumor patients.

The papers of HERMANN and OLESEN reflected an increasing interest in methods for measuring cerebral metabolism in vivo and in vitro. HERMANN pointed out the difficulties of studying glucose metabolism of gliomas, by examining jugular venous blood samples, because the contribution of the glioma is small compared to the remainder of the brain.

T. W. LANGFITT

Cerebral Hemodynamics and Metabolism Following Brain Trauma. Demonstration of Luxury Perfusion Following Brain-Stem Laceration

J. S. Meyer, A. Kondo, F. Nomura, K. Sakamoto, and T. Teraura

Wayne State University, Detroit, Michigan

Changes in cerebral blood flow and metabolism were correlated with the EEG in 64 experiments in 58 baboons during concussion, contusion or brain-stem laceration.

Anesthesia was maintained with sodium pentobarbital and ether. Cerebral venous outflow was measured with electromagnetic flowmeters about both internal jugular veins [6, 7]. Cerebral arteriovenous oxygen difference was monitored with the Guyton oxygen analyzer [3]. EEG, EKG, BP, end-tidal CO_2 and intracranial pressure (ICP) were recorded. Arterial and cerebral venous lactate, pyruvate, pH, PO_2 and PCO_2 were monitored.

Concussion and contusion were produced by a controlled blast of compressed air. Laceration was produced by thrusting a small scalpel into the brain-stem. Concussion was judged by cardio-respiratory and EEG changes.

Results and Comment

Results will be expressed in mean values. There were no measurable changes after subconcussive blows. In mild or severe concussion, all cardiorespiratory changes, as well as changes in EEG and cerebral oxygen consumption, were transient and there were no abnormalities of the brain at necropsy. After contusion and laceration, these changes were not always reversible and sometimes progressed.

In concussion and contusion, apnea was observed shortly after the blow, sometimes followed hyperventilation due to hypercarbia. Blood pressure fell with bradycardia immediately after the impact followed by transient hypertension.

Mild concussion increased CBF by 4.3% and $CMRO_2$ by 6.7% for about 1 min after the blow, associated in most cases with either low-voltage fast activity in the EEG or with brief convulsive movements of the extremities. Blood pressure, ICP and CVR transiently increased after the blow.

Severe concussion decreased CBF by 12% and transiently caused EEG slowing associated with decreased $CMRO_2$ by 9.7%. Blood pressure decreased by 3.6% and CVR increased by 10.8%.

Mild concussion due to a moderate impact produced signs of CNS excitation as indicated by the EEG and clinical signs. This was associated with increased cerebral metabolism compatible with the excitation theory of concussion [9]. On the other hand, in severe concussion due to a more severe impact, EEG slowing and a transient decrease in $CMRO_2$ indicated a transient traumatic neuronal paralysis [1].

Following contusion, CBF gradually decreased by 14%. The EEG showed prolonged slowing with or without changes in amplitude. Blood pressure showed little change but CVR increased by 15% as ICP increased. Cerebral lactate production increased while pyruvate change was minimal, suggesting excess lactate formation [4]. This view was supported by a decrease in CVpH and a decrease in $CVPO_2$ despite $CMRO_2$ reduction suggesting that cerebral contusion causes cerebral anoxia due to vascular injury.

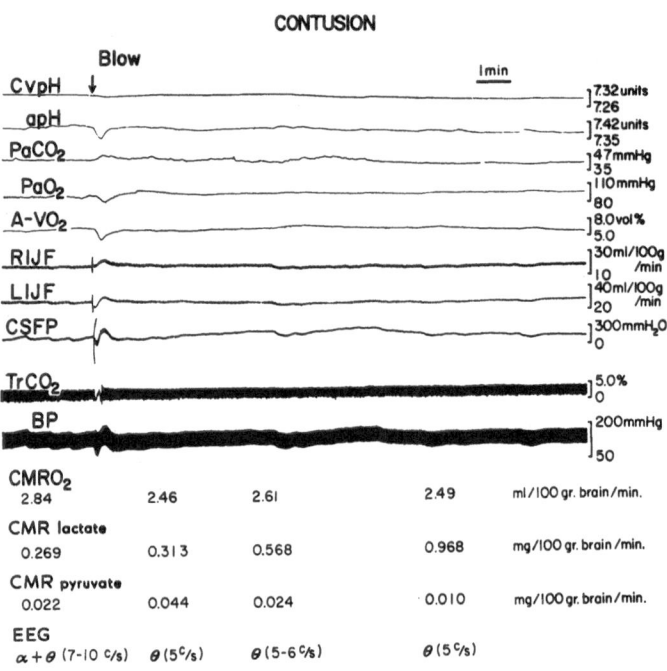

Fig. 1. Actual recording of cerebral hemodynamics and metabolism during cerebral contusion marked by arrow. CVpH = cerebral venous pH; apH = arterial pH; $PaCO_2$ = arterial PCO_2; PaO_2 = arterial PO_2; $A\text{-}VO_2$ = cerebral arterio-venous oxygen difference; RIJF = right internal jugular flow; LIJF = left internal jugular flow; CSFP = cerebrospinal fluid pressure; $TrCO_2$ = end-tidal PCO_2; BP = blood pressure; $CMRO_2$ = cerebral oxygen consumption; CMR lactate = cerebral lactate production; CMR pyruvate = cerebral pyruvate production. EEG changes are listed as predominant rhythms

Following brain-stem laceration, the BP and CVR increased and the EEG slowed and often became isoelectric. CBF decreased by 2%, yet $CMRO_2$ was decreased by 25% and hence the cerebral $A\text{-}VO_2$ difference decreased by 18%, associated with a remarkable increase of PVO_2 by 18%. This proved the existence of "luxury perfusion" [5, 8], at this time. Lactate production was profoundly and significantly increased within 2 min of the luxury perfusion without any notable change in pyruvate, indicating an excess of lactate formation. Hence, increased lactate in brain tissues [2] may have participated in the luxury perfusion, although decreased oxygen demand by cerebral tissues played an important part.

References

1. Denny-Brown, D.: Cerebral concussion. Physiol. Rev. 25, 296 (1954).
2. Gurdjian, E. S., J. E. Webster, and W. E. Stone: Experimental head injury with special reference to certain chemicals factor in acute trauma. Surg. Gynec. Obstet. 78, 618 (1944).
3. Guyton, A. C., R. J. Nichols, Jr., and C. Farisii: Arteriovenous oxygen difference recorder. J. appl. Physiol. 10, 158 (1957).

4. HUCKABEE, W. E.: Relationship of pyruvate and lactate during anaerobic metabolism. IV. Local tissue components of total body O_2 debt. Amer. J. Physiol. **196**, 253 (1959).
5. LASSEN, N. A.: The luxury perfusion syndrome and its possible relation to acute metabolic acidosis localized within the brain. Lancet II, 1113 (1966).
6. MEYER, J. S., S. ISHIKAWA, and T. K. LEE: Electromagnetic measurement of internal jugular venous flow in the monkey. J. Neurosurg. **21**, 524 (1964).
7. —, F. GOTOH, M. TOMITA, and M. AKIYAMA: New techniques for recording cerebral blood flow and metabolism in subjects with cerebrovascular disease. In: Cerebral Vascular Diseases, Fifth Princeton Conference. R. SIEKERT and J. WHISNANT (Eds.), p. 147. New York: Grune and Stratton 1966.
8. — The nature of high oxygen tensions in bordering zones of cerebral ischemia. Scand. J. clin. Lab. Invest. Suppl. **102**, p. XVI-A (1968).
9. WALKER, A. E., J. J. KOLLROS, and T. J. CASE: The physiologic basis of concussion. J. Neurosurg. **1**, 103 (1944).

Disturbance of Autoregulation Following Head Injury – Acute and Chronic Observations

A. R. TAYLOR

Royal Victoria Hospital, Belfast

The point of departure for this paper is the observation made a few years ago [10] that head-injured patients suffering from posttraumatic symptoms, have increased cerebral circulation times in the range of 5–25% above normal, with a mean increase of 13%. Similar increases in circulation time, with a reciprocal reduction in cerebral blood flow, have been observed in patients with atherosclerosis, subarachnoid haemorrhage, cerebral tumors and increased intracranial pressure. In these there is an obvious cause: narrowing or spasm of cerebral arteries, sluggish flow through a tumor bed or raised venous pressure. It is not immediately obvious why cerebral trauma should slow intracranial circulation. This paper examines possible pathological and biochemical causes and suggests a method of validating a likely hypothesis.

Pathology

The direct effects of cerebral trauma causing death, contusion, laceration and hemorrhage have been fully described by pathologists, but help little in the elucidation of the symptoms occurring in survivors. For this information we must look to post mortem material from patients who recovered from head injuries and died from other causes – this is scanty – and to histological findings in animals experimentally concussed. In deceleration concussion 3 types of lesion are constantly found: chromatolysis followed by neuronal death in the nuclei of the upper cervical cord, medulla and brain stem; axonal tearing accompanied by myelin distortion and microglial aggregation spread throughout the white matter; and multiple small ischaemic infarcts which are found characteristically in the subcortical areas, in the corpus callosum and basal nuclei.

Considering the possible effects of such lesions on blood flow, and taking them in order: – nuclear lesions in the medulla – brain stem axis could affect medullary respiratory centers [7], cerebral vaso motor centers [6] or midbrain CO_2 centers [9] – if such centers exist. Whether they do or not, which is still a matter of uncertainty, it seems entirely credible that there is a system of cell stations and feed-back circuits in the mid and hind brain which control the factors influencing blood flow and which would work less efficiently after deceleration trauma.

Diffuse white matter lesions are prima facie less likely to influence blood flow directly, but are more likely to contribute to the concomitant symptoms of intellectual disorganization which follow brain trauma. The effects of multiple small infarcts must be deduced from what is known of the behaviour of experimentally produced infarcts in animals [1]. In the area, and for a considerable distance around an infarct, cerebral blood flow reacts paradoxically to inhaled CO_2. This effect continues for about a week in normotensive cats and for at least 2 months in cats made hypertensive by renal surgery [4]. It is possible, then, that brains which have suffered anoxic insults from contusion, cerebral oedema, vascular occlusion (direct or indirect), or an obstructed airway may react paradoxically to the physiological variations of CO_2, O_2, blood pressure or other changes of internal milieu not yet investigated.

Biochemical Factors

Changes following trauma likely to influence blood flow in the short or long term are the presence of blood in the CSF causing an acid shift in CSF pH and hyperventilation [3], a rise in CSF lactic acid resulting from tissue necrosis [5], primary neurogenic hyperventilation causing respiratory alkalosis, inactivation of the NADP/NAD system reducing glucose oxidation with resulting tissue acidosis [2]. The changes in CSF pH caused by the above reactions cause alterations in CBF directly or through changes in ventilation.

Design of Investigation

From the above observations it would seem possible that the reversible slowing in cerebral circulation found after head injury is caused by an impaired capacity to react to physiological changes in internal milieu generally or regionally. Investigations are proceeding into the changes in cerebral circulation induced by changes in pCO_2 and systemic BP.

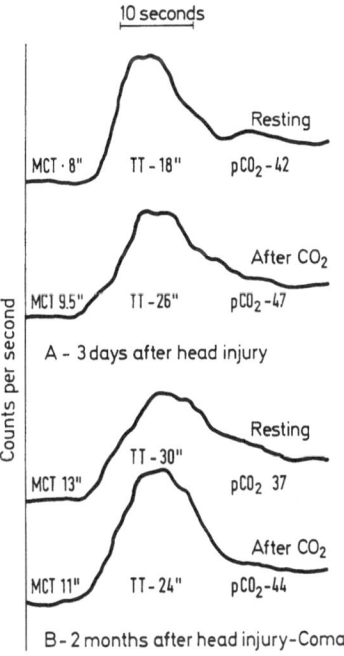

Fig. 1. A) showing paradoxical response. B) showing normal response to inhalation of CO_2

Following intravenous injection of a gamma-emitting isotope and its rapid deposition into the right heart by the blood pressure cuff technique described by OLDENDORF [8], gamma activity is monitored by a posterior mid line scintillation counter and passed to a chart recorder. A plot of the gamma activity against time is analyzed, and the arm to head and cerebral transit times noted. The mean circulation time (MCT) is calculated from the first derivative of the curve. Blood is taken for pCO_2 (ASTRUP), BP. and pulse rate noted. Following inhalation of 10% CO_2 in air, the examination is repeated and the nature of the reaction to CO_2 is immediately apparent from comparison of the curves. Insufficient numbers of observations have been made to date to permit a statistical analysis, but it is already clear that the acute cases can be categorized into those which react normally to CO_2, those which react little if at all, and those which react paradoxically. To date we have not found paradoxical response in patients more than 4 days after deceleration injury.

Fig. 1 illustrates a normal response to CO_2 in a patient in coma 2 months after head injury, and a paradoxical response 4 days after an acute deceleration injury.

References

1. BRAWLEY, B. W., D. E. STRANDNESS, and W. A. KELLY: The physiological response to therapy in experimental cerebral ischemia. Arch. Neurol. 17, 180 (1967).
2. CHASON, J. L., O. U. FERNANDO, V. R. HODGSON, L. M. THOMAS, and E. S. GURDJIAN: Experimental brain concussion: morphologic findings and a new cytologic hypothesis. J. Trauma 6, 767 (1966).
3. FROMAN, C., and A. C. SMITH: Metabolic acidosis of the cerebrospinal fluid associated with subarachnoid hemorrhage. Lancet I, 965 (1967).
4. HALSEY, J. H.: Personal communication.
5. KURZE, T., R. E. TRANQUADA, and K. BENEDICT: In: Head Injury. W. F. CAVENESS and A. E. WALKER (Eds.), p. 254. Philadelphia: J. B. Lippincott 1966.
6. LANGFITT, T. W., J. D. WEINSTEIN, and N. F. KASSELL: Vascular Factors in Head Injury. In: Head Injury. W. F. CAVENESS and A. E. WALKER (Eds.), p. 172. Philadelphia: J. B. Lippincott 1962.
7. MITCHELL, R. A.: The regulation of respiration in metabolic acidosis and alkalosis. In: Cerebrospinal Fluid and the Regulation of Ventilation, ed. by BROOKS, KAO, and LLOYD, Oxford, 1965.
8. OLDENDORF, W. H., and M. KITANO: Radioisotope measurement of blood brain turnover time as a clinical index of brain circulation. J. nucl. Med. 8, 570 (1967).
9. SHALIT, M. N., O. M. REINMUTH, S. SHIMOJYO, and P. SCHEINBERG: Carbon dioxide and cerebral circulatory control. III. The effect of brain stem lesions. Arch. Neurol. 17, 342 (1967).
10. TAYLOR, A. R., and T. K. BELL: Slowing of cerebral circulation after concussional head injury. A controlled trial. Lancet II, 178 (1966).

Loss of Autoregulation Produced by Cerebral Trauma

M. Reivich, W. J. S. Marshall, and N. Kassell

Spiller Neurological Unit and the Research Laboratories of the Department of Neurology, University of Pennsylvania, Philadelphia

Recent studies [3, 5] have demonstrated that blunt, non-necrotizing injury to the exposed brain followed by arterial hypertension caused acute brain swelling. It was postulated that cerebral autoregulation had been abolished or diminished by the trauma and that the subsequent increase in arterial pressure was transmitted to the capillaries and veins producing a large hydrostatic gradient across the capillary membrane with resultant cerebral edema. The present experiments were undertaken in order to investigate the autoregulatory ability of the brain following mechanical trauma to the cerebral cortex.

A series of 11 healthy adult cats weighing between 2.5 and 5.0 kg was studied in which a bilateral craniotomy was performed under light anesthesia with 70% nitrous oxide and 30% oxygen after an induction with 20 mg/kg body weight of thiopental. The animal was then paralyzed with gallamine, 8 mg i.v., and artifically respired with a Harvard small animal pump. Additional doses of 4 mg gallamine were given as required to maintain control of the animal's respiration. End tidal pCO_2 was continuously monitored with an infrared CO_2 analyzer and intermittent arterial samples were obtained for pCO_2, pO_2 and pH.

In order to avoid damage to the underlying brain during the craniotomy fluid was withdrawn from the cisterna magna allowing the brain to fall away from the overlying skull. The dura was opened and reflected and the brain covered with a thin polyethylene sheet in order to prevent drying or the escape of Krypton-85 from the brain during the blood flow studies. When the preparation was complete, 1 cc/kg body weight of a 3% solution of Evans blue dye in saline was injected intravenously. After a pause of 30 min the brain was inspected to see whether there was any blue staining. If this was present, it was presumed that the brain had been traumatized during preparation and the animal was discarded.

Regional cerebral blood flow was measured by 2 techniques. In 9 of the animals 2 end window Geiger-Müller tubes 0.5 inches in diameter were placed over homologous areas of the exposed cortex. Krypton-85 dissolved in saline was infused via a PE 10 polyethylene catheter placed in each common carotid artery and its cortical clearance was monitored from both hemispheres before and after mechanical trauma to the cerebral cortex at various levels of mean arterial pressure. In 5 animals regional cerebral blood flow was measured after cerebral trauma by an autoradiographic technique utilizing ^{14}C-antipyrine [4].

Arterial pressure was lowered by slowly bleeding the animal into an heparinized syringe and then raised by reinjecting the blood.

Supported in part by a U.S.P.H.S. Research Grant (NB-06314-04) from the National Institute of Neurological Diseases and Blindness. M. Reivich is recipient of Career Research Development Award 5 K03 HE 11896-04.

After control measurements were obtained one hemisphere was traumatized by means of a jet of compressed nitrogen at 50 lbs/square inch delivered from a fixed distance of 2 cm from the surface of the brain. The duration of the blast and the interval between blasts was controlled by a Tectronics pulse and wave form generator. The normal square wave form to the release of gas was modified to that of a sine wave in order to prevent sudden displacement of the whole brain and tearing of bridging veins. The brain was depressed approximately 4 mm with each impulse. The duration of each blast was $1/_2$ sec and the interval between blasts was 2 sec. 12 such blasts were administered.

During the control studies the mean $\pm$ S.E. arterial pCO_2 was 29.3 ± 0.6 mmHg while during the post trauma measurements the value was 28.9 ± 0.9 mmHg. Mean arterial pO_2 was 123 ± 3.8 mmHg during the control phase and 126 ± 4.9 mmHg following trauma. Arterial pH was $7.378 \pm .025$ and $7.370 \pm .032$ in the control and post trauma studies respectively. There were no significant differences between any of these measurements. During the control studies the autoregulatory ability of the cortex was demonstrated to be intact. Mean arterial pressure was changed from 75 to 187 mmHg without any significant effect on cortical blood flow. Following cerebral trauma mean arterial pressure was varied from 65 to 215 mmHg. Cortical blood flow passively followed the changes in mean arterial pressure indicating a loss of autoregulation. The regression lines and 95% confidence limits for the control and post trauma data are shown in Fig. 1. There is a highly significant difference between the slopes of these 2 lines ($p < .001$). The slope of the control data is not significantly different from zero. If the cortical blood flow data in the region of normal mean arterial pressure in these animals, i. e. about 120 mmHg, are examined; no significant difference is seen between the control and trauma studies. This suggests that a primary vasodilation does not occur after trauma under these conditions.

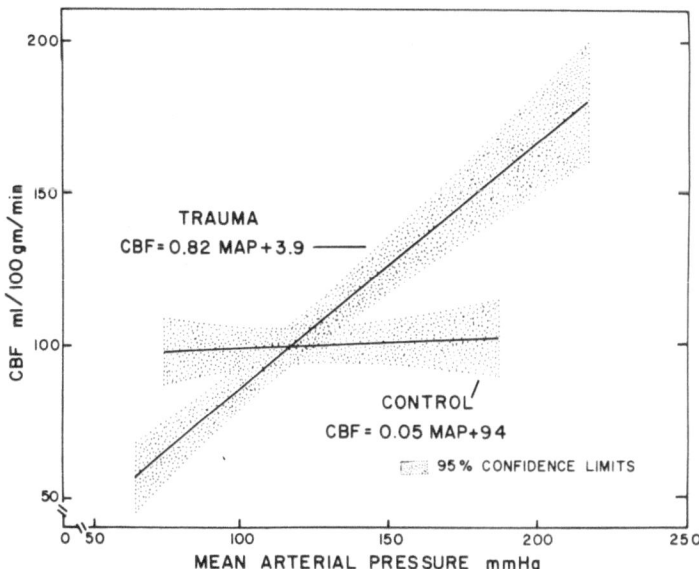

Fig. 1. The regression lines through the control and post trauma data are shown with their 95% confidence limits. There is a highly significant difference between the slopes of the 2 lines ($p < 0.001$)

A loss of autoregulation was also demonstrated by the autoradiographic regional cerebral blood flow studies. In addition these studies provided information regarding the distribution of these changes in autoregulation. Usually the loss of autoregulation was confined to the traumatized hemisphere but did extend to areas in that hemisphere not directly traumatized. These changes were most marked in the cortex and underlying white matter while more deeply placed structures such as the thalamus, hypothalamus and the basal ganglia did not show much change in flow. Structures in the posterior fossa also were unaffected. In 1 of the animals studied, loss of autoregulation was also seen in parts of the opposite non-traumatized hemisphere. Fig. 2 and 3 show autoradiograms from animals whose mean arterial pressure was 175 and 75 mmHg respectively. The areas of loss of autoregulation in the cortex and underlying white matter of the traumatized hemisphere are clearly seen.

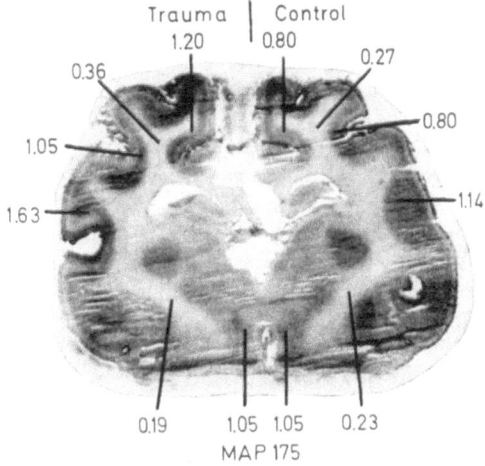

Fig. 2. Autoradiogram of a mid-coronal section of the brain in an animal whose mean arterial pressure (MAP) was elevated to 175 mmHg. In the traumatized hemisphere the flows in the cortex and subcortical white matter are higher than in the control hemisphere. The deep midline structures do not show this difference in flow

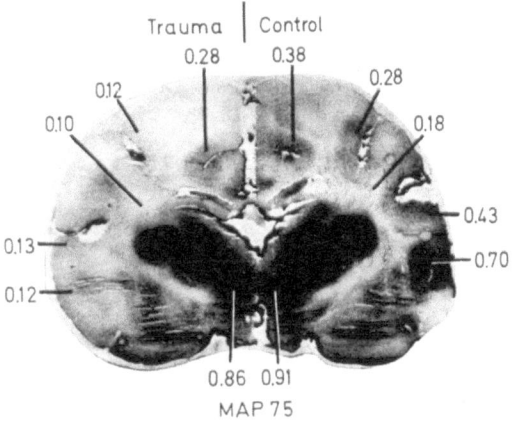

Fig. 3. Autoradiogram of a mid-coronal section of the brain in an animal whose mean arterial pressure (MAP) was reduced to 75 mmHg. Marked reductions in flow are present in the cortex and white matter of the traumatized hemisphere. These changes are also present in regions not directly traumatized in the same and contralateral hemispheres. The deep midline structures do not show these changes

These data demonstrate that a loss of autoregulation occurs following mechanical trauma to the exposed cerebral cortex [1, 2]. They also suggest that a primary vasodilation does not occur under these conditions. The changes in autoregulation are seen mainly in the cortex and underlying white matter of the traumatized area but also involve regions of the hemisphere not directly traumatized. Vessels of the deep midline structures and structures in the posterior fossa appear to retain their autoregulatory ability. Such changes in autoregulation may also occur in patients with closed head injuries or following manipulation of the brain during operative procedures. It would therefore appear important to maintain a normal cerebral perfusion pressure under these conditions in order to insure adequate blood flow to the involved regions.

References

1. Brock, M.: Regional cerebral blood flow (rCBF) changes following local brain compression in the cat. Scand. J. clin. Lab. Invest. Suppl. **102**, XIV:A (1968).
2. Freeman, J., and D. H. Ingvar: Elimination by hypoxia of cerebral blood flow autoregulation and EEG relationship. Exp. Brain Res. **5**, 61 (1968).
3. Langfitt, T. W., W. J. S. Marshall, N. F. Kassell, and H. S. Schutta: The pathophysiology of brain swelling produced by mechanical trauma and hypertension. Scand. J. clin. Lab. Invest. Suppl. **102**, p. XIV-B (1968).
4. Reivich, M., J. Jehle, L. Sokoloff, and S. S. Kety: Measurement of regional cerebral blood flow with ^{14}C-antipyrine in awake cats. J. appl. Physiol., 1969, in press.
5. Schutta, H. S., N. F. Kassell, and T. W. Langfitt: Brain swelling produced by injury and aggravated by arterial hypertension. Brain **91**, 281 (1968).

Cessation of Cerebral Blood Flow in Total Irreversible Loss of Brain Function

A. A. Hadjidimos, M. Brock, P. Baum and K. Schürmann

Departments of Neurosurgery and Internal Medicine I, University of Mainz

After the demonstration, by our group [3, 4], that in cases of so-called "cerebral death" there is *an absence of clearance* from the brain of ^{133}Xe selectively injected into the internal carotid artery, further attention was dedicated to establish possible sources of error in this technique.

As known, in cases of "cerebral death" with the angiographic phenomenon of "non-filling" of cerebral vessels, there is a reflux of contrast material into the external carotid territory following slug injection into the internal carotid artery. It is conceivable that, if the same takes place following an injection of ^{133}Xe-solution, the clearance of the tracer from the extracerebral tissue might simulate a more or less preserved cerebral circulation even in cases where a total cerebral circulatory arrest is present. In order to study this problem we performed rCBF measurements in a series of 5 patients in "coma dépassé". All of them presented clinical, electroencephalographic and angiographic signs of cerebral death. The heart was beating (sinus rhythm) in all patients, and ventilation was artificially maintained. The following technical procedures were used:

1. direct, selective tracer injection into the internal carotid artery, following *surgical exposure* of this vessel, with temporary occlusion of the external carotid artery during the injection of 4–7 mCi of ^{133}Xe;
2. percutaneous selective internal carotid tracer injection with and without digital compression of the external carotid in order to prevent reflux;
3. percutaneous selective ^{133}Xe-injection into the external carotid artery.

Our results indicate that only the selective tracer injection into the *surgically exposed* internal carotid artery under simultaneous temporary occlusion of the external carotid, allows to demonstrate in a conclusive way the total arrest of cerebral circulation in such patients. Fig. 1 shows the typical rCBF *plateau curves* recorded over the whole hemisphere. Three main aspects deserve special attention in these recordings:

a) the small amount of radioactivity reaching the brain (often less than 10% of full scale deflection = 50 c.p.s.; two-decades recording;
b) the fact that most of the tracer does not reach the cerebral convexity;
c) the total absence of isotope clearance for more than 30 min.

Percutaneous injections, even if they are performed selectively into the internal carotid with the catheter tip near to the siphon, and even if the external carotid is subject to digital compression, are always followed by extra-cerebral tracer contamination and clearance

(Fig. 2). Under such circumstances, the clearance curves recorded over the hemisphere have a form typical for extracranial tissues, namely a biexponential slope (during the first 2 min of clearance) with an initial, short-lasting fast component (slope equal to about 20–70 ml/100 g/min) of small relative weight, followed by a slow component (slope equal to about 0–10 ml/100 g/min) of larger relative weight depending on the (proximal or distal) location of the detectors. Thus, mean hemisphere rCBF under such conditions is about

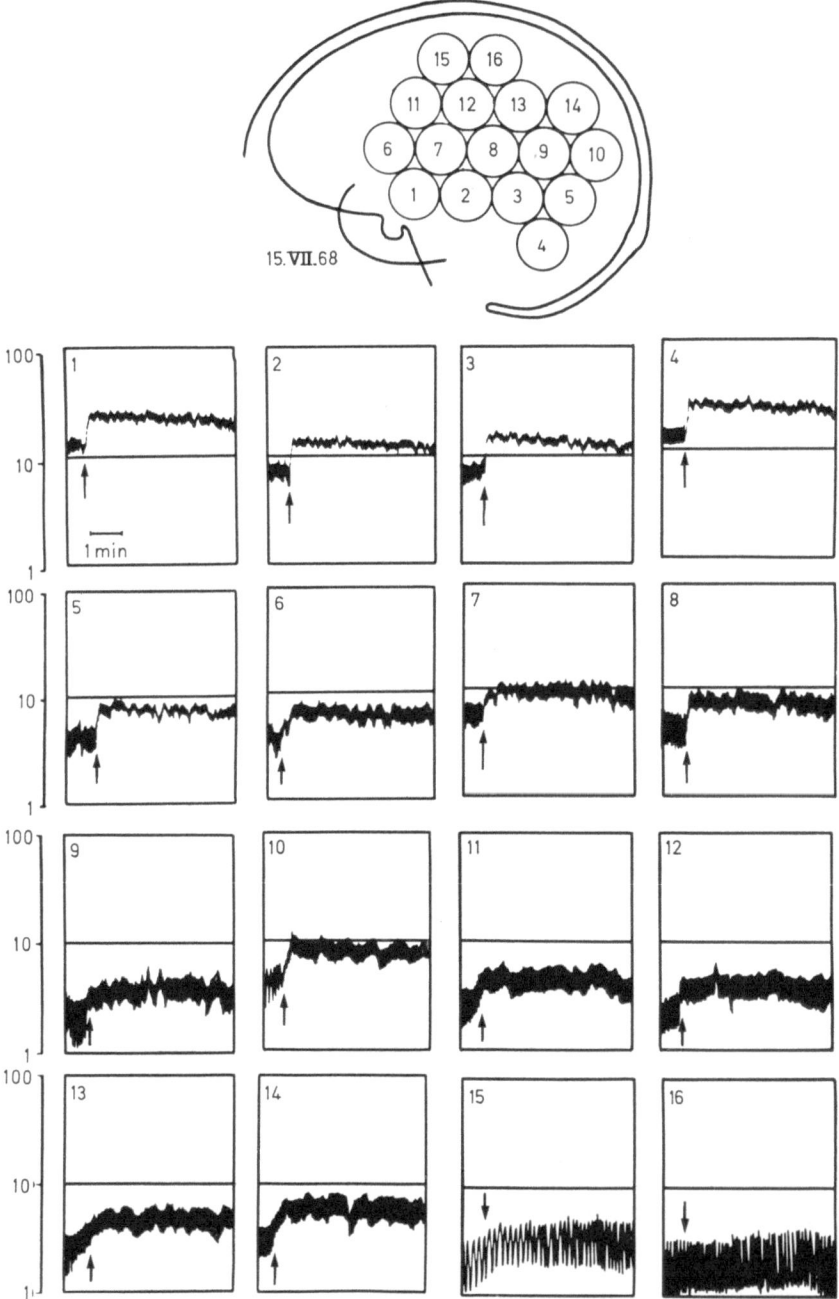

Fig. 1. Virtual absence of clearance from the brain of ^{133}Xe-solution forcefully injected into the surgically exposed internal carotid artery during simultaneous temporary occlusion of the external carotid (from Brock [3])

7–15 ml/100 g/min, while complete cerebral circulatory arrest should be expected. This kind of clearance curves can also be recorded in normals following a selective tracer injection into the external carotid artery ([5] and personal observations).

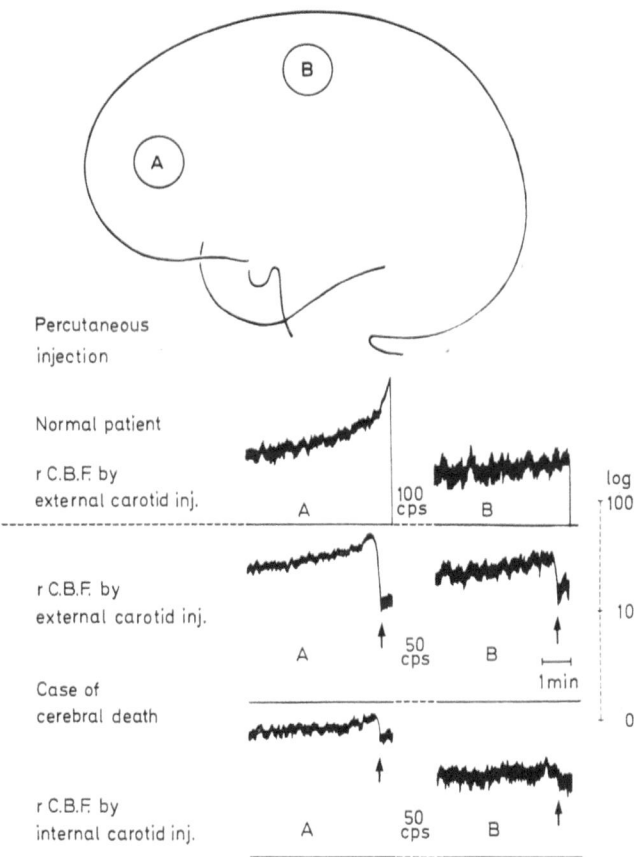

Fig. 2. ^{133}Xe clearance curves from 2 regions of the skull (A and B) following (1) percutaneous external carotid injection in the normal (upper curves), (2) following percutaneous external carotid injection in a case of cerebral death (middle curves), and (3) following percutaneous internal carotid injection in a case of cerebral death (lower curves)

In our material the shape of such curves can be triexponential during the first 2 min of clearance with high vascular peaks at the fronto-temporal region and may be as high as 20–30 ml/100 g/min (Fig. 2, upper part). In cases of cerebral death, as well as in normals, a rubber band placed around the head does not prevent the filling of the distal branches of the external carotid artery. Further, in contrast to the normal cerebral clearance curves, the external carotid flow can be increased by induced hypertension also in cases of cerebral death with high intracranial pressure.

Bès et al., have reported 2 cases of "coma dépassé" with angiographic arrest of cerebral circulation and reduced (7 and 11 ml/100 g/min) rCBF [2]. Furthermore, BALDY-MOULINIER and FRÈREBEAU, observed biexponential clearance curves of ^{133}Xe in the same type of patients [1]. According to our experience, these observations probably represent measurements of external carotid blood flow. It is evident that only by temporary *surgical* occlusion of the external carotid artery is it possible to avoid a tracer reflux into the external carotid territory and to force it into the cranium in cases of cerebral circulatory arrest with intracranial hypertension exceeding the systolic blood pressure.

If the above mentioned technical precautions are observed, cerebral blood flow studies can become a reliable criterium for establishing cerebral death in cases of total, irreversible loss of brain function.

References

1. Baldy-Moulinier, M., and Ph. Frèrebeau: Cerebral blood flow in cases of following coma severe head injury. Presented at the International CBF Symposium, Mainz, April 1969 (this volume, p. 216).

2. Bès, A., L. Arbus, Y. Lazorthes, M. Escande, M. Delpla, and J. P. Marc Vergne: Hemodynamic and metabolic studies in "coma dépassé". A search for a biological test of death of the brain. Presented at the International CBF Symposium, Mainz, April 1969 (this volume, p. 213).

3. Brock, M.: Contribution to the Discussion on Cerebral Death. In: H. Penin and C. Käufer (Eds.), Der Hirntod, p. 96–98. Stuttgart: Georg Thieme 1969.

4. — K. Schürmann, and A. Hadjidimos: Cerebral blood flow and cerebral death. Acta Neurochir., 20, 195 (1969).

5. Ueda, H., S. Hatano, T. Molde, and T. Gondaira: Discussion on "Compartmental analysis of the human brain blood flow". Acta neurol. scand. Suppl. 14, 89 (1965).

Hemodynamic and Metabolic Studies in "Coma Dépassé". A Search for a Biological Test of Death of the Brain

A. Bès, L. Arbus, Y. Lazorthes, M. Escande, M. Delpla, and J. P. Marc Vergne

Department of Neurology, University Hospital, Toulouse

In a previous study [1] of comatose patients we noted a striking difference with respect to cerebral hemodynamic and metabolic parameters between apoplectic patients in deep coma and patients in coma dépassé. Table 1 summarizes the results obtained in apoplectic patients in coma, classified according to the level of clinical severity: In the most grave cases (level C4 in our classification), where all reactions to pain as well as corneal reflexes had disappeared, it was found that cerebral metabolism was relatively unaltered. Cerebral blood flow (measured by the ^{85}Krypton inhalation modification of Kety's method) was lowered by about 50% (in average 26 ml/100 g/min as compared to 46 in normal subjects); cerebral oxygen consumption was proportionally decreased (1.96 ml/100 g/min as compared to 3.34). Cerebral arteriovenous difference of oxygen remained at values close to the normal (5.88 vol.%). Furthermore, in the most serious coma cases a moderate decrease of the arteriovenous glucose difference (—32%) and an increase in cerebral arteriovenous lactic acid difference was noted.

Table 1. *Cerebral arteriovenous differences in comatose patients (apoplectic comas)*

	NC	C 1	C 2	C 3	C 4
Art. O_2 (vol.%)	15.89	16.27	16.71	15.61	13.60
(A–V) O	7.23	6.86	6.84	6.38	5.88
(A–V) CO_2	7.60	6.68	6.44	5.98	6.19
CO_2/O_2	1.06	0.96	0.92	0.95	1.01
A Glucose (mg %)	119.3	175.1	183.6	190.5	230.3
(A–V) glucose	15.4	12.4	12.2	13.3	10.4
V Lact. (mg %)	13.3	14.6	15.9	17	16.7
(V–A) Lact.	0.33	1.19	1.84	1.32	1.03
(V–A) Pyr.	0.10	0.12	0.18	0.13	0.14

NC = stroke patients with normal consciousness.
C 1, C 2, C 3, C 4: depth of coma (cf. text).

The findings presented suggest, consequently, that in grave cases of coma due to apoplexy, there is a preservation of a basically normal metabolic pattern. In coma dépassé on the other hand, a completely different picture is seen, in that cerebral arteriovenous oxygen difference is usually extremely diminished to values of 1–2 vol.%. Table 2 shows a comparison between the values obtained in 10 cases of coma dépassé and those from 10 patients with severe apo-

214   A. Bès, L. Arbus, Y. Lazorthes, M. Escande, M. Delpla, and J. P. Marc Vergne:

plectic coma. The striking decrease of arteriovenous oxygen difference to values below 2 vol.% in 6 of the 10 coma dépassé cases should be noted. In 2 of the remaining cases, the difference was below 3 vol.%.

Table 2. *Cerebral arteriovenous differences in "coma dépassé"*

Diagnosis	Severe apoplectic coma (n = 10) (C 4)	"Coma dépassé"									
		trauma	trauma	trauma	cer. hem.	cer. hem.	cer. hem.	trauma	trauma	trauma	cardiac arrest
A O$_2$ (vol %)	13.60	17.68	16.22	13.51	13.94	15.71	18.48	18.05	28.99	21.78	19.75
(A–V) O$_2$	5.88	1.58	1.35	0.66	1.21	0.64	1.74	3.90	4.75–6.02	2.54	2.82
V CO$_2$ (vol. %)		38.25	46.48	40.29	30.15	47.97	46.6	48.73		41.51	54.35
(V–A) CO$_2$	6.19	1.57	1.44	0.97	2.10	0.81	2.5	4.19		3	3.33
QR	1.01	1	1	1	1.7	1.2	1.4	1.6		1.1	1.1
A glucose (cg %)	230.3		221.5	333.2	313.6	118.8	568		262	127.7	249.2
(A–V) glucose	10.4		14.3	32.1	7.1	11.6	9.8		27	8.9	4.4
V Lact. (mg %)	16.7		28.80	19.20	36.7	17.60	29.2		22	21.6	21.3
(V–A) Lact.	1.3		4	6.40	1.7	2	6.8		1.5	4.8	2.1
(V–A) Pyr.	0.14		0	0	0.06	0.19	0		0	0.40	0
PaCO$_2$	31.70		39	39.32	28.05	48.76				41.74	44
Mean CBF (ml/100 g/min)	26.1		11								6.9

With regard to the arteriovenous glucose difference the values were more scattered, even though a moderate decrease was observed in several cases. A marked increase in cerebral arteriovenous difference of lactic acid was noted, whereas no clearcut alteration in the difference for pyruvic acid was found.

In 2 patients we measured cerebral blood flow by the intracarotid injection technique using [133]Xenon and external counting. The average flow was extremely low, 7 and 11 ml/ 100 g/min, respectively. In the latter case, there was, as a matter of fact, no evidence of intracranial circulation on angiography. It is therefore possible that the measurement of blood flow is a more precise method than angiography to demonstrate persistence of a greatly reduced cerebral circulation. However, the possibility of reflux of isotope to the external carotid artery territory [2] must also be considered.

We conclude that the measurement of the cerebral arteriovenous oxygen difference constitutes a simple biological test of coma dépassé, that is of death of the brain. Indeed, one does not find a progressive decrease of this parameter when the level of coma deepens. As measured during normocapnia or slight hypercapnia, one finds what appears to be a sudden drop from 6 vol.% to values below 2 vol.% (cf. Kaasik [3]). The clinical application of this measurement would seem useful since the clinical signs associated with electrical silence of the EEG are not in all cases unequivocal for the diagnosis of brain death.

It should be observed that in our series a cerebral arteriovenous oxygen difference of below 2% was only found in cases of coma dépassé, suggesting that patients with such low values (at normal pCO_2) cannot survive. In this context, we should like to report that a cerebral arteriovenous oxygen difference equal to or superior to the normal value of 6 vol.% was found in 4 cases of toxic coma after barbiturate ingestion, a clinical condition resembling coma dépassé but with a completely different prognosis. We shall not suggest, however, that the biological test proposed is infallible: in 2 cases of our series of coma dépassé the arteriovenous difference remained above 2 vol.% even though the clinical picture was to all appearance similar to that of the other cases.

Finally, it could be added that the measurement of the cerebral arteriovenous oxygen difference might possibly be an earlier and more reliable test of impending cerebral circulatory arrest than cerebral arteriography.

Table 3. *Cerebral arteriovenous differences in four cases of deep barbiturate coma*

AO_2 (vol. %)	14.68	18.45	18.21	15.29
(A–V) O_2	5.20	10.69	9.77	7.18
V CO_2 (vol. %)	36.07	56.65	49.02	46.97
(V–A) CO_2	5.52	9.98	10.25	6.07
Pa CO_2 mmHg	22	34.5	24	36.5
Q R	1	0.93	1.05	0.84
A glucose (cg %)	365.5	192	405.5	327
(A–V) glucose	5.4	29.4	32.1	12.5
V – Lact. (mg %)	38.8	34.8	23.60	19.2
(V–A) Lact.	0.8	6.8	0	3.6
V – Pyr. (mg %)	0.90	0.65	0.80	0.62
(V–A) Pyr.	0	0	0	0
Mean CBF (ml/100 g/min)	21.4			33.9

References

1. GERAUD, J., A. BÈS, and M. ESCANDE: Hemodynamic and metabolic study of apoplectic comas. IVe International Symposium on Cerebral Circulation, Salzburg 1968, in press.
2. HADJIDIMOS, A. A., M. BROCK, P. BAUM and K. SCHÜRMANN: Total irreversible loss of brain function and cerebral blood flow. Presented at the International CBF Symposium, Mainz, April 1969 (this volume, p. 209).
3. KAASIK, A. E.: Reduction of cerebral arteriovenous oxygen difference in terminal phase of cerebral hemorrhage. Scand. J. clin. Lab. Invest. Suppl. 102, p. X:D (1968).

Cerebral Blood Flow in Cases of Coma Following Severe Head Injury

M. Baldy-Moulinier and Ph. Frèrebeau

Laboratory of Experimental Pathology and Neurosurgical Department, Centre Hospitalier Universitaire, Montpellier

For a long time it has been known that CBF is reduced in most types of coma [5]. A close correlation has been determined between reduction of CBF, decrease of oxygen consumption and the level of consciousness [4]. Later, the description of "luxury perfusion syndrome" of brain tissue [6] has shown that this assumption must be corrected in some conditions. Since experimental data support that cerebral trauma is often followed by loss of cerebral autoregulation [2, 8] a question must be raised: is the reduction of CBF that takes place in posttraumatic coma [1] present in all the cases? To answer this question rCBF has been measured in a series of 30 comatose patients by means of the ^{133}Xenon clearance method within 5 days of the head trauma. Mean rCBF was calculated by the compartmental method.

The patients were divided into 3 groups according to the cerebral angiograms.

I. The first group included 6 patients with normal angiograms. All showed clinical and electroencephalographic signs of coma and reacted to arousing stimulation. The average rCBF was 43 ml/100 g/min $\pm$ 8 SD with a mean arterial pCO_2 of 40 mmHg $\pm$ 2 SD When arterial hypertension was induced in 1 case, this was followed by an increase of rCBF in the 4 areas measured. Complete clinical recovery was observed in all patients but 2 who died from septicemia. In 2 cases the initial rCBF later diminished during a period of a transient fall in the level of consciousness. During recovery of consciousness, rCBF increased again to the initial level.

II. The second group included 15 comatose patients with symptoms of unconsciousness identical to those of the first group. They differed however from the first by showing clinical and angiographic signs of an intracranial space occupying lesion. These patients had an average rCBF of 27 ml/100 g/min $\pm$ 5 SD with an $ApCO_2$ of 27 mmHg $\pm$ 4 SD. Induced arterial hypertension increased rCBF in regions around the lesion. Hyperventilation was followed by an increase of rCBF in some cases.

In most of these patients, in which subdural hematomas, extradural hematomas or local brain contusion were found, the rCBF measurements were repeated after surgical operation. rCBF was generally decreased the first day after surgery and either raised or lowered 5 to 15 days later. Clinical improvement was observed when rCBF increased. Death or permanent coma occurred in the other cases.

III. The third group included 9 cases with severe coma, bilateral mydriasis and isoelectric EEG, and absence or disturbance of spontaneous respiration. In all of these patients intracranial or extracranial cerebral circulatory arrest was determined by angiography. Determination of rCBF was impossible in these cases. The curves recorded after injection of Xenon into the internal carotid artery generally showed an absence of the initial peak.

In some cases with intracranial circulatory arrest, a plateau, or a peak followed by a plateau, could be observed. In none of these cases were there any signs of reestablishment of the cerebral circulation and no recovery occurred after surgical treatment.

Comparison of the first 2 groups indicates a significant difference in the mean rCBF in patients with the same degree of unconsciousness. The higher rCBF observed in the first group suggests that a reactive hyperemia might have been present. A loss of autoregulation is indicated by the increase of rCBF during induced hypertension. In this group without signs of localized brain lesion, coma could be considered as induced by functional disturbance of the brain stem. This indicates that a functional disturbance of the brain stem may affect

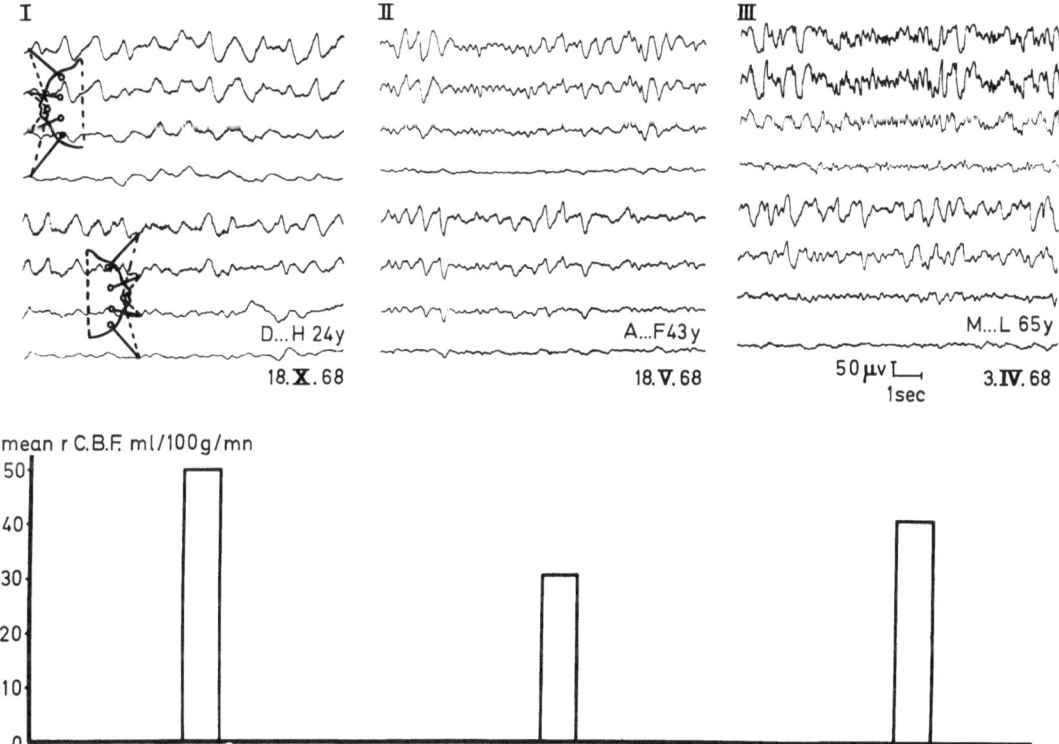

Fig. 1. EEG and mean rCBF in three different patients. I – Comatose patient after head injury without localized brain lesion and with uncoupling between CBF and EEG activity. II – Comatose patient after head injury with intracranial space occupying lesion. III – Brain tumor patient

cerebral blood flow differently from a localized lesion of the brain stem which may lead to a low CBF [3, 7]. The loss of autoregulation in this group was found to be transient and recovery was the rule although an aggravation with decrease of rCBF could be observed.

In the second group, in which loss of autoregulation was also observed, the reduction of CBF appeared to be aggravated by brain displacement and compression, or by intracranial hypertension produced by a space occupying lesion. However, comparing the rCBF values of the second group with those of 10 brain tumor patients without coma, which had a mean rCBF of 39 ml/100 g/min $\pm$ 7 SD, it can be observed that decrease of cerebral blood flow is less pronounced when the expansion of the mass is gradual. In fact, since rCBF did not immediately increase after surgical operation, it seems that compression of cerebral vessels is not the only factor decreasing CBF in the second group.

Correlation between EEG and the mean rCBF values (Fig. 1), demonstrated uncoupling between EEG activity and cerebral blood flow in 3 posttraumatic comatose patients without signs of localized brain lesion.

The present study demonstrates that rCBF measurements can be used to judge the prognosis in states of coma.

References

1. FAZEKAS, J. F., and A. N. BESSMAN: Coma mechanisms. Amer. J. Med. **15**, 804 (1953).
2. FOG, M.: Autoregulation of cerebral blood flow and its abolition by local hypoxia and/or trauma. Scand. J. clin. Lab. Invest. Suppl. **102**, p. V-B (1968).
3. INGVAR, D. H., E. HÄGGENDAL, J. J. NILSSON, P. SOURANDER, I. WICKBOM, and N. A. LASSEN: Cerebral circulation and metabolism in a comatose patient. Arch. Neurol. **11**, 13 (1964).
4. KETY, S. S.: The physiology of the human cerebral circulation. Anesthesiol. **10**, 610 (1949).
5. LASSEN, N. A.: Cerebral blood flow and oxygen consumption in Man. Physiol. Rev. **39**, 183–238 (1959).
6. — The luxury perfusion syndrome and its possible relation to acute metabolic acidosis localized within the brain. Lancet II, 1113 (1966).
7. REINMUTH, O. M., K. KOGURE, and S. SHIMOJYO: Total cerebral blood flow and metabolism in human brain stem disease. Neurology **18**, 280 (1968).
8. REIVICH, M., W. J. S. MARSHALL, and N. KASSELL: Loss of autoregulation produced by cerebral trauma. Presented at the International CBF Symposium, Mainz, April 1969 (this volume, p. 205).

Effect of High Concentrations of CO_2 on Cerebral Metabolism and EEG Activity

E. Betz, U. Knebel, L. Neumann, and H. Nguyen-Duong

Department of Applied Physiology, University of Tübingen

Wyke [1] demonstrated that the EEG-activity in artificially ventilated animals ceases during inhalation of 60–80% CO_2 in oxygen. In order to study the relations of the EEG and the energetic potential of the cerebral cortex we conducted the following experiments:

1. 10 cats anesthetized with barbiturates were artificially ventilated with a gas-mixture of O_2 and CO_2. Carbon dioxide was increased in the inspired air until the EEG disappeared. The electrical silence was maintained for $1^1/_2$ h. The behavior of the animals was observed when they woke up after the experiment.

2. In a second series of experiments in 5 cats the skull was opened on both sides of the sagittal sinus and local cortical blood flow, extracellular local cortical pH and local cortical pO_2 was recorded simultaneously. Arterial blood pressure, EEG and the ECG were controlled during the increased CO_2 concentration in the inspired air which was maintained high for 1 h.

3. During the CO_2 inhalation we excised small pieces of cortical tissue with a special steel punch, cooled in liquid air as described by Schmahl et al. in 1966, and analyzed the energy rich phosphates, lactate and pyruvate of the cortical tissue with enzymatic-optic methods.

Results

When the CO_2 concentration in the inspired air was increased to 70% or 80% the cortical blood flow was initially increased. During the inhalation period, CBF followed passively the variations of arterial blood pressure. P_{art}, however, reacted different in the various animals. In most cases we were able to maintain the arterial blood pressure values higher than 80 mmHg for more than $1^1/_2$ h when the CO_2 in the inspired air was higher than 70%. In some other experiments, however, the P_{art} decreased transiently below this critical level, and the CBF fell also.

The simultaneously measured cortical pO_2 was usually high because of the high O_2 content in the inspired gas mixture. In the experiments in which the blood pressure decreased below a mean pressure of 80 mmHg the cortical pO_2 also sometimes decreased below the initial value. These cases were excluded from the analysis. In all experiments the A-V difference (between arterial blood and the blood in the sagittal sinus) became reduced after about 15 min of CO_2 inhalation. At the same time EEG activity also ceased. It therefore can be concluded that the oxygen consumption of the cortex decreases during the isoelectric phase of the EEG. During the exposure to CO_2 the EEG frequency shifts towards slow frequencies. The EEG disappeared completely between 50 and 70% CO_2 in the inspired air. The cortical pH at which the EEG disappeared was between 6.5 and 6.3. This means that there is not a very strong correlation between extracellular pH and the disappearance of the EEG. During inhalation of 70% CO_2 the cortical pH was 6.3 in the mean of 6 experiments.

The heart rate of the cats became slower during the CO_2 inhalation. In the electrocardiogram QRS and the repolarization phase were prolonged. In 4 of 20 experiments the ST segment was elevated and T became negative. When the pCO_2 was normalized at the end of the exposure severe disturbances in the repolarization were observed in nearly all animals. Fig. 1 gives an example of the course of the EEG and the ECG during an exposure to CO_2. The phase of the normalization was the most critical one during the experiments. In most cases the ECG was normalized again after 5–20 min, when the animals recieved normal air or oxygen. In 4 experiments the cats reacted with a cardiac arrest of several minutes duration when CO_2 was brought back to normal values.

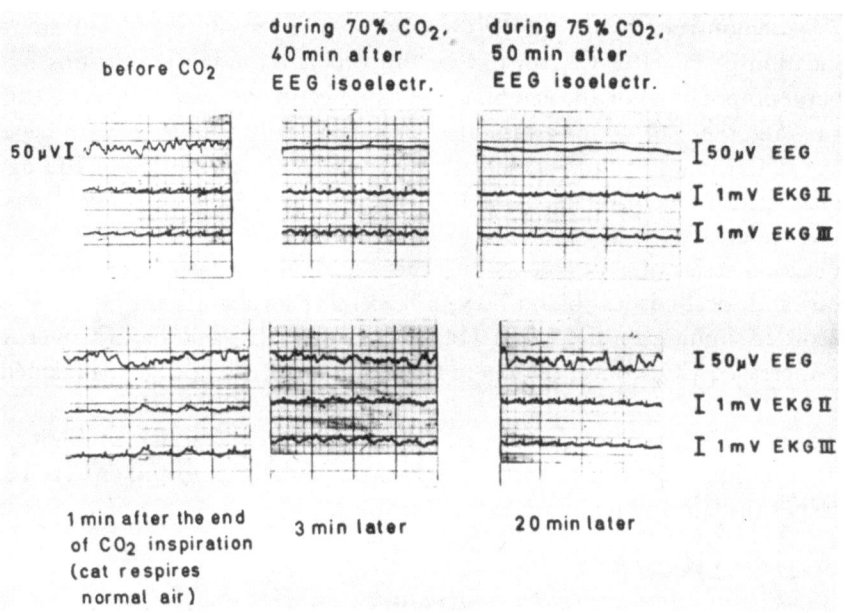

Fig. 1. Reactions during a short-term exposure to high CO_2 in the inspired air

The EEG normalized very rapidly after the end of exposure to CO_2. The energy rich phosphates, lactate and pyruvate were analyzed at different time intervals after the onset of the CO_2 inhalation. The values 1 h after the onset of the inhalation are shown in Figure 2f Initially, ATP increased somewhat but after 1 h of exposure to high CO_2 ATP and Cr P were not significantly altered, when compared with controls. The quotients ATP/ADP and Cr P/Cr did not show the same tendency as the absolute values. The apparent creatinkinase equilibrium was not altered during the CO_2 exposure. Lact/Pyr was insignificantly lower than the control values.

In 10 animals no cortical tissue was excised. 8 animals in this group survived a $1^1/_2$ h cessation of the EEG. The anesthetized state lasted some hours longer than that of the controls. The behaviour of these animals showed no striking alteration when they woke up. In 1 animal with a 10 min cardiac arrest we observed signs of ataxia and the brain showed histologically diffuse destructions of nerve cells in the cerebellar and cerebral cortex.

In conclusion, it can be said that during an increasing cortical acidosis the EEG frequency spectrum shifts independently of the tissue content of energy rich substrates. Initially the cortical ATP increases similarly as in deep anesthesia.

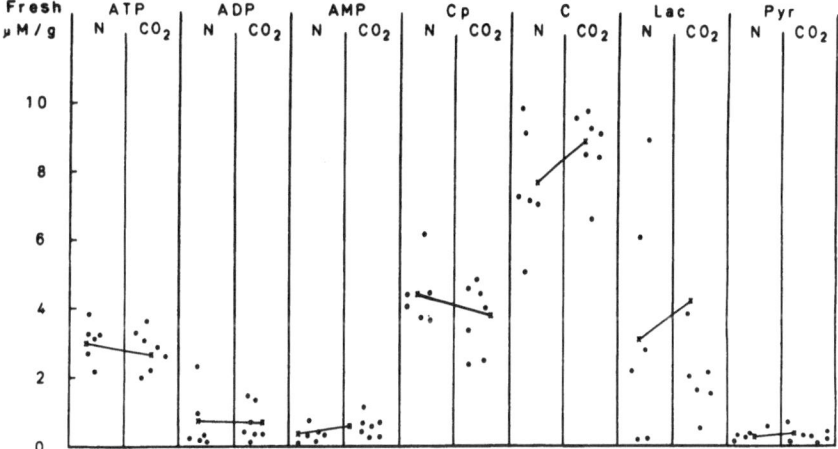

Fig. 2. Energy-rich phosphates and lactate/pyruvate in cats. N = normals; CO_2 = exposure for 1 h to 70 % CO_2 with arrest of the EEG

Because of the differences in the relations between ECG and the pH of the myocardium, the heart had to be supervised very carefully in these experiments in order to prevent cardiac arrest and secondary cerebral anoxia.

Reference

1. WYKE, B.: Brain functions and metabolic disorders. London: Butterworths 1963.

Cerebral Vasomotor Control and CSF pH in Metabolic and Respiratory Coma

C. Fieschi, A. Agnoli, N. Battistini, M. Nardini, and M. Prencipe

Department of Neurology and Psychiatry, University of Genoa

Regional CBF was measured in the resting state and during vasomotor tests, along with CSF acid-base status, in 15 patients comatose due to diffuse brain lesions (Table 1). Fig. 1 illustrates 1 patient, showing the data obtained in a typical study. Flow was measured in 5 regions with ^{85}Kr, analyzing the entire clearance curves; the arterial blood was repeatedly sampled, and CSF was sampled at the end and/or at the beginning of the study, for pCO_2, pH and pO_2 determinations. Clearance curves were obtained in the resting state, and during different stimuli. In this case (hepatic coma with a slight metabolic acidosis and marked respiratory alkalosis; flow slightly reduced) the stimuli were mannitol infusion followed by an induced increase in arterial pressure. Both stimuli revealed a diffuse loss of autoregulation. In some other cases CO_2 inhalation or hyperventilation have been tested.

The observations in our case material will be analyzed in the light of the 2 main theories of the mechanisms of regulation of cerebral blood flow in states of coma. On the one hand, the "classical" concept assumes that the cerebral perfusion rate is a function of the metabolic

F.C. n. 228, hepatic coma

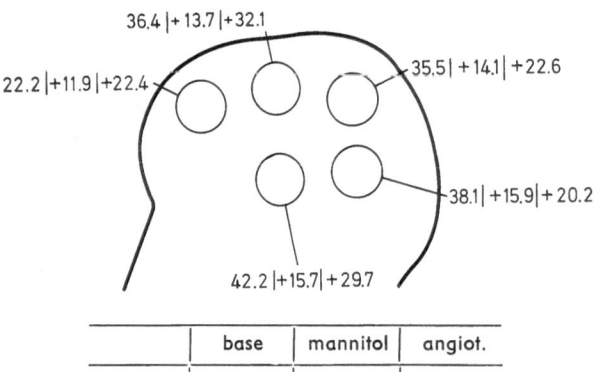

	base	mannitol	angiot.
MABP	98	97	108
aPCO$_2$	21.8	20.4	21.8
apH	7.51	7.48	7.46
CSF pH		7.43	

Fig. 1. The numbers represent the mean regional blood flow (^{85}Kr intracarotid injection, compartmental analysis) in ml/100 g/min, and its absolute changes during functional tests (in this case mannitol and angiotensin infusion). Changes in MABP, aPCO$_2$, apH during the tests, and CSF pH in the resting state are indicated at the bottom of the figure

Supported by CNR contract No. 115/2422.

Table 1

	Number of patients	Arterial blood						CSF			CBF	MABP	CVR	Regulation
		Ht	pH	pCO_2	pO_2	SO_2%	HCO^-_3	pH	pCO_2	HCO^-_3				
Respiratory acidosis	5	46.0	7.294	75.2	41.6	71.4	36.0	7.234	80.6	32.1	60.91	97.4	1.6	Impaired
(treated)	4	46.0	7.427	41.0	63.5	91.7	26.6	7.317	52.0	25.0	35.62	92.7	2.6	Impaired
Metabolic acidosis	4	41.0	7.285	29.0	82.7	94.6	14.1	7.173	31.8	11.2	45.17	99.0	2.2	Impaired
Metabolic alkalosis	2	43.5	7.545	56.5	61.0	86.4	48.3	7.498	59.0	42.4	27.22	80.0	3.0	Impaired
Hepatic coma	2	45.0	7.510	26.0	65.0	92.2	21.0	7.385	34.7	19.4	42.70	95.5	2.3	Impaired
Post-anoxic coma	2	41.5	7.380	41.0	57.5	95.7	25.9	7.320	50.5	25.0	33.60	81.5	2.4	Impaired

A.C. n. 250, metabolic acidosis

	base	angiot.
MABP	77.0	90.0
aPO2	98.0	89.0
aPCO2	22.5	20.7
apH	7.22	7.26
CSF pH	7.14	

$51.3 | +3.3$
$69.3 | -9.3$
$47.1 | +9.2$
$55.4 | +4.8$
$61.3 | -4.3$

Fig. 2. For explanation see Fig. 1

B.A. n. 222, metabolic acidosis

	base	angiot.
MABP	92	113
PaO2	84	87
PaCO2	30	31
aPH	7.27	7.26
CSF pH	7.195	

$43.2 | +69$
$41.5 | +11.7$
$41.2 | +15.6$
$41.4 | +6.8$
$37.3 | +5.2$

Fig. 3. For explanation see Fig. 1

T.B. n. 201, respiratory acidosis

	base A	hypert. A	base B
aPCO2	73	79	39
aPO2	41	44	65
CSF pH	7.26		MABP 100
MABP	107	133	CSF pH 7.31

$68.7 | +9.8$ 30.4
$66.6 | +14.3$ 31.1
$68.1 | +20.6$ 31
$61.9 | +20.0$ 32
$61.2 | +12.5$ 27.8

Fig. 4. For explanation see Fig. 1

and functional activity of the brain. Therefore, CBF should be reduced in coma, this reduction being related to the depth of coma rather than to its etiology [3]. On the other hand, the modern theory attributes to the concentration of H^+ in the ECF of the brain the role of the main "signal" or determinant of the tone of the arteriolar smooth muscle [1, 4]. Therefore a reduction in functional and metabolic activity influences the CBF through a reduction in hydrogen ions concentration of the ECF.

This however is a final common pathway, that may be independently affected by several other factors. Any time a disturbing factor of systemic origin affects the acid-base status of the brain – monitored, in clinical cases, by the CSF pH [5] – the relationship between reduction in flow and degree of impairment of the state of consciousness may be lost.

The first known example of such uncoupling was reported more than 20 years ago by Kety et al. [2], who in severe acidosis (diabetic) coma (arterial pH < 7.00) recorded a CBF 20% higher than normal. This uncoupling was unexplained until the "Severinghaus electrode" theory of CBF control was advanced.

Recently however, a new kind of "uncoupling" has been noted, this time between CBF and CSF pH. It follows that no simple theory seems to hold entirely, and the problem of cerebral vasomotor control is still a matter of debate. Some examples chosen from our case material clarify our point of view on this matter: Case no. 250 (Fig. 2) is a patient with diabetic coma and severe metabolic acidosis, partially compensated by a respiratory alkalosis. The rCBF is diffusely increased: in this case the flow is uncoupled from the state of consciousness (and, as a matter of fact, also from arterial pCO_2) and follows the CSF pH. A moderate increase in pressure obtained by intravenous infusion of angiotensin is accompanied by some further decrease in a pCO_2 and hydrogen ion concentration. Nevertheless, the flow increases showing a definite impairment of autoregulation in 2 regions.

In case no. 222 (Fig. 3) the clinical conditions are worse: the EEG and state of consciousness are more severely impaired and the CSF pH, which was 7.195 at the time of blood flow measurements, was lower (7.08) the day before, when treatment was started. Despite this low pH the CBF is moderately decreased, being thus dissociated from CSF pH.

We submit that such a dissociation between low CSF pH and low flow is an indication of a severe encephalopathy with brain edema. Vasoparalysis is present in this case despite no obvious vasodilatation as shown by the resting CBF values. The vasoparalysis is revealed by the diffuse impairment of autoregulation. Case 201 (Fig. 4) is a respiratory encephalopathy (chronic hypoxic hypercapnia). With moderately reduced CSF pH, drowsiness and diffuse EEG abnormalities, the flow is increased (following CSF pH rather than cerebral metabolic demands) and autoregulation is lost. After treatment, when the state of consciousness, EEG and clinical conditions have returned to normal, the flow is reduced: again, flow dissociates from the state of consciousness and is related to CSF pH (and to pCO_2, in this case).

Case 195 (Fig. 5) is another patient with a respiratory encephalopathy. The hypoxic hypercapnia is more pronounced, as well as the neurological and EEG abnormalities, and CSF pH is more acidotic than in the previous case. In contrast, CBF is within normal limits, and does not follow the CSF acidosis. Autoregulation has not been tested but it was probably impaired, since we found it so when, after treatment, the clinical and acid-base status were almost normalized. In summary, these results seem to indicate that:

1. The CSF pH does reflect an important system controlling the steady state cerebral blood flow level. Actually, when in a metabolic or respiratory acidosis the CSF pH is not sufficiently protected and shifts towards acidotic values, the CBF tends to increase above normal in

presence of a disturbed state of consciousness and reduced cerebral metabolic demands (uncoupling between CBF and neuronal activity).

2. The same conditions inducing a CSF acidosis however predispose to brain edema (BE), with increased extravascular resistances. When these are increased markedly, in the presence of an impaired autoregulation, the CBF decreases below normal reaching very low values (uncoupling between CBF and CSF pH in cases of BE).

3. For practical purposes any attempt to interpret rCBF findings in pathological terms must take into account the fact that an important component of CVR in disease may depend on extravascular factors: namely BP and CSF pressure.

4. Even if CBF measurements may be interfered with by these factors, the cerebrovascular reactivity tested by changing the arterial pCO_2 level, the arterial pressure or the intracranial pressure with osmotic agents reveals the existence of a vasoparalysis and indicates the degree of impairment of the vasomotor regulatory mechanisms.

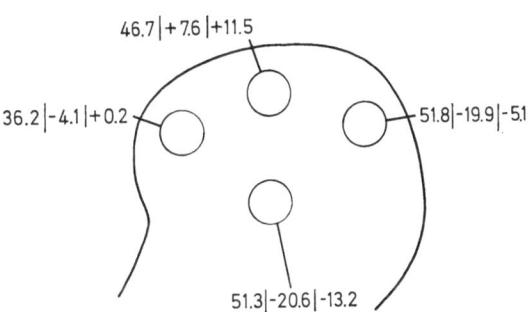

R.A. n. 195, respiratory acidosis

	base 1	base 2	angiot.
MABP	95	133	93
aPCO$_2$	71.0	36.0	36.0
aPO$_2$	39.0	67.0	67.0
apH	7.27	7.45	7.45
CSF pH	7.21	7.34	

Fig. 5. For explanation see Fig. 1

References

1. BETZ, E., and D. HEUSER: Cerebral cortical blood flow during changes of acid-base equilibrium of the brain. J. appl. Physiol. **23**, 726 (1967).
2. KETY, S. S., B. D. POLIS, C. S. NADLER, and C. F. SCHMIDT: The blood flow and oxygen consumption of the human brain in diabetic acidosis and coma. J. clin. Lab. Invest. **27**, 500 (1948).
3. — The cerebral circulation. In: Handbook of physiology, Neurophysiology III, 1751 (1960).
4. LASSEN, N.: Brain extracellular pH: the main factor controlling cerebral blood flow. Scand. J. clin. Lab. Invest. **22**, 247 (1968).
5. POSNER, J. B., A. G. SWANSON, and F. PLUM: Acid base balance in cerebrospinal fluid. Arch. Neurol. **12**, 479 (1965).

CBF and Metabolism in Uremic and Hepatic Precomatose States

U. Gottstein, K. Held, and W. Berghoff

Department of Internal Medicine, University of Kiel

Psychiatric and neurological disturbances are often the leading symptoms in metabolic diseases. Frequently the reason cannot be well defined, as is the case in Wernicke's encephalopathy, in which the glucose metabolism is reduced due to a disturbance of oxydative decarboxylation of pyruvate and α-ketoglutarate as a consequence of thiamine deficiency.

This review of 6 cases from the literature [3, 4, 13] and 3 measurements of our own shows equally the strong reduction of the cerebral oxygen and glucose consumption.

In uremia and hepatic precoma the relations are not that clear and the reason for mental disturbance is not yet well known. Therefore we have done measurements of the cerebral metabolism to have a second look on the results of earlier authors, who had to work with nonspecific methods of glucose determination. Our measurements were made with the specific hexokinase-glucose-6-phosphate-dehydrogenase reaction for determination of glucose and with lactate-dehydrogenase for lactate and pyruvate [7].

In uremics (Fig. 1) we found a statistically significant decrease of cerebral O_2- and glucose consumption while the CBF lactate- and pyruvate delivery did not show any significant change.

With these findings the results of the earlier authors could be confirmed (Fig. 1). In all 44 cases reported the CBF was normal (60.6 ml/100 g/min), while the O_2 uptake was significantly decreased to 2.4 ml/100 g/min as well as the glucose uptake to 3.5 mg/100 g/min.

Correlating the results to the mental state, there is a good relation: O_2 and glucose uptake are normal in patients without psychiatric or neurological symptoms, however, the values are reduced in uremic precoma and coma.

In order to look for the cause of this correlation we brought the results into relation to urea-N and arterial pH. While there was no correlation to the urea-N there was a definite one to the pH. The more acid the pH, the lower were the cerebral O_2 and glucose-uptake values. Of interest in this connection is a result of Kety [9], who found nearly identical results in diabetic acidosis. Kety also found a decrease of oxygen-uptake with the decrease of pH. As in our measurements his O_2-uptake values were 2.0 ml/100 g/min at a pH of 7.0–7.2.

Corresponding to this interesting result there is the clinical observation that in uremia nearly only acidotic patients become comatose, and that the mental state improves when acidosis has been treated, even if the urea-N can not be lowered. Merrill et al. [10] have dialyzed chronic uremic patients against a urea-N containing solution, so that the BUN was not changed. Nevertheless the psychiatric symptoms of uremia disappeared.

The explanation for these phenomena is possibly in the extreme susceptibility to acid of the hexokinase, which is necessary for the phosphorylation of glucose. At a pH of 7.25 the hexokinase of the erythrocytes, for example, has only 30% of its normal activity [2].

The results of previous studies could also be confirmed in 12 patients with hepatic insufficiency. FAZEKAS et al. [5] first reported that in patients with liver cirrhosis cerebral O_2-uptake was diminished. When psychiatric symptoms or coma developed O_2-uptake was especially severely reduced. ERBSLÖH [3], SCHEINBERG [12] and our group also found a significantly reduced O_2 and glucose consumption in patients with psychiatric and neurological symptoms during hepatic precoma. These 3 teams found a disproportional severely reduced glucose-uptake, i. e. the quotient of glucose consumption to oxygen consumption decreased in a statistically significant way. In the measurements of SCHEINBERG it decreased to 1.39, in those of ERBSLÖH and ours to 1.07.

The explanation for these phenomena cannot be given at present. We are now controlling these results also in animal experiments and we are searching for biochemical explanations, for example a cerebral utilization of noncarbohydrates.

Naturally one might think that the reason for cerebral metabolic disturbance in such cases is an ammonia-intoxication, but our measurements did not show any correlation between the blood-ammonia level and the values of O_2 and glucose consumption or the mental state. We know, however, that several toxic phenol bodies are increased in the blood in cases of liver insufficiency and that as an expression of a disturbance of the oxydative metabolism of pyruvate acetoin and 2.3-butylenglycol are increased [14]. Pyruvate and α-ketoglutarate were also found to be severely increased in the spinal fluid of patients with hepatic coma. Thus, it can be discussed whether intracerebral pyruvate-metabolism is also disturbed. The cause for such a disturbance is unknown. GEIGER and MAGNES [6], on the basis of animal experiments, believe that the normal liver liberates a substance which is necessary for a normal cerebral metabolism.

Our results remind us that the brain has, as we all know, a good protective mechanism against many circulatory disturbances, but that it is quite vulnerable as concerns its metabolism.

Table 1. *Brain Metabolism in Uremia*

CBF ml/100 g/ min	O_2-consumption ml/100 g/ min	Glucose consumption mg/100 g/ min	Urea N mg %	Hb g %	BP mmHg	n	Authors
57.0	2.2	—	159	9.7	100	7	HEYMAN et al.
45.0	2.1	—	158	10.2	166	9	HEYMAN et al.
74.0	2.6	3.5	?	?	117	7	SCHEINBERG et al.
78.0	2.5	3.8	?	?	?	10	BIANCHI PORRO et al.
49.6	2.5	3.3	140	8.3	126	11	GOTTSTEIN et al.
60.6 *55–62*	*2.4* *3.1–3.9*	*3.5* *5.3–6.2*				*44*	Total

References

1. BIANCHI PORRO, G., A. T. MAIOLA e M. DELLA GRAZIA: Effetti del trattamento emodialitico cronico sul metabolismo cerebrale. Gaz. Sanitaria **38**, 10 (1967).
2. BOCK, H. E., H. D. WALLER, W. KAUFMANN u. G. W. LÖHR: Experimenteller Beitrag über die Entstehung renaler Anaemien. Dtsch. Arch. Klin. Med. **211**, 73 (1965).
3. ERBSLÖH, F., A. BERNSMEIER u. H. R. HILLESHEIM: Der Glukoseverbrauch des Gehirns und seine Abhängigkeit von der Leber. Z. ges. Neurol. **196**, 611 (1958).

4. Fazekas, J. F., and A. N. Bessman: Coma mechanisms. Amer. J. Med. **15**, 804 (1953).

5. —, H. E. Ticktin, W. E. Ehrmantraut, and R. W. Alman: Cerebral metabolism in hepatic insufficiency. Amer. J. Med. **21**, 843 (1956).

6. Geiger, A., J. Magnes, R. M. Taylor, and M. Veralli: Effect of blood constituents on uptake of glucose and on metabolic rate of the brain in perfusion experiments. Amer. J. Physiol. **177**, 138 (1954).

7. Gottstein, U., A. Bernsmeier u. I. Seldmeyer: Der Kohlenhydratstoffwechsel des menschlichen Gehirns. Klin. Wschr. **41**, 943 (1963).

8. Heyman, A., J. L. Patterson, and R. W. Jones: Cerebral circulation and metabolism in uremia. Circulation **3**, 558 (1951).

9. Kety, S. S., L. D. Polis, C. S. Nadler, and C. F. Schmidt: The blood flow and oxygen consumption of the human brain in diabetic acidosis and coma. J. clin. Lab. Invest. **27**, 500 (1948).

10. Merrill, J. P., E. Sabbaga, L. Henderson, W. Welzant, and C. Crone: Trans. Amer. Soc. Artif. Int. Org. **8**, 252 (1962).

11. Scheinberg, P.: Effects of uremia on cerebral blood flow and metabolism. Neurology **4**, 101 (1954).

12. — Brain metabolism in liver disease. Wld. Neurology **2**, 288 (1961).

13. Shimojyo, S., P. Scheinberg, and O. Reinmuth: Cerebral blood flow and metabolism in the Wernicke-Korsakoff Syndrome. J. clin. Lab. Invest. **46**, 849 (1967).

14. Thölen, H., u. F. Bigler: Zur Pathogenese und Therapie des endogenen Lebercomas. Schweiz. med. Wschr. **93**, 1836 (1963).

Energy Metabolism, Lactate/Pyruvate Ratio and Extracellular Space in Cortex and White Matter Adjacent to and Distant from a Local Freezing Lesion

M. Samii, H. J. Reulen, F. Fenske, U. Hase, and K. Schürmann

Department of Neurosurgery, University of Mainz

The present investigation was undertaken to study the distribution of pathological fluid accumulation and the local energy metabolism in the cortex and white matter following a localized traumatic lesion of the cortex. A systematic examination of the changes in the tissue following a trauma is the basis for further studies on the relationship between cortical blood supply, local tissue metabolism and local function.

A localized cold lesion was inflicted to the right temporoparietal cortex of dogs under general anaesthesia. Edematous areas were outlined by vital staining with Trypan blue or Geigy blue. 24 h later, water and sodium contents, as well as CrP, ATP, glucose, pyruvate and lactate were measured in the control hemisphere and in the stained edematous areas, as well as in the unstained distant areas of the experimental hemisphere.

In the animals studied the lesion was surrounded by a faintly bluish stained area. The subjacent white matter was deeply stained, in the manner Klatzo et al. have described. In the bluish cortex adjacent to the lesion, there was a slight but significant increase in the water and sodium content. The water content increased by approximately 1% and the sodium content by 52 mEq/kg dry weight as compared with the control hemisphere. In the subjacent stained white matter, water and sodium content increased considerably more than in the cortex. The increase in the water content amounted to approximately 8% and the sodium content to about 170 mEq/kg dry weight, as compared to the control hemisphere and the unstained distant areas of the injured hemisphere.

In order to determine whether the pathological fluid accumulation involved the extra- or intracellular space, the extracellular space in the brain tissue was measured in these studies. The measurement was performed by means of a newly developed method, using the combination of an extracellular tag administered intravenously and a ventriculo-subarachnoido-cisternal perfusion. Sodium-S^{35}-thiosulfate was used as the extracellular tag.

The distribution of thiosulfate is constant in all areas after 180 min of perfusion. It is interesting to note that, using this new method, the extracellular space amounts to about 10% in the normal cortex and to about 8% in the normal white matter, as has been shown also recently by Oldendorff and Davson. Thus, it is considerably larger than was formerly assumed on the basis of electron-microscopical investigations.

In the perifocal bluish cortex there is a definite increase in the extracellular space as opposed to a slight decrease in the distant cortex. Accordingly, the pathological fluid accumulation is mainly extracellular in the perifocal cortex and intracellular in the distant areas. In the white matter, there is enlargement of the extracellular space in both the stained and unstained areas, so that here it is chiefly a matter of extracellular edema.

The question of whether the oxygen supply in edematous regions, that is, regions with increased water content, is sufficient, is also of special interest. Since at present there is no method available for the determination of oxygen consumption in individual cortical areas, we have performed an analysis of the energy-rich phosphates and of the lactate/pyruvate ratio in the edematous tissue, since these metabolites immediately show charateristic changes with the onset of tissue hypoxia.

A comparison of the control cortex with the edematous perifocal cortex showed that the latter is characterized by a marked reduction of CrP, ATP and glucose. The changes in lactate content and in the lactate/pyruvate ratio were not uniform. Compared with the control cortex, there was an increase in the lactate content in some animals, whereas a decrease was registered in others. On the average, however, there was a slight increase in the lactate level and the lactate/pyruvate ratio.

These changes are characteristic for oxygen deficiency in edematous tissue. The reduction of glucose and the accumulation of lactate must be seen as a cellular attempt to compensate for the oxygen deficiency through an intensified anaerobic production of energy. Since glycolysis cannot provide the entire energy requirement, the cells use up their energy compounds, chiefly CrP and ATP.

The question of the differing lactate contents in the edematous cortex arises. We suppose that such a varying lactate level is due to the fact that tissue biopsies were taken at varying distances from the lesion. Perhaps we will arrive at an explanation if we perform studies of the present type together with measurements of cortical blood flow.

Regional Cerebral Blood Flow in Chronic Alcoholism

D. H. Ingvar, J. Risberg, S. Cronqvist, U. Zätterström, L. Gustavsson, and K. Ljungberg

Departments of Clinical Neurophysiology, Neuroradiology, and Psychiatry I, University Hospital, Lund

Introduction

Several years ago Battey et al. [1] and Fazekas et al. [3] demonstrated with the Kety method that the cerebral oxygen uptake and blood flow is reduced in demented chronic alcoholics. This finding suggests similarities between this disorder and organic dementia of the senile and presenile type in which there is also a reduced cerebral metabolism and blood flow ([4] and others) which is proportional to the intellectual impairment [7, 9, 11].

However, it is not known if a reduced CBF is always present in chronic alcoholism, whether the reduction is general or focal, or whether the flow decrease correlates to the mental symptoms. These problems were the object of the present study with the ^{133}Xenon clearance rCBF technique.

Materials and Methods

21 men, 29–66 years old (mean 50.4), were studied. They had all abused alcohol for at least 5 years. All, except 3 cases of delirium, underwent the rCBF measurement (with some sedative medication) in a "dry" state. Apart from the alcoholism, the histories of the patients were in general negative. There were no severe head injuries. 7 patients showed, however, signs of hepatic insufficiency. One of them suffered from hepatic cirrhosis. There were no reports of apoplexy, and all patients showed normal neurological findings. Many of them had had epileptic attacks of the abstinence type. The EEG showed mild diffuse slowing in 3 patients, and in 4 some low voltage 3–6 cps episodes with predominance on the left side. No paroxysmal changes were recorded. Carotid angiography (in conjunction with the rCBF study) showed normal findings. Pneumoencephalography revealed slight cerebral atrophy in 7 cases.

All patients presented the typical clinical picture of chronic alcoholism with a more or less reduced intellectual capacity, as well as affective disturbance. They were submitted to psychological tests of verbal performance, memory, spatial orientation, persistence, cognitive adaptability, etc. For the present purpose the test results were summarized into 4 classes: 1. Normal findings. 2. Memory disturbance, or defective Gestalt perception, or reduced

Supported by the Swedish Medical Research Council (Contracts B 6821 × 8404 and B 6921 × 8405 A), Riksbankens Jubileumsfond, and the Wallenberg Foundation, Stockholm. Dr. B. Lizelius-Lanke, head of the Alcohol Clinic, Malmö General Hospital, is thanked for valuable cooperation.

psychomotor and/or cognitive reduction. 3. Moderate general cognitive reduction. 4. Severe general cognitive reduction. These groups are denoted by a, b, c, and d in Table 2.

Table 1. *Mean rCBF variables in chronic alcoholism* (21 cases)

	n	f_g	f_w	W_g	$f_{10'}$
Alcoholics	21	61.1 ± 13.3	16.7 ± 2.7	48.5 ± 5.1	44.4 ± 10.5
Healthy controls	7	79.9 ± 8.3	20.9 ± 2.1	49.2 ± 2.0	49.8 ± 4.1
Signific. *t*-test		$p < 0.001$	$p < 0.001$	n. s.	$p < 0.01$

Table 2. *Clinical correlations (18 patients)*

18 alcoholics with	Psychological defects				Cerebral atrophy	Abnormal EEG	Hepatic insuffic.
	a	b	c	d			
Normal or high f_g	—	2	4	2	4	3	1
Subnormal f_g	1	2	3	4	3	4	6
Totals	17				7	7	7

One patient died of coronary thrombosis about 2 months following the rCBF study. Autopsy of the brain did not reveal any significant abnormalities and the cortical morphology was well preserved, also in the temporal region where the lowest f_g was found. This patient showed a significant reduction of mean f_g (58.7 ml/100 g/min), but the W_g was normal (51.3%), as well as the f_w and $f_{10'}$ (17.7 and 44.6 ml/100 g/min, respectively).

Measurements of rCBF were made on the left side (except in 1 case) with the [133]Xenon residue detection method, using an 8 detector unit with a magnetic core memory [8]. Calculations were made for each region of noncompartmental 10 min flow ($f_{10'}$), grey matter flow (f_g), white matter flow (f_w), and relative weight of grey matter (W_g). The mean arterial pCO_2 was 36.5 mmHg. Only 1 patient (the one with hepatic cirrhosis) showed an excessively low value (26.0 mmHg).

Results

The group as a whole showed a significant reduction of f_g and f_w ($p < 0.001$), and a less significant $f_{10'}$, reduction ($p < 0.01$). The total mean W_g was within normal limits (Table 1). Very marked variations were found in the flow values with some "hyperemic" cases with a mean f_g for the 3 highest of 94.1, 83.5, and 76.0 ml/100 g/min (corresponding f_{10}: 67.8, 64.5, and 56.4). The 3 lowest had f_g:s of 47.5, 41.1, and 38.8 ml/100 g/min (corresponding $f_{10'}$:s: 27.6, 33.8, and 27.8). The range of variation thus comprised values from almost half of the normal flow, and up to about a 35% increase.

The intra-individual variation of the rCBF values were in general small, i. e. the deviations of the regional values from the mean of a given patient were less than about 10% in the vast majority of the measurements. In 10 cases a "focal" reduction of f_g (i. e. a deviation of more than 15% from the patient's mean) was recorded in the temporal region. In view of the fact that isotope recirculation in this region, according to observations in our laboratory, may cause about a 10% understimation of the rCBF, no significance can at present be attributed to this finding. It might be mentioned though, that a memory deficit was recorded in at least 5 of the 10 cases with a temporal "focus".

A summary of the clinical correlations, except for the delirium cases, is shown in Table 2. The main finding was a lack of correlation between the psychological deficits and the rCBF variables (only f_g shown in Table 2). Thus, all 8 patients with normal or high ("hyperemic") f_g showed psychological deficits, which in 6 was of the general type. 9 of the 10 patients with subnormal f_g showed abnormalities in the tests, and in 5 the deficits were of the general type.

Cerebral atrophy (7 cases), and EEG abnormalities (7 cases) were also about equally distributed in the 2 groups, while signs of hepatic insufficiency (7 cases) were all found in the group with subnormal f_g, except in 1 case.

The 3 delirium patients (mean age 39.3 years) all showed classical symptoms with confusion and hallucinations. They were studied in the predelirium stage and had all been given adequate medication (Hemineurine), a fact probably contributing to their low mean rCBF values: f_g 52.2, f_w 14.8, f_{10} 37.3 ml/100 g/min, and W_g 47.2%). EISENBERG [2] has recently found similar low CBF values in delirium.

Discussion

The markedly wide distribution of the rCBF values recorded in the present series of alcoholics cannot, it seems, be explained as due to age differences, since this variable was relatively uniform with only 2 patients under the age of 45 years (both delirium cases). Nor does arterial pCO_2 variations explain it. The 3 "hyperemic" cases, as well as 2 of those with the lowest f_g:s had a normal pCO_2. The two patients with cirrhosis had, however, both the lowest mean f_g (38.8 ml/100 g/min) and pCO_2 (26.0 mmHg).

The most important finding of the present study is that no correlation was found between the rCBF variables and the psychological defects in chronic alcoholism. Large such defects of the classical type were, in fact, found amongst the patients in which the rCBF was considerably elevated, as well as in those with a marked reduction. This places dementia due to alcohol abuse in another cathegory than organic dementia of the senile or presenile type in which CBF (ref. above) and rCBF ($f_{10'}$; [7, 11]) correlates well to the intellectual impairment. Another difference is that the W_g variable which showed a high correlation (r 0.95) to the psychological defects in senile dementia (OBRIST et al. 1969) did not relate to the defects found in the present series of alcoholics.

The presence of reduced function of the central nervous system in alcoholism without a proportional reduction of the cerebral blood flow – and especially the finding of cerebral "hyperemia" in some cases – suggest that chronic alcohol abuse may upset the normal metabolic regulation of the cerebral blood flow, in other words, that luxury alcohol consumption may, for as yet unknown reasons, in certain patients produce cerebral "luxury perfusion" [10]. The high perfusion in some alcoholics studied in a "dry" state may be related to the increased rCBF found in animals during acute alcohol intoxication [5]. A high flow was not, however, found in the 3 delirium cases studied a short period following cessation of alcohol intake, but the reduced rCBF which they showed may be explained by the medication given.

In conclusion, the present study raises 2 important questions. First, do the psychological defects in chronic alcoholism correlate better to $CMRO_2$ or other cerebral metabolic indices, than to mean rCBF variables? Second, what is the relation in time of the "hyperemic" state to the state with reduced rCBF, i. e. does the high flow state precede, or perhaps interchange with, the reduced flow state? These problems will be dealt with in our further studies which will include longitudinal observations of some of the patients presented here.

Summary

Twenty one chronic alcoholics were studied with an 8 detector rCBF technique. The mean grey matter flow was high in 3 cases, about normal in 5, and subnormal in 13. Three of the subnormal cases were studied under medication in predelirium. 7 patients showed signs of hepatic insufficiency and 6 of those had subnormal grey matter flow. All patients, except 1, showed psychological defects, which in 11 was considerable. These defects were found at all flow levels, and rCBF thus did not correlate to the type or degree of intellectual impairment. Alcohol dementia consequently differs from senile and presenile dementia in which the cerebral blood flow is reduced in proportion to the psychological deficit. The occurence of markedly high as well as low rCBF in mentally reduced alcoholics indicates that the normal metabolic regulation of the cerebral blood flow is deficient in this disorder.

References

1. BATTEY, L. L., A. HEYMAN, and J. L. PATTERSON, JR.: Effects of ethyl alcohol on cerebral blood flow and oxygen consumption. J. Amer. Med. Ass. **152**, 6 (1953).
2. EISENBERG, S.: Cerebral blood flow and metabolism in patients with delirium tremens. Clin. Res. **16**, 71 (1968).
3. FAZEKAS, J. F., S. N. ALBERT, and R. W. ALMAN: Influence of chlorpromazine and alcohol on cerebral hemodynamics and metabolism. Amer. J. Med. Sci. 230, 128 (1955).
4. FREYHAN, F. A., R. B. WOODFORD, and S. S. KETY: Cerebral blood flow and metabolism in psychosis of senility. J. nerv. ment. Dis. **113**, 449 (1951).
5. HADJI-DIMO, A., R. EKBERG, and D. H. INGVAR: Effects of ethanol on EEG and cortical blood flow in the cat. Quart. J. Stud. Alcohol **29**, 828 (1968).
6. INGVAR, D. H., S. CRONQVIST, R. EKBERG, J. RISBERG, and K. HØEDT-RASMUSSEN: Normal values of regional cerebral blood flow in man, including flow and weight estimates of grey and white matter. Acta neurol. scand. **41**, Suppl. **14**, 72 (1965).
7. —, W. OBRIST, E. CHIVIAN, S. CRONQVIST, J. RISBERG, L. GUSTAVSSON, M. HÄGERDAL, and G. WITT-BOM-CIGEN: General and regional abnormalities of cerebral blood flow in senile and "presenile" dementia. Scand. J. clin. Lab. Invest. Suppl. **102**, XII-B (1968).
8. —, T. LUNDMARK, J. RISBERG, E. VON SABSAY, U. BURKLINT, and S. SUNDELIN: Recording of multiple clearance curves by means of a magnetic core memory. Scand. J. clin. Lab. Invest. Suppl. **102**, p. XI-H (1968).
9. LASSEN, N. A., O. MUNCK, and E. R. TOTTEY: Mental function and cerebral oxygen consumption in organic dementia. Arch. Neurol. Psychiat. **77**, 126 (1957).
10. — The luxury perfusion syndrome of the brain: A condition of relative cerebral hyperemia occuring in a variety of acute brain disorders. Lancet II, 1066 (1966).
11. OBRIST, W. D., E. CHIVIAN, S. CRONQVIST, and D. H. INGVAR: Regional cerebral blood flow in senile and presenile dementia. Neurology, 1969, in press.

Cerebral Blood Flow and Metabolism in Senile Dementia

S. Hoyer

Department of Pathological Chemistry and General Neurochemistry,
University of Heidelberg

CBF was determined in 20 patients by the Kety-Schmidt technique [7] modified according to Bernsmeier and Siemons [3]. The average age of the patients was 73 years; all were hospitalized because of severe and progressive senile dementia. We measured the uptake of oxygen and glucose, the output of CO_2 and lactate and, in the amino acid auto-analyzer[1], the cerebral arteriovenous differences of taurine, urea, aspartic acid, threonine-aspargine, serine, glutamic acid, proline, citrulline, glycine, alanine, valine, cystine, iso-leucine, leucine, tyrosine, phenylalanine, ammonia, ornithine, lysine, and histidine by the method of Moore and Stein [9]. The results of our studies were compared with those from 15 healthy volunteers whose average age was 25 years. Statistical calculations were carried out by the Wilcoxon-test ($2 \alpha = 0.05$).

Under physiological conditions the human brain gains energy almost exclusively from the oxydation of glucose [4]. About 67% of the glucose is metabolized into amino acids in the Krebs cycle which is extended by the metabolites glutamic acid and γ-aminobutyric acid. About 33% of the glucose is oxidized directly to CO_2. The brain consequently synthesizes the amino acids it constantly requires from glucose. These are only results of experiments in animals [1, 2, 10].

The results of our studies of CBF and metabolism in 15 volunteers agreed with the results of other authors. In regard to the amino acids we found a small uptake and a small release of amino acids. Statistically, however, there were no differences between the concentrations of amino acids in arterial and venous blood. The small uptake of ammonia is, however, remarkable. We found that the normal human brain releases minute amounts of amino acids [6]. That means that under normal conditions only a small exchange of amino acids occurs between brain and blood. That is, the amino acids taken up from the blood are not metabolized for energy.

The ratio between glucose minus lactate and oxygen may be used to differentiate 2 types of cerebral metabolic disorders. Normally the ratio is 1.34, theoretically 1.33 because 1 ml oxygen oxidizes 1.33 mg glucose. Numbers smaller than 1.34 signify disorder type I a hypoglykoxidosis, numbers larger than 1.34 signfy disorder type II a hypoxydosis. These 2 types show remarkable differences in their amino acid metabolism which was investigated in patients suffering from different diseases by Knauff et al. [8] (Table 1).

In hypoglycoxidosis the cerebral release of amino acids seemed to be normal. We found, however, a cerebral output of ammonia. Probably the release of amino acids has a pathological significance. Such a metabolic disorder represents no rigid entity, however, since in some patients we found an uptake as well as an ouput of amino acids from the brain.

Supported by the Strebel-Stiftung.

On the other hand the type of metabolic disorder with alterations suggestive of cerebral hypoxidosis seems to form an entity, since in this group we found a release of amino acids from the brain in all except 1 patient. It is possible that the glucose uptake was responsible for the increased release of amino acids; for amino acids constituted in the brain under these conditions cannot be used by the cerebrum.

Table 1

	Normals $n = 15$	I $n = 8$	II $n = 12$	
CBF	52.9	40.2[a]	42.7	ml/100 g/min
O_2-uptake	3.54	2.92	2.48[a]	ml/100 g/min
CO_2-output	3.77	2.78	2.63	ml/100 g/min
Glucose-uptake	4.97	2.53[a]	5.64	mg/100g/min
Lactate-output	0.36	0.40	−0.31[a]	mg/100g/min
Aminoacid avD	−0.033	−0.030	−0.143[a]	mmol/l
Ammonia avD	+0.008	−0.004[a]	−0.013[a]	mmol/l
RQ	1.06	0.97	1.05	
GOQ	1.34	0.71	3.09[a]	
Age	25	72	74	years

I = Hypoglycoxidosis; II = Hypoxidosis   + = uptake
[a] = Statistically significant differences   − = output (except lactate in II).

Contrary to the results of KNAUFF et al., who found especially a cerebral uptake of glutamic acid we were able to show changes in all amino acids which can form glucose or originate from glucose.

We do not know what biochemical defects are responsible for these different metabolic disorders.

Although the clinical findings in all 20 patients were characterized by a chronic and slowly progressive senile dementia, the metabolic conditions of each varied considerably, a fact which in our opinion carries certain therapeutical implications.

References

1. BARKULIS, S. S., A. GEIGER, Y. KAWAKITA, and V. AGUILAR: A study on the incorporation of ^{14}C derived from glucose into the free amino acids of the brain cortex. J. Neurochem. 5, 339 (1960).
2. BERL, S., G. TAKAGAKI, D. D. CLARKE, and H. WAELSCH: Metabolic compartments in vivo. Ammonia and glutamic acid metabolism in brain and liver. J. biol. Chem. 237, 2562 (1962).
3. BERNSMEIER, A., u. K. SIEMONS: Die Messung der Hirndurchblutung mit der Stickoxydulmethode. Pflügers Arch. ges. Physiol. 258, 149 (1953).
4. GOTTSTEIN, U., A. BERNSMEIER u. I. SEDLMEYER: Der Kohlenhydratstoffwechsel des menschlichen Gehirns. I. Untersuchungen mit substratspezifischen enzymatischen Methoden bei normaler Hirndurchblutung. Klin. Wschr. 41, 943 (1963).
5. HOYER, S., u. K. BECKER: Hirndurchblutung und Hirnstoffwechselbefunde bei neuropsychiatrisch Kranken. Nervenarzt 37, 322 (1966).
6. — in preparation.
7. KETY, S. S., and C. F. SCHMIDT: The nitrous oxide method for the quantitative determination of cerebral blood flow in man. Theory, procedure and normal values. J. clin. Lab. Invest. 27, 476 (1948).
8. KNAUFF, H. G., U. GOTTSTEIN u. B. MILLER: Untersuchungen über den Austausch von freien Aminosäuren und Harnstoff zwischen Blut und Zentralnervensystem. Klin. Wschr. 42, 27 (1964).
9. MOORE, S., and W. H. STEIN: An improved Ninhydrin System. J. biol. Chem. 211, 908 (1954).
10. SACKS, W.: Cerebral metabolism of doubly labeled glucose in humans in vivo. J. appl. Physiol. 20, 117 (1965).

Comments to Chapter VI

Trauma, Coma, Alcoholism and Dementia

The symptoms of the post-concussional syndrome are well-known to the clinician. Its pathogenesis and the changes in cerebral hemodynamics under these conditions, however, are still lacking explanation. It is therefore of great interest that alterations in local circulation have been observed within traumatized areas by means of diverse experimental procedures. J. S. Meyer et al. found a slight increase in blood flow following light concussion in the monkey, a reversible decrease of cerebral blood flow and oxygen consumption following severe concussion and irreversible changes in cases of contusion. The results of Taylor are in good agreement with these data. He demonstrated a slowing of circulation time and necrosis of ganglion cells after severe head trauma, probably as a consequence of ischemia and edema. Reivich et al. found neither hyperemia nor ischemia following a commotion in the cat as long as blood pressure was normal. However there was a loss of autoregulation, that is, blood flow in the traumatized cortex and in the subjacent white matter passively followed changes in blood pressure. This finding seems important, since patients with commotion frequently present a derangement in blood pressure regulation with tendency to an orthostatic pressure drop, with the danger of impending brain ischemia. A new finding contained in Reivich's observations, is the vasoparalysis without vasodilatation in the presence of normotension. Further studies, specially during CO_2-breathing, will have to show if this is a total or partial loss of autoregulation. The latter seems more probable.

An important point to be taken into account when considering the studies of cerebral concussion and cerebral contusion upon cerebral blood flow is the difference in diagnostic criteria. Further, attention has to be given to the differences in the severity of trauma and in localization of the traumatized areas. This was shown by the studies of Baldy-Moulinier et al. In post-traumatic patients who were unconscious but had a normal angiogram, cerebral blood flow was normal, a finding which is in good agreement with the results of Reivich. Such patients recuperated. Others, with intracranial expanding processes in the angiogram and a low cerebral blood flow had a bad prognosis.

Hadjidimos et al. brought a valuable contribution to the problem of *determination of cerebral death*: following an injection of ^{133}Xenon into the internal carotid after temporary occlusion of the surgically exposed external carotid (to prevent reflux) there is an absence of isotope clearance in cases of arrest of cerebral circulation. This permits an additional safety in the diagnosis of cerebral death, which can not be conclusively made only by angiography and EEG, as also stressed by Bès et al. The possible absence of correlation between EEG and recuperation capacity of ganglion cells is shown by the interesting studies of Betz et al. These authors observed a complete arrest of EEG without changes in cortical energy metabolism following ventilation with 50–70% CO_2 in the cat. Important and surprising is the fact that the animals awoke from anesthesia and survived normally even after 1 h of EEG silence.

The papers by FIESCHI et al., and by GOTTSTEIN et al. referred to patients with *non-traumatic comatose states*. FIESCHI stressed the importance of acidosis, specially of cerebrospinal fluid, for the regulation of cerebral blood flow, as exemplified by patients with repiratory insufficiency and with severe metabolic acidosis. Although most of them had an increased circulation, in the majority of severely sick patients cerebral blood flow was lowered despite CSF acidosis. The low CBF values are attributed to brain edema by the authors. Similarly to the cases of cerebral trauma, all comatose patients studied had a loss of autoregulation.

Thus, all previous works have shown disturbances in brain function to take place as well in the presence of increased as of normal or decreased CBF. Therefore it is evident that alterations in brain metabolism play a prominent role in the genesis of comatose states. GOTTSTEIN et al. called attention to the disturbances of cerebral metabolism in patients with Wernicke encephalopathy, uremia and hepatic insufficiency. In all these conditions no specific circulatory alteration takes place, but there is a reduction of cerebral oxygen and glucose uptake. Thus, in these internal diseases coma and precoma are not directly attributable to cerebral blood flow disturbances, but to alterations in cerebral metabolism. When ischemia and brain edema are added, the consequence is a further derangement of substrate supply to the nerve cells, as shown by the studies of SAMII et al.: the energy rich phosphates are reduced, while lactate and pyruvate are increased in brain regions injured by cold edema.

Finally INGVAR et al., as well as HOYER, called attention to the interesting alterations of CBF and metabolism accompanying chronic derangements in brain function such as observed in cases of chronic alcoholism and senile dementia.

U. GOTTSTEIN

Effect of Blood Pressure Alterations on CBF during General Anesthesia in Man

A. L. Smith, J. L. Neigh, J. C. Hoffman, and H. Wollman

Department of Anesthesia, University of Pennsylvania, Philadelphia

Cerebral autoregulation can be disturbed by pathology such as stroke [4], or stresses such as hypercarbia [5] or hypoxia [3]. However the effects of general anesthetics on autoregulation have not been quantitated. The present study was designed to investigate this problem.

Informed, human volunteers were anesthetized lightly with 70% N_2O or deeply with 20% cyclopropane and paralyzed as necessary with d-tubocurarine. Ventilation was controlled to maintain normocarbia, and CBF was determined by sampling arterial and jugular bulb blood during inhalation of ^{85}Kr. $CMRO_2$ was calculated as the product of CBF and the manometrically determined arteriovenous oxygen content difference. The arteriovenous oxygen saturation difference was determined 3 times during each flow measurement to document steady state conditions. Arterial and venous pressures were obtained from calibrated transducers. Following measurements at normotension, blood pressure was raised by an intravenous infusion of angiotensin or phenylephrine or lowered with a veratrum alkaloid (Unitensin). Measurements were made in each subject during the infusion of 2 drugs and the order of the drugs was alternated. Angiotensin [1] and veratrum [2] have previously been shown to have no direct effect on cerebral vessels. Phenylephrine was chosen because it seemed desirable to study the effects of hypertension produced by 2 different drugs.

The results for the N_2O study are summarized in Table 1. The results of the 3 studies with phenylephrine and 4 studies with angiotensin are similar and have been pooled. Fig. 1 shows the data expressed as mean change in cerebral perfusion pressure and mean change in CBF. Mean perfusion pressure was increased 41.8 torr in the 7 studies with angiotensin and phenylephrine and there was no significant change in CBF. The slight fall in mean CBF from

Table 1. *Autoregulation during nitrous oxide anesthesia*

	PaCO$_2$ torr	PVO$_2$ torr	Perfusion pressure torr	CBF ml/ 100 g/min	CVR torr/ml/ 100 g/min
Normotension	36.5	38.3	83.5	44.5	1.95
Hypertension	36.0	37.8	125.3	45.9	2.92
p (7 studies)	NS	NS	< .001	NS	< .01
Normotension	36.1	38.9	86.6	40.0	2.17
Hypotension	35.8	38.9	49.1	35.0	1.41
p (5 studies)	NS	NS	< .001	NS	< .05

p was obtained from paired *t*-test. NS = not significant.

40.0–35.0 ml/100 g/min, when perfusion pressure was lowered from 86.6–49.1 torr, is not statistically significant. Thus, autoregulation was unimpaired during nitrous oxide anesthesia. There were no significant differences in CMRO₂ among any of the groups.

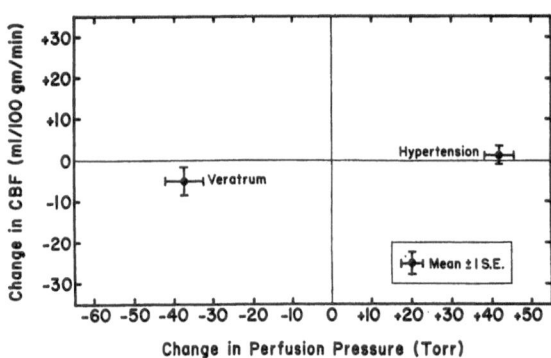

Fig. 1. Mean change in CBF as a function of mean change in cerebral perfusion pressure under nitrous oxide anesthesia

The results of the cyclopropane study are summarized in Table 2 and Fig. 2. When mean perfusion pressure was increased 31.1 torr with angiotensin and 36.0 torr with phenylephrine, there were no significant changes in CBF. There was greater variability of CBF with cyclopropane than with nitrous oxide among subjects and between experimental conditions. This has been observed in previous studies in this laboratory and is due in part to the greater measurement error of the [85]Kr technique with high CBF's. When perfusion pressure was lowered from a mean of 87.5–39.1 torr, the mean CBF fell from 67.3–42.8 ml/100 g/min. The decrease in CBF occurred in every subject and there was a tendency for the higher control flows to decrease more. The CMRO₂ was not changed significantly by any of the drugs.

Table 2. *Autoregulation during cyclopropane anesthesia*

	PaCO₂ torr	PVO₂ torr	Perfusion pressure torr	CBF ml/ 100 g/min	CVR torr/ml/ 100 g/min
Normotension	38.2	60.2	90.7	74.5	1.57
Angiotensin	37.6	71.8	121.8	66.1	1.99
p (5 studies)	NS	NS	< .001	NS	< .05
Normotension	37.4	56.3	93.4	60.4	1.84
Phenylephrine	35.9	53.7	129.4	66.0	2.13
p (6 studies)	NS	NS	< .01	NS	< .05
Normotension	38.6	55.3	87.5	67.3	1.42
Veratrum	38.5	44.2	39.1	42.8	0.93
p (5 studies)	NS	< .01	< .001	< .05	< .05

p was obtained from paired *t*-test. NS = not significant.

With cyclopropane, the cerebral vessels seem able to constrict in response to an increased perfusion pressure and maintain CBF unchanged. However, unlike the case of nitrous oxide, in the face of low perfusion pressures, the flow decreases. This fall in CBF cannot be entirely due to the fact that the mean perfusion pressure fell 10.9 torr more in the cyclopropane studies. The control CVR, which was 2.17 with nitrous oxide and normotension, was only 1.42 torr ml/100 g/min with cyclopropane during normotension. CVR fell farther to a mean of 0.93 at low perfusion pressures with cyclopropane, probably representing maximal cerebral vaso-dilation.

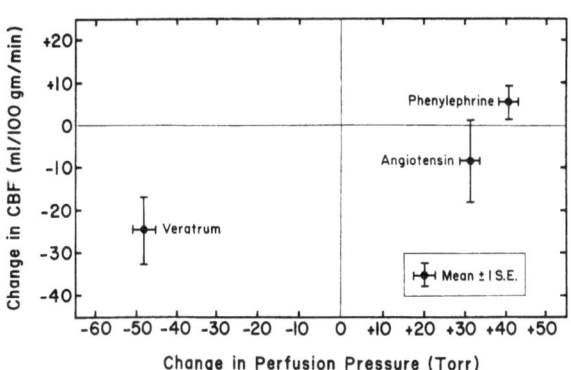

Fig. 2. Mean change in CBF as a function of mean change in cerebral perfusion pressure under cyclopropane anesthesia

We conclude that during either light anesthesia with nitrous oxide or deep anesthesia with cyclopropane the ability of the cerebral vessels to constrict and maintain a constant flow in response to increased perfusion pressure is intact. The normal CBF obtained with nitrous oxide can be maintained in the face of a fall in perfusion pressure to 49.1 torr. But at low perfusion pressure, the cerebral vasculature could not dilate enough to sustain the supranormal flows typical of deep cyclopropane anesthesia even though the very low CVR value indicated a maximal autoregulatory dilatation.

References

1. AGNOLI, A., N. BATTISTINI, L. BOZZAO, and C. FIESCHI: Drug Action on Regional CBF in Cases of Acute Cerebro-Vascular Involvement. Acta neurol. scand. Suppl. 14, 142 (1965).
2. FREIS, E. D.: The Hemodynamic Effects of Hypotensive Drugs in Man. Veratrum Viride. J. clin. Lab. Invest. 28, 353 (1949).
3. HÄGGENDAL, E., and E. JOHANSSON: Effects of Arterial Carbon Dioxide Tension and Oxygen Saturation on Cerebral Blood Flow Autoregulation in Dogs. Acta physiol. scand. 66, Suppl. 258, 27 (1965).
4. HØEDT-RASMUSSEN, K., E. SKINHØJ, O. PAULSON, J. EWALD, J. BJERRUM, A. FAHRENKRUG, and N. LAS-SEN: Regional Cerebral Blood Flow in Acute Apoplexy. Arch. Neurol. 17, 271 (1967).
5. RAPELA, C. F., P. P. MACHOWICZ, and G. FREEMAN: Effect of CO_2 on Autoregulation of Cerebral Blood Flow. Fed. Proc. 22, 344 (1963).

Effects of General Anesthetics in Man on the Ratio of Cerebral Blood Flow to Cerebral Oxygen Consumption

H. Wollman, A. L. Smith, and S. C. Alexander

Department of Anesthesia, University of Pennsylvania, Philadelphia

The effects of several general anesthetics on cerebral blood flow are shown in Fig. 1. The pattern which appears to be consistent for all of the general anesthetics studied in man in these laboratories is one of decreased cerebral blood flow during light anesthesia and increasing cerebral perfusion as anesthesia deepens.

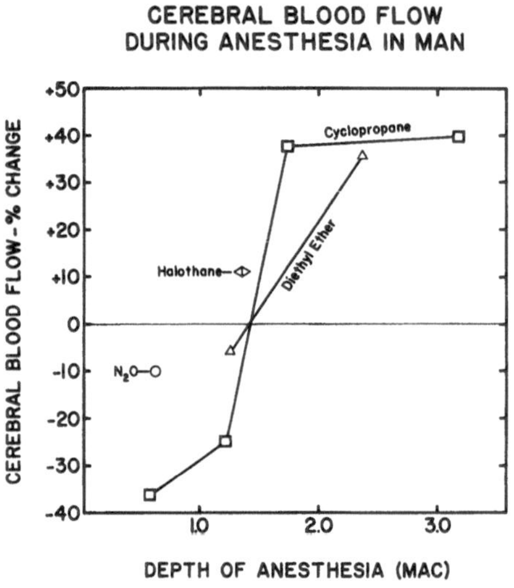

Fig. 1. The figure shows the per cent change in cerebral blood flow from normal awake control values as a function of the depth of anesthesia. All measurements were made during normocarbia, with adequate arterial oxygenation. Each point represents the mean of at least 6 determinations. The measure of depth of anesthesia chosen was MAC, the minimum alveolar concentration which permits skin incision without movement in 50 % of patients. It is necessary to use such a generalized concept for depth of anesthesia in order to compare all of the anesthetic agents

The actions of the general anesthetics on cerebral oxygen consumption in man are illustrated in Fig. 2. $CMRO_2$ is below normal at all depths of anesthesia, but appears to be least depressed at moderate depths. $CMRO_2$ is most decreased when anesthesia is either light or very deep.

Further understanding of the data results from consideration of the ratio of cerebral blood flow to cerebral oxygen consumption (Fig. 3). In awake normocarbic man the ratio of

Supported (in part) by U.S.P.H.S. research grants GM-09070-06, GM-15430-02, 5-Tl-GM 215-11, and HE 06352.

cerebral blood flow to cerebral oxygen consumption is 14.9 ml blood/cc oxygen. The same ratio holds for anesthetized normocarbic man up to 1 MAC. Beyond 1 MAC, the ratio of cerebral blood flow to cerebral metabolic rate increases approximately linearly with depth of anesthesia, reaching 34 at 3 MAC.

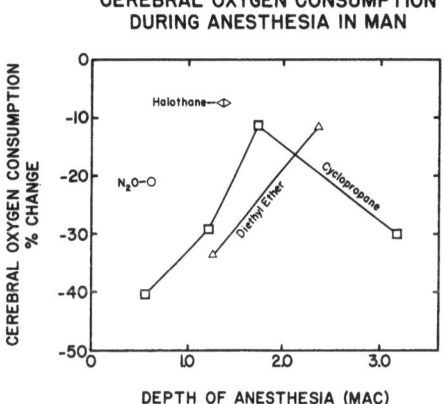

Fig. 2. The figure shows cerebral oxygen consumption during anesthesia in adequately oxygenated and normocarbic man. The per cent change in cerebral oxygen consumption is compared to the awake control, and is plotted as a function of the depth of anesthesia (MAC). Each point represents the mean of at least 6 determinations

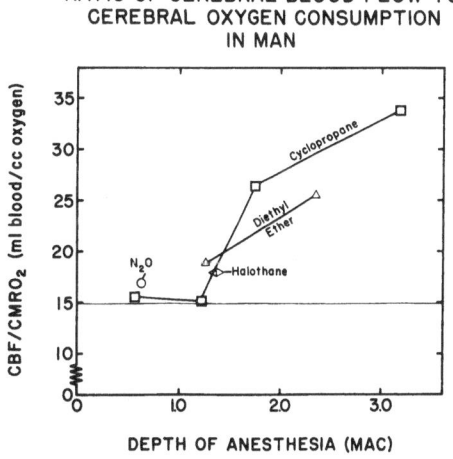

Fig. 3. The figure shows the ration of CBF to CMRO₂ as a function of depth of anesthesia (MAC). All values were obtained during normocarbia in adequately oxygenated man. Each point represents the mean of at least 6 determinations. In awake normocarbic man the ratio CBF/CMRO₂ is 14.9 ml blood/cc. O₂, and a light horizontal line indicating this value is shown on the figure

If cerebral oxygen supply is increased as anesthesia deepens, new questions are raised. Do general anesthetics block certain metabolic pathways necessitating higher tissue pO_2 to drive oxidative metabolism? Or, on the other hand, do these data suggest failure of processes governing cerebral vascular diameter and tissue pO_2 during deep anesthesia? If such failure of oxygen regulation occurs one must then speculate on the efficiency of autoregulation to blood pressure changes, and perhaps even impairment of response to carbon dioxide during deep anesthesia.

Effect of Hyper- and Hypoventilation on CBF during Anesthesia

H. Kreuscher and J. Grote

Department of Anaesthesiology and Physiology, University of Mainz

Alexander and Wollman [1, 2, 4, 8], Harper [5] and McDowall and other investigators have published observations on CBF during anaesthesia in man and dog, when arterial pCO_2 was altered by hyper- or hypoventilation.

Nevertheless, there were discussions about whether or not the CBF increases when the pCO_2 rises, and whether or not the CBF decreases when the pCO_2 falls during deep anaesthesia. The opinion was advanced that clinical anaesthesia will inhibit the pCO_2-depending regulation of CBF.

The behaviour of CBF under anaesthesia is of great importance for the clinical work of anaesthetists, who are familiar with the common technique of hyperventilating the patient so as to keep him "anaesthetized" with a minimal amount of anaesthetic agent.

We found that patients receiving this kind of anaesthesia often have prolonged recovery periods.

Could the decrease in pCO_2 by means of intensive and long lasting hyperventilation cause a severe decrease in CBF in a brain, the oxygen requirement of which is not markedly diminished by deep anaesthesia? In this case the $AVDO_2$ would increase. The pCO_2 in the brain capillaries could reach values too low to supply sufficient oxygen to the brain cells.

We have measured CBF under light N_2O and under Halothane anaesthesia in dogs, using the dye dilution technique. We induced anaesthesia in our experimental animals by i.v. application of the very short acting Epontol. After endotracheal intubation and complete relaxation the animals were ventilated artificially. Body temperature was maintained within the normal range. During determination of the normal individual CBF control value, arterial pCO_2 was kept at 40 mmHg. Afterwards we induced hypo- or hyperventilation and used a gas mixture of nitrous oxide and oxygen with or without 1% Halothane as the anesthetic agents. In no case did the mean arterial blood pressure fall below 80 mmHg. The pO_2 was kept in the normal range, that means it did not decrease below 70 mmHg in the arterial, and not below 30 mmHg in the cerebral venous blood.

The results here presented are to be understood as preliminary (Fig. 1). On the Y-axis of the graph we plotted the changes of CBF in % of the individual normal control value, and on the X-axis the corresponding arterial pCO_2. The unbroken line shows the behaviour of CBF under light N_2O-anaesthesia as a function of pCO_2. The course of this curve is in close conformity to that of the Philadelphia group (Alexander, Cohen, Wollman et al.) and of Harper from Glasgow, published in 1964 and 1965, respectively. The crossmarks on the graph are the corresponding values measured during halothane anaesthesia.

It is easy to recognize, that changes of pCO_2 by hyper- or hypoventilation are followed by changes of CBF under light (N_2O) and under deep (halothane) anaesthesia.

We know that these changes are not dependent only upon the oxygen requirement of the brain cells, because the $AVDO_2$ increases markedly in subjects undergoing hyperventilation under light anaesthesia. Under deep anaesthesia the $AVDO_2$ increases less probably due to decreased oxygen consumption of the brain secondary to the effect of the anaesthetic agent.

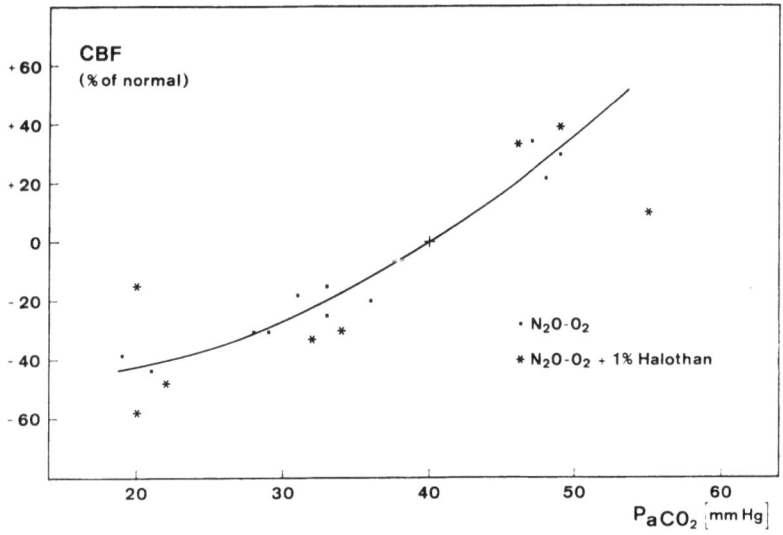

Fig. 1. CBF as a function of $aPCO_2$ during light N_2O or N_2O-halothane anaesthesia in dogs. The curve represents the approximated course during N_2O-anaesthesia. ($+$: normal value; $\blacksquare$, $*$: mean values taken from 3–4 single measurements)

This would confirm the clinical observations, mentioned above, that patients who have been hyperventilated over long periods of time occasionally have a prolonged stage of recovery. It is possible that those patients suffer from cerebral disturbances due to local hypoxia.

For the practice of clinical anaesthesia we would like to recommend normoventilation for patients under anaesthesia and normothermia.

References

1. ALEXANDER, S. C., H. WOLLMAN, P. J. COHEN, P. E. CHASE, E. MELMAN, and R. D. DRIPPS: Cerebral blood flow and metabolism during halothane anesthesia in man. Fed. Proc. **22**, 187 (1963).
2. — — —, and M. BEHAR: Cerebrovascular response to $aPCO_2$ during halothane anesthesia in man. J. appl. Physiol. **19**, 561 (1964).
3. BETZ, E., H. OEHMIG u. W. WÜNNENBERG: Die Wirkung verschiedener Narkotika auf die lokale Gehirndurchblutung der Katze. Z. Kreisl.-Forsch. **54**, 5 (1965).
4. COHEN, P. J., H. WOLLMAN, S. C. ALEXANDER, P. E. CHASE, and M. G. BEHAR: Cerebral carbohydrate metabolism in man during halothane anesthesia. Anesthesiology **25**, 185 (1964).
5. HARPER, A. M.: The inter-relationship between $aPCO_2$ and blood pressure in the regulation of blood flow through the cerebral cortex. Acta neurol. scand. Suppl. **14**, 94 (1965).
6. KIENLE, G.: Hydrodynamik, Elektrolyt- und Säure-Basen-Haushalt im Liquor und Nervensystem. Stuttgart: Thieme 1967.
7. KREUSCHER, H.: Die Hirndurchblutung unter Neuroleptanaesthesie. Anaesthesie und Wiederbelebung, vol. 21. Berlin-Heidelberg-New York: Springer 1967.
8. WOLLMAN, H., S. C. ALEXANDER, P. J. COHEN, P. E. CHASE, E. MELMAN, and M. G. BEHAR: Cerebral circulation of man during halothane Anesthesia. anesthesiology **25**, 180 (1964).

Cerebral Blood Flow and Oxygen Consumption in Man during Electroencephalographic Seizure Patterns Associated with Ethrane Anesthesia

H. Wollman, A. L. Smith, J. L. Neigh, and J. C. Hoffman

Department of Anesthesia, University of Pennsylvania, Philadelphia

Abnormal electroencephalographic patterns occurred in some patients during clinical testing of a new general anesthetic, Ohio 347 (Ethrane; 2-chloro-1,1,2-trifluoroethyl difluoromethyl ether). Despite these patterns, recovery from general anesthesia was prompt, and no mental or psychological sequelae were detected. In view of the other advantages of the agent, the decision was made to continue clinical testing and attempt to define the cerebral effects of the drug more quantitatively. With this in mind the following studies were undertaken.

Methods

6 young, normal, informed, human volunteers were studied. Control measurements were made while the subjects were awake breathing 100% oxygen. Then general anesthesia was induced with Ethrane in oxygen and the trachea was intubated. Ventilation was controlled with a Bird respirator in a non-rebreathing system, with enough carbon dioxide added to the inspired gases to maintain normocarbia. D-tubocurarine was injected intravenously as necessary for muscle paralysis to permit controlled ventilation. When a stable and moderately deep level of Ethrane anesthesia had been attained, electroencephalographic patterns were recorded which consisted of frequent (about 1 per sec) sharp spikes separated by periods of electrical silence or relatively normal EEG. The duration of the spikes was 150 msec and their amplitude up to 200 μv. This pattern was a stable one and often continued for 1 h or more. It was at this time that all measurements were repeated (steady state Ethrane anesthesia).

Cerebral blood flow was measured by intermittent sampling of arterial and jugular-bulb blood during ^{85}Krypton inhalation. Arterial and jugular bulb pO_2, pCO_2, and pH were measured with blood-gas electrodes. Arterial and jugular bulb oxygen content was measured manometrically (Van Slyke) and A-V difference used to calculate $CMRO_2$. Body temperature was monitored with a rectal thermistor probe, and arterial and jugular blood pressures were measured with calibrated strain gauges.

In 3 of the 6 volunteers after measurements had been made during steady state Ethrane anesthesia, grand mal convulsions spontaneously occurred. Since these did not last for a

Supported (in part) by U.S.P.H.S. research grants GM-09070-06, GM-15430-02, 5-Tl-GM215-11, and HE 06352. The authors wish to acknowledge the laboratory assistance of Mrs. Linda Lovette, Miss Marilyn Shaw, Miss Sarah Childs, and Mrs. Cheryl McIlvaine.

sufficiently long time to permit measurements of cerebral blood flow, only arteriovenous differences for oxygen and arterial and jugular blood gases could be measured. This was done during the convulsions, in the period of electrical silence following the convulsions, and later when the electroencephalogram had reverted to its previous pattern.

Results

Steady State Ethrane Anesthesia

The experimental conditions are defined in Table 1. During anesthesia body temperature, arterial pCO_2, pO_2, and pH were unchanged from the control measurements. The cerebral perfusion pressure was decreased, and this was the result of depression of arterial blood pressure by Ethrane. The experimental results are given in Table 2. During Ethrane anesthesia cerebral blood flow was unchanged from the control value. This was true in spite of the decreased cerebral perfusion pressure, and could be attributed to decreased cerebral vascular resistance. $CMRO_2$ during Ethrane anesthesia was decreased to half of the control value. This resulted in an increase of jugular venous pO_2 from the control value of 38.0 to 60.7 torr, and a doubling of the ratio of cerebral blood flow to cerebral metabolic rate.

Table 1. *Effects of Ethrane anesthesia on cerebral blood flow and metabolism in man (6 subjects). Experimental conditions*

	Control Mean	S.E.	Ethrane Mean	S.E.
Temperature (° C)	37.0	0.1	36.8	0.2
aPCO$_2$ (torr)	36.5	1.6	35.9	1.5
aPO$_2$ (torr)	602.8	10.7	589.1	30.2
apH	7.462	0.013	7.445	0.013
Cerebral Perfusion Pressure (torr)	86.8	4.3	49.8	3.9

Table 2. *Effects of Ethrane anesthesia on cerebral blood flow and metabolism in man (6 subjects). Results*

	Control Mean	S.E.	Ethrane Mean	S.E.
Cerebral Blood Flow (ml/100 g/min)	41.0	5.6	40.0	5.6
Cerebral Vascular Resistance (torr/ml/100 g/min)	2.32	0.33	1.34	0.18
CMRO$_2$ (cc/100 g/min)	3.06	0.42	1.53	0.15
vPO$_2$ (torr)	38.0	3.4	60.7	3.1
CBF/CMRO$_2$	13.4	—	26.1	—

Grand Mal Convoulsions

In the 3 subjects who underwent spontaneous grand mal convulsions, during the convulsion arteriovenous oxygen differences were approximately double the low values observed during Ethrane anesthesia. However, this doubling merely brought the A-V oxygen difference and the jugular venous pO_2 back to a normal level.

Discussion and Conclusion

These results indicate that during moderately deep Ethrane anesthesia cerebral blood flow remains normal in spite of decreased cerebral perfusion pressure. The effects of Ethrane on cerebral oxygen consumption were marked. When the EEG showed frequent spikes (steady state Ethrane anesthesia), cerebral oxygenation appeared to be increased due to the maintenance of normal flow and the halving of metabolic rate. It was only during the grand mal convulsion that cerebral metabolism per ml of blood increased to the normal awake level. This increase during the grand mal convulsion brought the jugular venous pO_2 back to normal levels but not lower. Thus there was no evidence to indicate that cerebral oxygenation was impaired during the grand mal convulsions occurring with Ethrane anesthesia.

An additional interesting speculation concerns the termination of the grand mal seizure patterns. If cerebral oxygenation remains adequate during the seizures it is difficult to implicate a hypermetabolic state and the utilization of all available substrate as the reason for termination of these seizures.

Effects of Hyperventilation on Focal Brain Damage Following Middle Cerebral Artery Occlusion

N. Battistini, M. Casacchia, A. Bartolini, G. Bava, and C. Fieschi

Department of Neurology and Psychiatry, University of Genoa

Several experimental and clinical data were presented in the last Lund Symposium on CBF-CSF, demonstrating the paradoxical effects of CO_2 on occluded vascular beds [6]. Since the "counter-steal" or "inverse-steal" phenomenon has many physiological and clinical implications, we made the present experiments to investigate the effects of hypocapnia on CBF in focal cerebral ischaemia.

In 12 adult cats under general anesthesia, artificially ventilated with a gas mixture of nitrous oxide and oxygen, the right middle cerebral artery (MCA) was occluded at its origin, as described by Sundt and Waltz [5]. The experiments lasted 12 h, after which local CBF was measured with the ^{14}C-labeled-antipyrine technique of Reivich et al. [3]. Body temperature, arterial pressure, EEG and end-tidal CO_2 were continuously monitored. Arterial pH, pO_2, pCO_2 and Hb were measured every 2 h. The extent of the tissue damage was evaluated on fresh frozen tissue slices by means of histo-enzymatic methods.

2 different groups of animals were studied:

1. Normo- or slightly hypoventilated animals; $aPCO_2$ was kept between 35 and 40 mmHg throughout the experiment.

2. Hyperventilated animals; $aPCO_2$ was lowered to 20–25 mmHg for 6 h, starting 30 min after middle cerebral artery occlusion.

Both groups were again divided in 2 subgroups of animals in which the MABP was kept at a normal or a high level.

In the normoventilated normotensive group (Fig. 1) the autoradiography showed a large area of non-flow involving 62% of the hemisphere, that is the entire territory of distribution of the occluded middle cerebral artery. All other groups of cats showed lesions which were considerably less extensive. In the hyperventilated normotensive group, the extension of the ischaemic lesion involved an average of 15% of the hemisphere. In normocapnic hypertensive group only a diffuse reduction in flow was recorded, without any area where flow was entirely arrested. In the last group of hypocapnic hypertensive animals, the average extension of the ischaemic lesion involved 22% of the hemisphere, slightly more than in hypocapnic normotensive group, and definitely less than in normocapnic normotensive group (Fig. 2).

The extension of tissue damage shown by histo-enzymatic methods is comparable to the extension of the area of non-flow shown by autoradiography.

Supported by US-PHS Grant 05017NB.

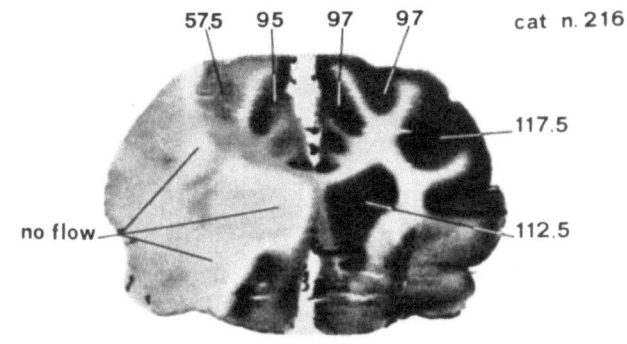

Normoventilated normotensive

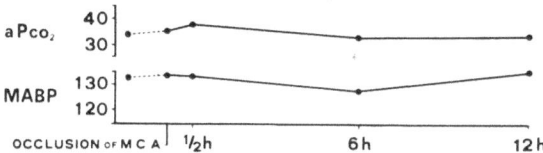

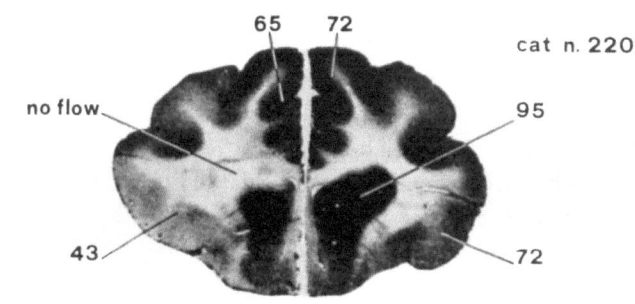

Hyperventilated normotensive

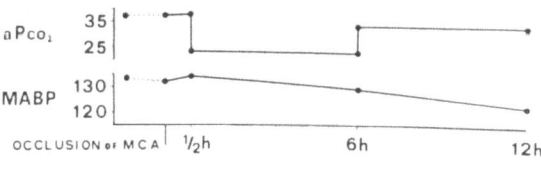

Fig. 1 a

These data confirm previous observations [4] that hyperventilation considerably reduces the extension of the brain damage produced by the occlusion of the MCA. We may add now that this reduction of brain damage is accompanied by a less extensive reduction of blood flow in the territory of the occluded artery, and that this protective effect takes place also when the hyperventilation starts 30 min after the occlusion. This effect may be due to the combination of several factors, namely a counter-steal phenomenon, the attenuation by respiratory alkalosis of the posthypoxic metabolic acidosis of the nervous tissue, improvement of brain edema, and decrease of intracranial pressure.

However the protective effect of the hyperventilation is observed only in normotensive animals. Actually, in normocapnic animals, the arterial hypertension induced before the MCA occlusion facilitates the collateral circulation and produces only a diffuse reduction in the

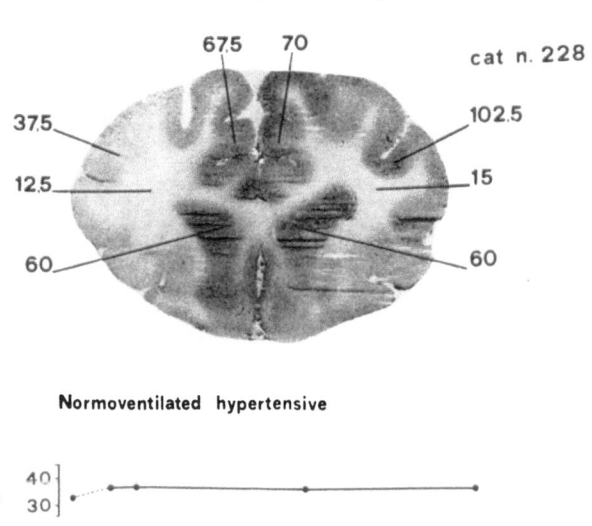

Normoventilated hypertensive

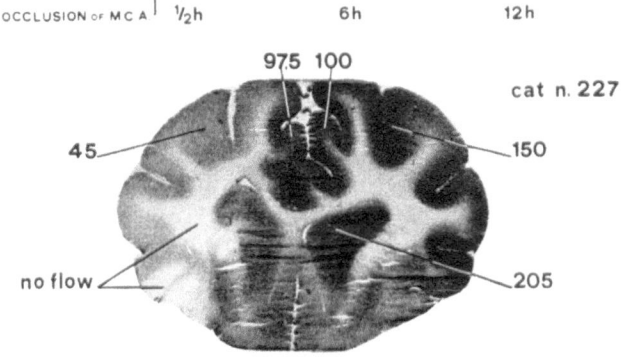

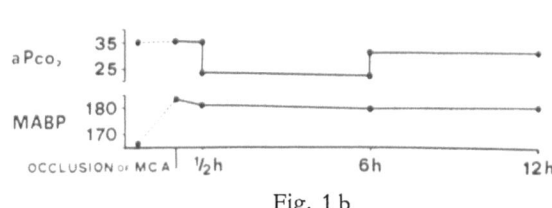

Hyperventilated hypertensive

Fig. 1 b

Fig. 1. ^{14}C-antipyrine autoradiography of coronal sections of brain in 4 cats with right MCA occlusion. Effects on CBF of arterial hypertension and hyperventilation are shown. The figure represents the values of regional blood flow in ml/100 g/min

blood flow and no areas of complete ischaemia, while when these hypertensive animals are hyperventilated a definite although small ischaemic lesion is produced. In this last group, during the initial 30 min after the occlusion of the MCA the cats are in the same experimental conditions of those of the normocapnic hypertensive group (that is, a diffuse reduction in flow). The subsequent hyperventilation apparently induces a further reduction in blood flow to the point where an ischaemic lesion is produced. The area of total ischaemia does not return to the previous conditions even when aPCO$_2$ is restored to the resting values.

A "classical" and easily understood interpretation is that hyperventilation, initiated half an hour after the arterial occlusion, produces a vasoconstriction of the leptomeningeal

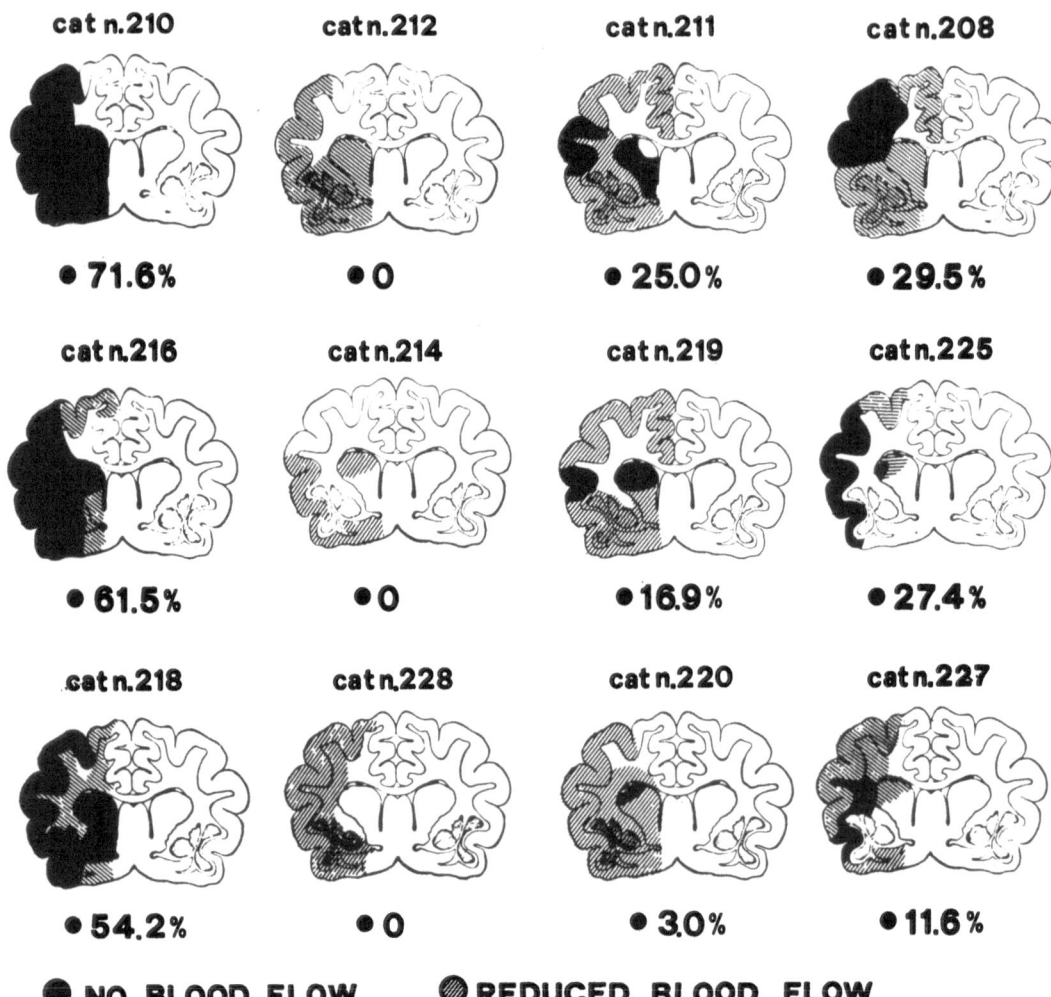

NORMOVENTILATED   **HYPERVENTILATED**

NORMOTENSIVE   HYPERTENSIVE   NORMOTENSIVE   HYPERTENSIVE

cat n.210   cat n.212   cat n.211   cat n.208

● 71.6%   ● 0   ● 25.0%   ● 29.5%

cat n.216   cat n.214   cat n.219   cat n.225

● 61.5%   ● 0   ● 16.9%   ● 27.4%

cat n.218   cat n.228   cat n.220   cat n.227

● 54.2%   ● 0   ● 3.0%   ● 11.6%

● NO BLOOD FLOW   ◐ REDUCED BLOOD FLOW

Fig. 2. The figure summarizes the data observed in the 4 groups of cats. Areas of non-flow (black) or reduced flow (cross hatched) are drawn in coronal schematic sections of the brain. ● indicates the % of the non-flow area of the entire right hemisphere. In the cross hatched area, blood flow is reduced below 30 % of the opposite (control) hemisphere

arteries supplying the collateral circulation to the ischaemic area. These anastomotic arteries seem indeed to be normally reacting to CO_2 in hypertensive cats, while they are not reacting in normotensive animals, whose brain ischaemia results in a more extensive lesion, with subsequent metabolic acidosis. Other interpretations however may be advanced, based on factors other than pure vasomotor responses, namely extramural sources of resistance and brain edema favoured by the loss of autoregulation [2].

In order to test these hypotheses, determinations of brain edema, total CO_2, and lactate/pyruvate ratio in brains of similar groups of animals are now in progress.

Clinical implications from our findings are as follows: In a general way, patients whose condition suggests extensive tissue damage should be treated with passive hyperventilation whenever possible. On the other hand, in cases with less severe lesions this treatment may be contra-indicated, as is clearly shown in the case presented by BROCK et al. [1].

The situation may be different in cases without permanent arterial occlusion: in these cases a combined hyperventilation and controlled hypertension might prove useful.

In individual cases the best way of choosing the correct treatment remains that of measuring cerebral blood flow in the damaged area and its responses to induced changes in $aPCO_2$ and arterial blood pressure.

References

1. BROCK, M., A. A. HADJIDIMOS, and K. SCHÜRMANN: Possible adverse effects of hyperventilation on rCBF during the acute phase of total proximal occlusion of a main cerebral artery. Presented at the International CBF Symposium, Mainz, April 1969 (this volume, p. 254).
2. LANGFITT, T. W., J. D. WEINSTEIN, N. F. KASSELL, and H. M. SHAPIRO: Cerebrovascular dilation and compression with intracranial Hypertension. Presented at the International CBF Symposium, Mainz, April 1969 (this volume, p. 177).
3. REIVICH, M., J. JAHYLE, L. SOKOLOFF, and S. S. KETY: Autoradiographic technique for measuring cerebral blood flow using C14 antipyrine. Circulat. Res., 1969, in press.
4. SOLOWAY, M., W. NADEL, M. S. ALBIN, and R. J. WHITE: The effect of hyperventilation on subsequent cerebral infarction. Anesthesiology 29, 975 (1968).
5. SUNDT, T. M., JR., and A. G. WALTZ: Experimental cerebral infarction; retro-orbital, extradural approach for occluding the middle cerebral artery. Mayo Clin. Proc. 41, 159 (1966).
6. SYMON, L., and R. WÜLLENWEBER: Discussion and Comments to section XIII. Scand. J. clin. Lab. Invest. Suppl. 102, p. XIII-F (1968).

Possible Adverse Effects of Hyperventilation on rCBF during the Acute Phase of Total Proximal Occlusion of a Main Cerebral Artery

M. Brock, A. A. Hadjidimos, and K. Schürmann

Neurosurgical Department, University of Mainz

Recent studies seem to favor a beneficial effect of hyperventilation (HV) in cases of cerebrovascular occlusion [1, 4] and brain trauma [2, 5]. On the other hand, it seems that, at least under certain circumstances, HV may exert an adverse effect upon cerebral blood flow and metabolism [3]. Since this question is not settled yet, it seems worthwhile to report briefly the present case of reversible total occlusion of the middle cerebral artery closely followed by successive rCBF studies (^{133}Xe-gamma-clearance method), in which, during the acute phase, HV caused a latent ischemia to become manifest.

The 9 years-old girl was first seen a few hours after the sudden onset of a complete, flaccid, right-sided hemiplegia with mild motor aphasia. She was well oriented and presented no clinical signs of increased intracranial pressure.

The *first rCBF study*, 36 h after the onset of hemiplegia, showed no clearcut focal rCBF anomalies during the resting state (Fig. 1 A). During HV a definite ischemia developed in the middle cerebral artery (m.c.a.) territory (Fig. 1 B). No hypertension (HT) was induced for fear that the patient might have an intracranial bleeding. Angiography following the rCBF study (Fig. 1, 1st angiogram) showed a complete proximal occlusion of the m.c.a.

As shown by the clearance curves of this study, reproduced in Fig. 2, there was no significant change of tracer arrival (peak height), between the first and the second studies, at the various regions involved.

The *second rCBF study*, $5^1/_2$ days after the onset of the illness, showed a clearcut relative hyperemia in the territory of the anterior cerebral artery (a.c.a.) and a relative ischemia in the m.c.a. region during the resting state (Fig. 1 C). At this stage HV reduced interregional rCBF differences and appeared to improve blood flow to the ischemic territory (Fig. 1 D). Autoregulation was partly impaired, since rCBF of the proximal ischemic areas could be increased by HT (Fig. 1 E). Hypoventilation (hV) did not alter the rCBF significantly (Fig. 1 F).

Clinically, the patient's condition had remained unchanged as compared to the previous study.

A *third rCBF study* was performed $19^1/_2$ days after the onset of the illness, while the patient was beginning to regain mobility of the proximal musculature of both paralyzed limbs.

At this stage, although angiography showed that the m.c.a. had again become patent (Fig. 1, 2nd angiogram), there was a clearcut relative ischemia in the m.c.a. territory, and a relative hyperemia in the a.c.a. region. Neither HV, HT or hV did change this status to a significant degree (Fig. 1 G–H).

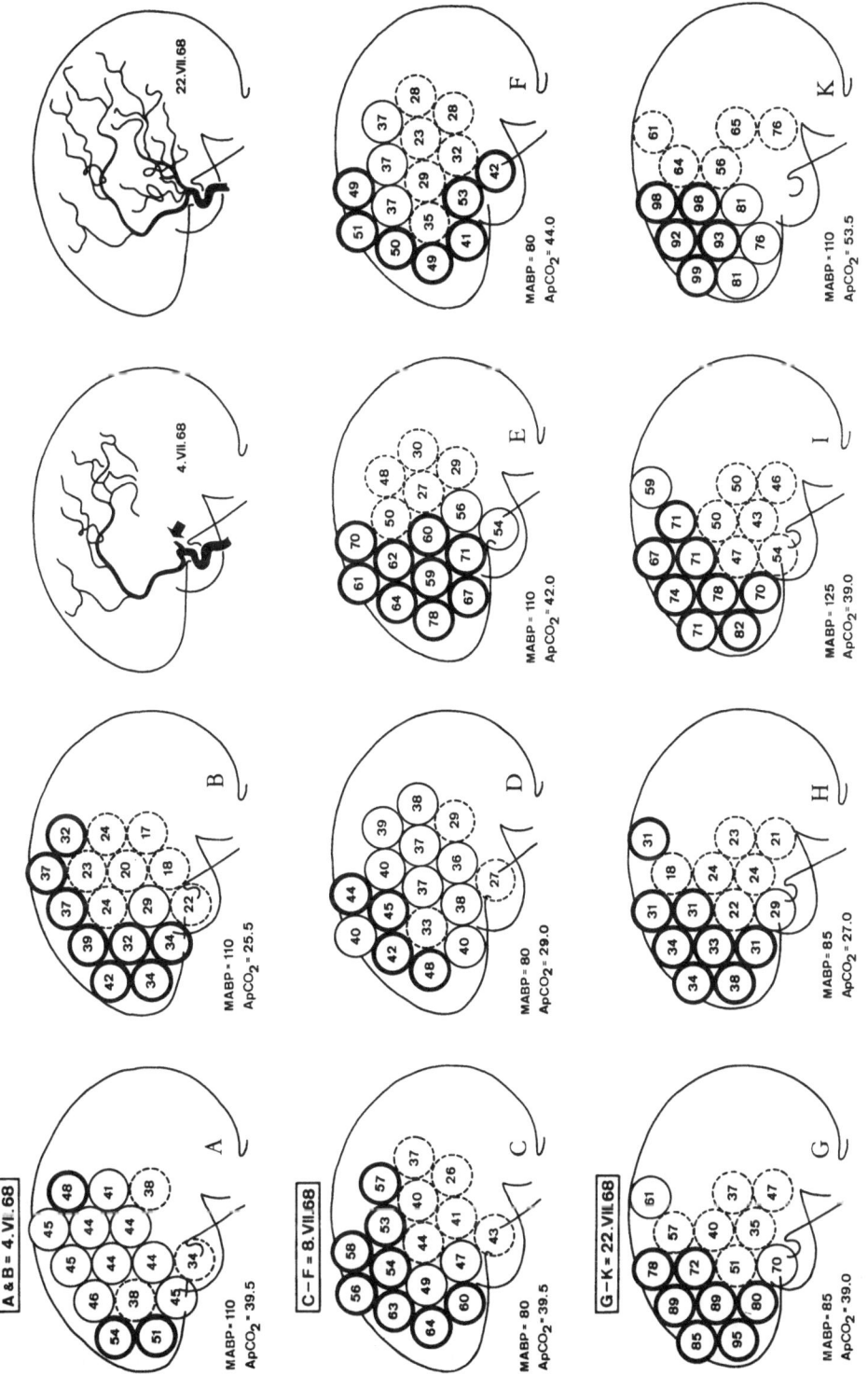

Fig. 1. rCBF values and angiographic findings of the case reported in the present paper

The patient eventually recovered completely from her hemiplegia and returned to school after a few months.

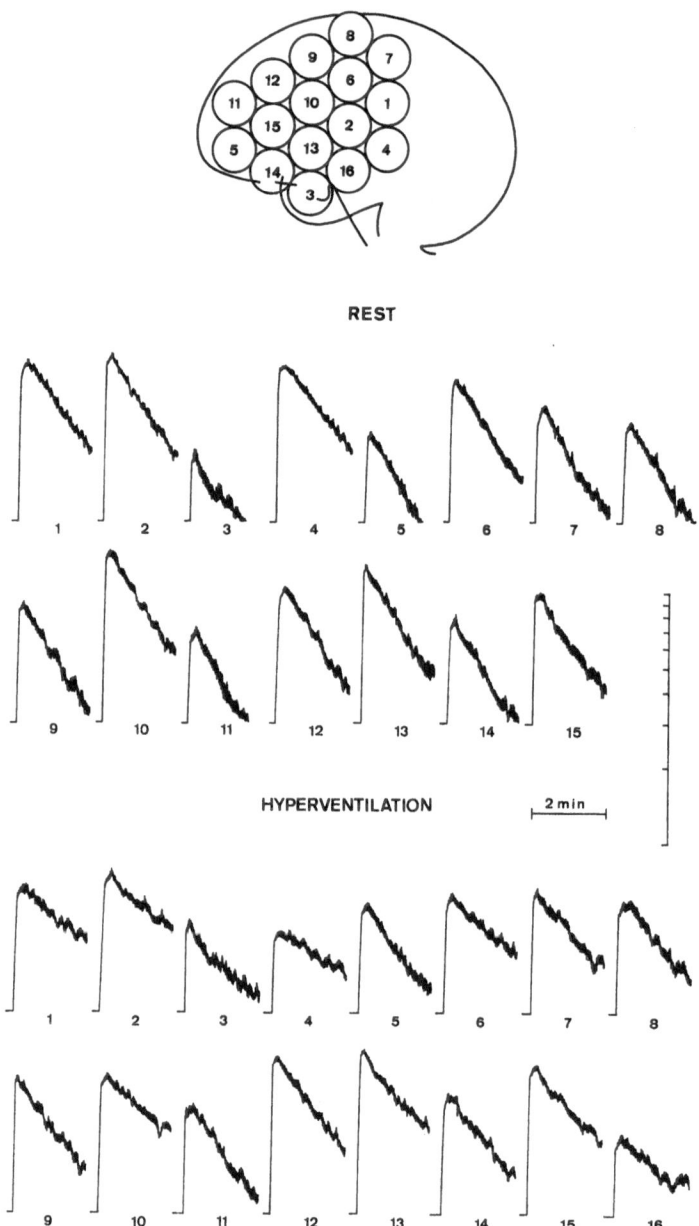

Fig. 2. Channel location and curve morphology of the rCBF study performed in the acute phase, 36 h after onset of the illness

It is suggested that, at least during the acute phase of the occlusion of a main cerebral artery, although the collaterals to the resulting ischemic territory are dilated by ensuing tissue acidosis and provide a sufficient collateral blood supply, these collateral (convexity) vessels still seem to retain a certain degree of reactivity to hypocapnia. Under such conditions it is understandable that hyperventilation could have a deleterious effect on rCBF in the affected area, as suggested in Fig. 3.

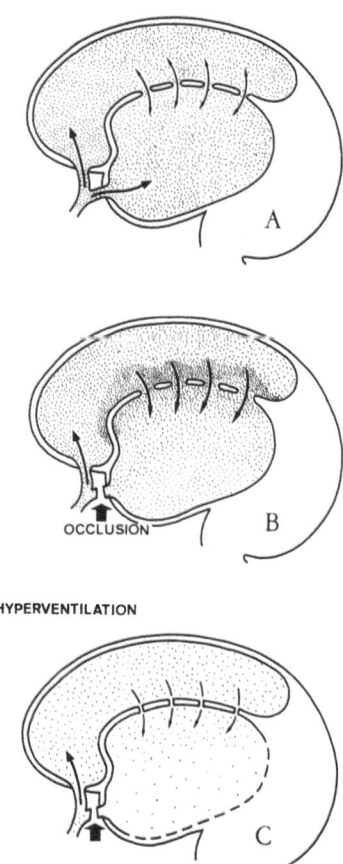

Fig. 3. Tentative schematic explanation for the effect of hyperventilation on rCBF during the acute phase of total proximal occlusion of a major cerebral vessel. The vessels communicating the anterior cerebral artery territory with the middle cerebral artery territory in the normal state (A) dilate in the presence of occlusion of the middle cerebral artery (B) but are still able to react to hyperventilation (C), rendering the distal territories ischemic.

References

1. BATTISTINI, N., M. CASACCHIA, A. BARTOLINI, G. BAVA, and C. FIESCHI: Effects of hyperventilation on focal brain damage following middle cerebral artery occlusion. Presented at the International CBF Symposium, Mainz, April 1969 (this volume, p. 249).
2. GORDON, E., and M. ROSSANDA: Artificial hyperventilation in the treatment of patients with severe brain lesions. Presented to the International CBF Symposium, Mainz, April 1969 (this volume, p. 258).
3. MEYER, J. S., T. SAWADA, M. TOYODA, and A. KITAMURA: Anaerobic cerebral metabolism induced by ischemia. Relation at hyperventilation and cerebrovascular disease in man. In: Cerebrovascular Diseases, p. 111. Grune & Stratton, Inc. 1968.
4. SOLOWAY, M., W. NADEL, M. S. ALBIN, and R. J. WHITE: The effect of hyperventilation on subsequent cerebral infarction. Anesthesiology 29, 975 (1968).
5. VAPALAHTI, M., H. TROUPP, and O. HEISKANEN: Extremely severe brain injuries treated with hyperventilation and ventricular drainage. Presented at the International CBF Symposium, Mainz, April 1969 (this volume, p. 266).

Artificial Hyperventilation in the Treatment of Patients with Severe Brain Lesions

E. Gordon and M. Rossanda

Department of Neuroanaesthesia, Karolinska Sjukhuset, Stockholm

Artificial respiration has been previously recommended in cases of severe brain lesion [1, 3, 5]. Lassen [4] suggested the use of moderate hyperventilation both for the correction of cerebral acidosis and for the reduction of intracranial pressure. Gordon and Rossanda [2] found not only clinical amelioration after the use of artificial ventilation in patients with brain lesions, but also a parallel regress of CSF acidosis.

Encouraged by these favorable initial results we have, over the past 2 years, treated patients with severe brain damage routinely with moderate hyperventilation.

The clinical course in 102 patients, treated at the Neurosurgical and Neurotraumatological Clinics in the Karolinska Sjukhuset, were studied. There were 30 females and 72 males. Age varied between 5 and 77 years with a mean value of 38.6 years. The diagnoses included 16 tumors, 16 cerebrovascular lesions (mostly arterial aneurysms) and 70 severe brain contusions. The indication for treatment with hyperventilation was either a certain degree of lowered consciousness and/or some type of abnormal breathing – and treatment was begun as soon as the patient arrived in the clinic or, if the patient's status was normal on admission, when the clinical condition deteriorated so that one of the above mentioned criteria was fulfilled.

Of the 102 patients 14 had a normal conscious level before the institution of treatment, 4 were somnolent, 39 had light coma and 45 deep coma. Breathing was normal in 26 cases, 15 patients hyperventilated spontaneously, 17 had hypoventilation, 12 irregular breathing (mostly of Cheyne-Stokes type) and 32 had apnoea.

86 of the 102 patients underwent one or more neurosurgical operations and most of them were treated with routine doses of cortisone. 84 patients were tracheotomized within the first 2 days, while 18 were treated by artificial ventilation through an endotracheal tube. Artificial ventilation was carried out with the Engström respirator using a gas mixture containing 30–50% O_2 and 50–70% air. The minute volume was adjusted so that the arterial pCO_2 remained between 25–35 mmHg. This treatment was continued in the patients with fatal outcome until cardiac standstill and in the unconscious stabilized group until normal spontaneous respiration could be maintained and the patient showed no improvement after at least a week. In the other 2 groups of patients respirator treatment was maintained until the patients' consciousness and breathing became normal with normal blood gas analyses. The period of respirator treatment varied from 6 h to 41 days (mean 10 days).

During respirator treatment arterial or capillary blood acid-base balance, and when possible oxygen tension and saturation studies, were done daily.

24 patients made a complete recovery, 9 patients recovered consciousness but showed more or less pronounced neurological sequelae, 15 patients remained comatose during

their stay at our clinic. 54 patients died after a period varying from some hours to 42 days. The majority of deaths (40 cases) occurred within 7 days. Postmortem examination of the brain was made in every case and in the majority of cases showed severe damage in vital regions.

The result of the treatment of patients with severe brain lesions is undoubtedly influenced by several factors. One of the most important is the site and severity of the *primary brain damage*, which results in an immediate and more or less extensive irreversible necrosis of the affected neurons. This damage can hardly be influenced by any therapeutic measures. This primary lesion, however, which in itself may not always be incompatible with normal or useful cerebral function, always initiates a chain of *secondary changes* – a sort of vicious circle – which, if not treated, usually results in more extensive irreversible cerebral damage often incompatible with life. The measures which one can take to avoid these *secondary changes* are therefore of cardinal importance in the final clinical outcome.

In this series of 102 patients treated with artificial hyperventilation it is extremely difficult to evaluate the effect of this treatment on the rate of recovery. However, the careful study of the clinical course of the patients who recovered or became much improved, gives strong evidence that the institution of artificial hyperventilation played a decisive role in the favourable clinical outcome in most of these cases thus altering a clinical course which before the onset of treatment showed a progressive deterioration. Relief from the very exhausting and oxygen-consuming respiratory overwork, shown by the patients with spontaneous hyperventilation, certainly played a great part in ameliorating the patient's clinical condition by allowing a greater part of the available oxygen to be utilized by the brain cells. The effectiveness of artificial hyperventilation in facilitating the removal of accumulated CO_2 from the brain cells and thus raising intracerebral pH is difficult to evaluate in this series. Obviously patients with severe primary brain lesions in vital areas may die or remain unconscious in spite of vigorous and early therapeutic measures. However, these results suggest that early respirator treatment will benefit patients with respiratory disturbances, whether leading to hypo- or hyperventilation, and patients with depression of consciousness of both minor and major degree.

References

1. Bozza-Marrubini, M., G. Marrubini, and R. Ghezzi: Mortality from respiratory complication during coma. Results of resuscitation treatment. Symposium Anaesthesiologiae Internationale, Prague, Aug. 1965.
2. Gordon, E., and M. Rossanda: The importance of the cerebrospinal fluid acid-base status in the treatment of unconscious patients with brain lesions. Acta anaesth. scand. 12, 51 (1968).
3. Huang, C. T., A. W. Cook, and H. A. Lyons: Severe craniocerebral trauma and respiratory abnormalities. Arch. Neurol. 9, 545 (1963).
4. Lassen, N. A.: The luxury-perfusion syndrome and its possible relation to acute metabolic acidosis localised within the brain. Lancet II, 1113 (1966).
5. Rossanda, M., G. Di Giugno, S. Corona, N. Bettinazzi, and G. Mangione: Oxygen supply to the brain and respirator treatment in severe comatose states. Acta anaesth. scand. Suppl. 23, 766 (1966).

Clinical Results of Respirator Treatment in Unconscious Patients with Brain Lesions

M. Rossanda, M. Bozza-Marrubini, and A. Beduschi

Department of Anaesthesiology and Intensive Care, Ospedale Maggiore, Milano,
and Department of Neurosurgery, Ospedale Civico, Palermo

In 1965 it was first noticed that the use of artificial ventilation in unconscious patients with brain lesions was followed by unexpected recoveries [4] and a decrease in death rate [1]. Since then unconscious patients have been subjected to respirator treatment as frequently as possible. This has not meant a systematic treatment of all the patients who might have benefited from it. Shortage of respirators and personel often caused the indication to be limited to the most severely ill patients, that is, to patients with extensor hypertonus or associated extracerebral lesions. A statistical comparison between treated and untreated patients could not be made, on account of the obvious inhomogeneity of the material.

Only treated patients are presented here (33 cases). The group includes: head injuries (14 cases), cerebral tumours after operation (9 cases), brain lesions following cardiac arrest (4 cases), encephalitis (5 cases) and cerebral embolism (1 case). The patients' age ranged from 18 months to 67 years. All patients were deeply unconscious. In 70% of the cases severe signs of brain stem involvement were present. It must be emphasized, however, that the patients with persistent bilateral pupillary areflexia, absence of motor reaction and apnea were not taken into consideration for therapeutical respirator treatment.

Respirator treatment was started within 6 h from the onset of coma in 14 cases. During treatment the trend was to avoid heavy sedation and to give the patients "as much air as they wanted" – which usually meant 14–18 l per min. The air was always enriched with oxygen in variable amounts. The duration of artificial respiration ranged from 2 to 44 days (average 12 days). Blood gas studies could not be extended to all the patients treated: in all cases of this group, however, arterial pCO_2 was checked daily, at least during the first days of artificial ventilation. In 25 cases arterial pO_2 could also be recorded. The relations between mortality rate and 1) blood gas levels during ventilation, 2) the time interval between onset of coma and start of the artificial ventilation were also analysed.

Results

The overall mortality was 36%. All patients who survived also recovered consciousness. Age seemed to influence survival: the death-rate was 66% in patients over 50, and 25% in those under this age (chi square: 3.27; p between 5 and 10%).

Mortality related to arterial pCO_2: The range of arterial pCO_2 was calculated from the daily variations. In patients with arterial pCO_2 ranging from 20–28 mmHg (8 cases) the mortality was 25%, in those with a range of 29–35 (15 cases) 26.6%, and in those with a range 36–46 (10 cases) 60%.

The difference between the 2 first groups and the third group is barely significant (chi square $= 3.46$; p between 5 and 10%).

Mortality related to arterial pO_2: The death-rate was found to be 81.8% for the patients with arterial pO_2 lower than 90 mmHg (11 cases), 7.1% for the patients with arterial pO_2 above 90 mmHg (14 cases). This difference is highly significant (chi square $= 11.37$; $p < 1\%$). The only patient who died having a high pO_2 had been treated late (36 h after a head injury). In the group with lower pO_2, there was a high percentage of old patients. One may wonder if the difficulty in maintaining a high arterial pO_2 is a factor in the high death-rate of elderly subjects in coma.

In this group the relation of pO_2 and pCO_2 to mortality are rather difficult to distinguish from one another, since most patients that recovered had both a high pO_2 and a low pCO_2 (Fig. 1). The separation between death and recovery, however, seems more clearcut for pO_2 than for pCO_2. The present data do not allow any conclusion about the relation between pCO_2 and recovery: this point would require a special experimental scheme, and also a more homogeneous group of patients. The relation between pO_2 and mortality, instead, is too impressive to be accidental. This result may give further support to other observations concerning the relation between severe brain lesions, arterial hypoxemia and mortality of unconscious patients [2, 3, 5].

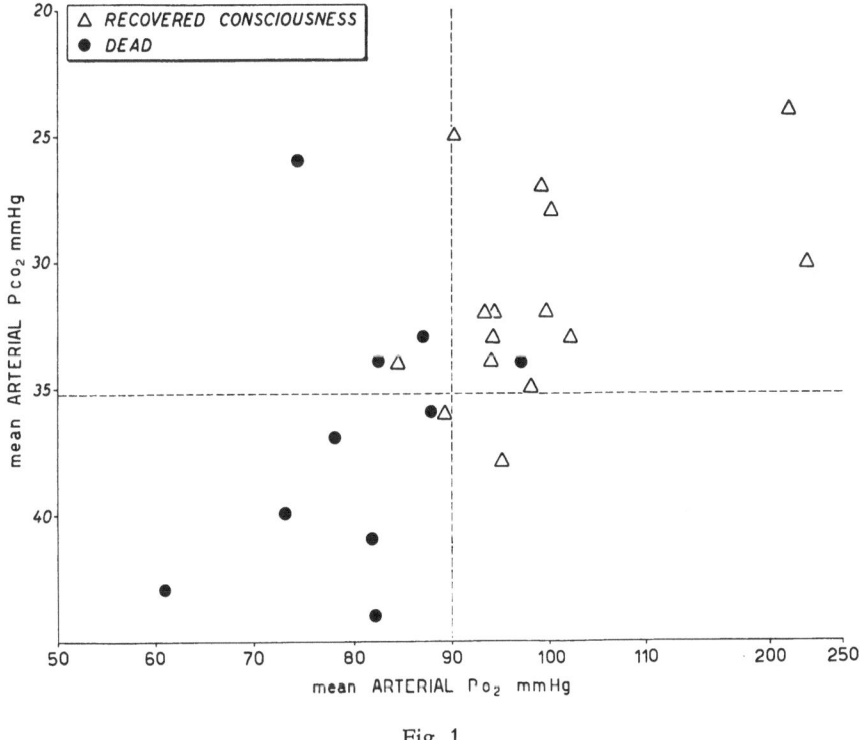

Fig. 1

Mortality according to the interval between onset of coma and start of respirator treatment: When the treatment was started within 6 h mortality was 21.4%; from 6–24 h 33%, and after 24 h 53.4%. The differences in death-rate, are not statistically significant (chi square $= 2.83$, $p = 10\%$). However, all 3 patients who died having been treated early had a low pO_2 and a high pCO_2 during the treatment.

We believe that a beneficial effect may be obtained only if artificial ventilation is started early after the onset of unconsciousness and is maintained with high oxygen concentrations and with high volumes, which usually do not reduce arterial pCO$_2$ to excessively low level. As to the possible advantages of a definite hypocarbia, the data at present available are not sufficient for any conclusion.

References

1. BOZZA-MARRUBINI, M.: Acquisizioni recenti e problemi insoluti della rianimazione neurologica. Report at a meeting in the Institute of Neurology, University of Genoa, December 1966.
2. FROWEIN, R. A., u. A. KARIMI-NEJAD: Sauerstoffversorgung des Hirngewebes nach schweren Hirnschädigungen. Acta Neurochir. 19, 3 (1968).
3. HUANG, C. T., A. W. COCK, and H. A. LYONS: Severe craniocerebral trauma and respiratory abnormalities. Arch. Neurol. 2, 545 (1963).
4. ROSSANDA, M., G. DI GIUGNO, S. CORONA, N. BETTINAZZI, and G. MANGIONE: Oxygen supply to the brain and respirator treatment in severe comatose states. Acta anaesth. scand. Suppl. 23, 766 (1966).
5. SIRIS, J. H., E. I. HENRY, and D. S. CUKIER: Occult hypoxemia complicating acute and subacute intracranial lesions. N.Y. St. J. Med. 62, 1444 (1962).

Cerebrospinal Fluid and Cerebral Venous pO_2 in Unconscious Patients with Brain Lesions

M. ROSSANDA and E. GORDON

Department of Anaesthesiology, Section of Neuroanaesthesia, Karolinska Hospital, Stockholm

The measurement of CSF pO_2 has been proposed by several authors [1, 2, 3, 6, 7] as a test of oxygen availability to the brain. The validity of this test is not yet perfectly established. However, since this is at present the only direct way of approaching the problem of cerebral hypoxia in clinical practice, it deserves further exploration.

Methods and Materials

The oxygen tension in CSF, arterial blood and, in some cases, in blood drawn from the bulb of the internal jugular vein, was measured in 22 patients without apparent brain lesions and subjected to neuroradiological examinations, in 34 patients with various brain lesions, but not unconscious, and in 20 comatose patients with severe brain lesions. No patients with total areflexia or shock were studied.

The oxygen tension was measured with the Radiometer oxygen electrode and apparatus, the electrode was calibrated with water equilibrated with air at 37 °C. The value was then corrected to the actual body temperature with the aid of SEVERINGHAUS' nomogram. CSF was collected in air-tight glass syringes, the first ml was discarded with the air filling the deadspace of the syringe. Only clear fluids were examined.

The subjects breathing air have been separated from those breathing oxygen-rich mixtures.

Results

In the group of patients breathing air, CSF pO_2 was found significantly lower in coma (26.2 $\pm$ 4.32 mmHg) than in patients with no brain lesion (43.0 $\pm$ 10.79) and with mild brain lesions (39.3 $\pm$ 9.45) ($t = 3.79$ and 4.60, $p < 0.01$). In this group also arterial pO_2 was significantly lower in comatose patients (79.0 $\pm$ 11.62) than in patients with mild brain lesions (89.1 $\pm$ 12.90) ($t = 4.37$, $p < 0.05$).

In the group of patients breathing oxygen-rich mixtures, the comatose subjects had again a significantly lower CSF pO_2 (29.7 $\pm$ 8.13) than those with no brain lesion (58.6 $\pm$ 15.21) and those with mild brain lesions (41.2 $\pm$ 14.14) ($t = 4.30$, $p < 0.01$, and 2.09, $p < 0.05$). In this group arterial pO_2 did not differ significantly in patients with no brain lesions (187 $\pm$ 45.7), mild brain lesions (212 $\pm$ 60.2) and coma (173 $\pm$ 70.3).

In a similar series of patients it had been found previously that jugular venous pO_2 was a little higher in comatose patients than in patients with milder brain lesions (mean value

40 mmHg against 39 during air breathing and 41.4 against 37.3 under O_2-rich mixture; differences not statistically significant).

A high oxygen content of the cerebral venous blood in cases of severe brain lesions may be the consequence of the hyperemia of the injured brain, where the oxygen consumption would be primarily depressed. In this case, however, one should expect to find a high CSF pO_2, as recorded by BLOOR et al. [1] in dogs poisoned with potassium cyanide. The finding of low CSF pO_2 is then discordant with this view.

In order to understand better this point, in 22 cases with simultaneous sampling of arterial blood, jugular venous blood and CSF, the mean capillary pO_2 was calculated from the arterial and jugular pO_2, the hemoglobin content and the pH, using SEVERINGHAUS' nomograms [4]. In each case the calculated mean capillary pO_2 was compared with the respective CSF pO_2. The mean capillary pO_2 (cap pO_2) was 54.3 ± 6.62 mmHg in 5 control patients and 55.2 ± 4.92 in comatose patients under air-breathing. The difference is not significant. Also in 6 patients with O_2-rich mixtures the mean-capillary pO_2, while increased, was not appreciably different in unconscious and not unconscious subjects (Fig. 1). The average difference between mean capillary pO_2 and CSF pO_2 was instead significantly lower in the control patients (18.1 ± 9.69 mmHg) than in the unconscious patients (30.0 ± 6.27) ($F = 12.37$ $p < 0.05$) (Fig. 1). The possible differences between lumbar and ventricular fluids do not seem to play a role in this case, the mean values for controls and comatose patient's being respectively 18.5 and 30.3 in case of lumbar samples, and 17.5 and 28.5 in case of ventricular samples.

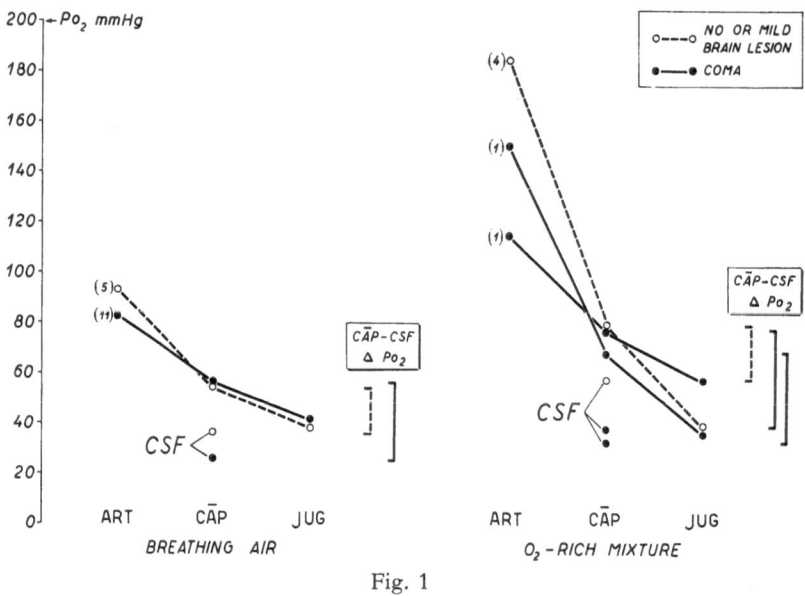

Fig. 1

Comments

In this group of patients CSF pO_2 did not seem to depend passively on arterial pO_2 changes. A low CSF pO_2 has been consistently found in comatose patients and also in some of the patients with no or mild brain lesions. We are bound to believe that there is some correlation between CSF pO_2 and cerebral tissue pO_2 in man.

We think therefore that the brains of the unconscious patients in this group were hypoxic in spite of the normal venous and calculated mean capillary pO_2. The increased difference between blood and CSF pO$_2$ may be explained by an alteration of the normal perfusion patterns and/or an obstacle to gas diffusion, such as brain edema.

The increased oxygen gradient between blood and CSF may be correlated to the increased gradient of carbon dioxide that has been found in very similar experimental conditions [5]. The increased pO$_2$ gradient in severely injured brains may also help to explain the strong relationship recorded between the mortality rate of the unconscious patients and their arterial pO$_2$ [8].

References

1. BLOOR, B. M., J. FRICKER, F. HOLLINGER, H. NISHIOKA, and J. McCUTCHEN: A study of cerebrospinal fluid oxygen tension. Arch. Neurol. **4**, 37 (1961).
2. DUNKIN, R. S., and S. BONDURANT: The determinants of cerebrospinal fluid PO$_2$. Ann. intern. Med. **64**, 71 (1966).
3. GÄNSHIRT, H.: Der Sauerstoffdruck im Liquor cerebrospinalis. Med. Wschr. **116**, 953 (1966).
4. GLEICHMANN, U., D. H. INGVAR, D. W. LÜBBERS, B. K. SIESJÖ, and G. THEWS: Tissue PO$_2$ and PCO$_2$ of the cerebral cortex, related to blood gas tension. Acta physiol. scand. **55**, 127 (1962).
5. GORDON, E., and M. ROSSANDA: The importance of the cerebrospinal fluid acid-base status in the treatment of unconscious patients with brain lesions. Acta anaesth. scand. **12**, 51 (1968).
6. JARNUM, S., I. LORENZEN, and E. SKINHØJ: Cisternal fluid oxygen tension in man. Neurology **14**, 703 (1964).
7. MOLLARET, P., J. J. PECIDALE, M. C. BLAYE et G. POYART: La pression partielle de l'oxygène du liquide céphalo-rachidien, nouvelle constante biologique. C. R. Acad. Sc. Paris **259**, 4857 (1964).
8. ROSSANDA, M., M. BOZZA-MARRUBINI, and A. BEDUSCHI: Clinical results of respirator treatment in unconscious patients with brain lesions. Presented at the International CBF Symposium, Mainz, April 1969 (this volume, p. 260).

Extremely Severe Brain Injuries Treated with Hyperventilation and Ventricular Drainage

M. Vapalahti, H. Troupp, and O. Heiskanen

Department of Neurosurgery, Central University Hospital, Helsinki

Since 1964 our department has been using Lundberg's method [2] of continuous recording of intraventricular pressure for unconscious brain-injured patients. We have found that the intraventricular pressure is a good prognostic guide as to survival [5]; in our previous series all patients with a prolonged intraventricular pressure over 60 mmHg died in spite of all efforts.

Hyperventilation was used by Lundberg [3] in terminating plateau-waves in patients with brain tumours and Rossanda [4] suggested hyperventilation as a routine treatment in severe brain injuries. Between January 1st, 1967, and March 31st, 1969, we have had 37 patients with severe brain injuries in whom the intraventricular pressure was recorded for at least the first few days after injury. All patients were deeply unconscious responding to pain only, and all except 1 child with prolonged intubation had undergone tracheotomy. 26 of these patients had pressures below 60 mmHg and 25 survived. None of these 26 had hyperventilation (Table 1).

Table 1. *Fate of 37 patients with severe brain injuries*

Intraventricular pressure	Recovered	Vegetative survival	Died	Total
less than 15 mmHg	3	4	—	7
15–30 mmHg	5	6	—	11
30–60 mmHg	2	5	1	8
above 60 mmHg	1	2	8	11

11 patients had a prolonged intraventricular pressure of over 60 mmHg. Hyperventilation was used in 9 of these patients, and 3 out of 9 survived; the case reports of these survivors are as follows:

Patient 1. An 18-year-old boy suffered a generalized brain injury in a traffic accident. On admission he was unconscious and had extension rigidity. An intraventricular pressure recording was started and showed very high plateau-waves. Hyperventilation was started, but as this did not keep the pressure down, we started ventricular drainage. 10 days after the injury a Holter shunt was performed. However, the patient did not recover consciousness and died 11 months later from pneumonia.

Patient 2. A 14-year-old boy suffered multiple fractures of the legs and a generalized brain injury. He also had extension rigidity and hemiballism. During the first day after the injury, he was breathing normally, and his acid-base balance in CSF was normal. He developed high plateau-waves during the first few days after the injury, even after the artificial hyperventilation had brought arterial pCO_2 down to 20 mmHg. At that time CSF bicarbonate was 15 mEq/l. Ventricular drainage was started and set to permit overflow if the intraventricular pressure rose above 40 mmHg. On the 17th day of the injury the intraventricular pressure was 20 mmHg even with the patient breathing spontaneously and the recording was stopped. This boy made an amazingly good recovery and seems normal 10 months after the injury.

Patient 3. An 8-year-old boy suffered a generalized brain injury when he was run down by a car. He had a CSF pH of 7.10 and his high intraventricular pressure could not be kept down by hyperventilation alone, so ventricular drainage was started. The pressure was normal with the patient breathing spontaneously on the 13th day after the injury. This boy is still unconscious 3 months after the injury and does not seem to be recovering.

Except for one patient, whose pressure recording was possibly terminated too early to gauge his condition, all patients with an intraventricular pressure mainly under 60 mmHg survived. These were all treated in the conventional manner, with tracheostomy, dehydrating agents, oxygen and/or corticosteroids. The treatment of severe brain injuries by routine hyperventilation demands much time and effort from the medical and the nursing staff. We think it fruitless to treat brain-injuried patients with hyperventilation without accurate knowledge of the intracranial pressure [1]: our series suggests that pressures below 60 mmHg responded to current conventional treatment and that pressures above 60 mmHg responded in a reasonable manner in only 1 patient; the intensive therapy outlined in the case reports seems worth while for children and adolescents only. Unless it can be shown, therefore, that the *quality* of survival can be improved by hyperventilation – a controlled trial might give such an information – this costly method used routinely would be better replaced by routine continuous pressure recording.

References

1. LANGFITT, T. W., J. D. WEINSTEIN, and N. F. KASSELL: Vascular Factors in Head Injury. In: Head Injury, W. F. CAVENESS and A. E. WALKER (Eds.), p. 172. Philadelphia: J. B. Lippincott 1966.
2. LUNDBERG, N.: Continuous recording and control of ventricular fluid pressure in neurosurgical practice. Acta Psych. Neur. Scand. Suppl. **149**, (1960).
3. —, A. KJÄLLQUIST, and C. BIEN: Reduction of increased intracranial pressure by hyperventilation. Acta psychiat. scand. Suppl. **139** (1959).
4. ROSSANDA, M.: Prolonged hyperventilation in treatment of unconscious patients with severe brain injuries. Scand. J. clin. Lab. Invest. Suppl. **102**, p. XIII-E (1968).
5. TROUPP, H.: Intraventricular pressure in patients with severe brain injuries. J. Trauma **7**, 875 (1967).

The Effect of Hyperbaric Oxygen on Intracranial Pressure and Cerebral Blood Flow in Experimental Cerebral Edema

J. D. Miller, I. Mc. A. Ledingham, and W. B. Jennett

Institute of Neurological Sciences and Hyperbaric Unit, Glasgow

Previous experimental work [1, 2, 10] has suggested that hyperbaric oxygen (OHP) will reduce the mortality in animals with cerebral edema and compression. Clinical experience with OHP in patients with head injuries has however been less promising [3].

In previous studies we have shown that OHP will reduce raised intracranial pressure (ICP) when arterial pCO_2 and blood pressure are kept constant, and that the decrease in ICP is accompanied by an increase in cerebral arterio-venous oxygen content difference [7, 8]. It was also found that reduction in ICP could be obtained only when the cerebral vessels were reactive, as indicated by a rise in ICP when CO_2 was added to the inspired gases. Since the changes in ICP were extremely rapid, and since OHP is known to cause vasoconstriction, it appeared likely that the fall in ICP was mediated by a reduction in CBF; the present study was undertaken to test this hypothesis.

Methods

Cerebral edema with raised ICP was produced in 7 anaesthetised, ventilated dogs by the application of liquid nitrogen to the intact dura over the cerebral hemisphere [9]; intracranial pressure was measured from the extradural space overlying the lesion and rCBF from the contralateral hemisphere by counting β emissions through the intact but devascularised dura [8] following intracarotid injection of ^{85}Kr, and using the stochastic analysis to obtain mean rCBF over a 10 min period. Cerebral A-V oxygen content and pCO_2 difference were also measured. All animals were studied at an ambient pressure of 2 ATA in a walk-in hyperbaric chamber; during control periods they breathed an air-equivalent mixture of 10% O_2 and 90% N_2 which was changed to 100% O_2 for 1 h during each run of OHP. Ventilation was adjusted to maintain normocapnia, and at intervals 5% CO_2 was added to the inspired gases to test cerebrovascular reactivity.

Results

In 5 dogs OHP caused a mean reduction of 30% in ICP, and of 19% in CBF (12 runs) (Table 1). There was no significant change of arterial pCO_2 or BP. The arterial pO_2 showed a 10-fold rise, while cerebral venous pO_2 rose from 47.4–81.4 mmHg on changing from

Supported by the Medical Research Council. JDM was in receipt of a MRC Fellowship.

air to OHP. There was a rise in cerebral A-V pCO_2 difference and a small increase in A-V O_2 content difference. Cerebrovascular resistance increased significantly with OHP, confirming that OHP was causing cerebral vasoconstriction despite the fact that arterial pCO_2 had not altered. In these 5 dogs addition of CO_2 to the inspired gases invariably produced a rise in ICP and/or CBF throughout the experiments which lasted 6–8 h.

Table 1. *Effects of hyperbaric oxygen in experimental cerebral edema. 5 dogs : 12 runs : $M \pm SE$*

	Air			OHP	
Intracranial Pressure (mmHg)	43	$\pm$ 4	< 0.001	30	$\pm$ 4
Arterial Blood Pressure (mmHg)	119	$\pm$ 3	NS	116	$\pm$ 2
Cerebral Perfusion Pressure (mmHg)	76	$\pm$ 4	< 0.001	86	$\pm$ 5
Arterial pCO_2 (mmHg)	40.5	$\pm$ 1.1	NS	40.4	$\pm$ 0.8
Arterial pO_2 (mmHg)	105	$\pm$ 4	< 0.001	1196	$\pm$ 29
Cerebral Venous pO_2 (mmHg)	47.5	$\pm$ 2.8	< 0.001	81.4	$\pm$ 9.0
Regional Cerebral Blood Flow (ml/g/min)	0.94	$\pm$ 0.13	< 0.001	0.76	$\pm$ 0.11
C (a-v) O_2 (ml/100 ml)	5.2	$\pm$ 0.7	NS	5.7	$\pm$ 0.5
P (a-v) CO_2 (mmHg)	10.9	$\pm$ 1.1	< 0.01	14.4	$\pm$ 1.1
rCVR (mmHg/ml/100 g/min)	0.91	$\pm$ 0.09	< 0.02	1.29	$\pm$ 0.23

In the 2 remaining dogs larger brain lesions were made, ICP rose to more than 70 mmHg within 2 h of making the lesion, and the response to CO_2 was lost. OHP did not influence either ICP or CBF which was much reduced.

Comments

Under normal conditions OHP is known to reduce CBF.

LAMBERTSEN et al. [5] attributed this reduction to a fall in arterial pCO_2, but JACOBSON et al. [4] were able to demonstrate cerebral vasoconstriction in dogs when pCO_2 was controlled. This study was carried out under conditions of increased ICP, but ZWETNOW et al. [11] have shown that CBF does not become reduced until cerebral perfusion pressure falls below 40 mmHg. Only in the 2 non-reactive dogs was the perfusion pressure in this low range. The results in the 5 reactive dogs support the concept that OHP can reduce CBF by direct cerebral vasoconstriction since arterial pCO_2 and BP were not changed. Despite the reduction in CBF cerebral oxygenation is improved, as indicated by a rise in cerebral venous pO_2.

LANGFITT et al. [6] have demonstrated the loss of cerebrovascular reactivity to CO_2 that occurs with increasing ICP. This loss of reactivity appears to apply also to OHP. Furthermore, it has been shown that this "vasomotor paralysis" may occur focally in relation to many types of brain lesion. Thus, in considering the reactivity of the cerebral circulation as a whole in relation to such lesions (whether ischaemic or traumatic), and therefore the usefulness of hyperbaric oxygen therapy, 2 dimensions have to be considered; the volume of brain damaged by the lesion, and the time from the occurrence of the lesion, during which ICP may have been rising. Differences in these factors may explain the widely varying results so far obtained in OHP therapy of head-injured patients and indicate possible avenues of progress.

References

1. Coe, J. E., and T. M. Hayes: Treatment of experimental brain injury by hyperbaric oxygenation. Preliminary Report. Amer. Surg. **32**, 493 (1966).
2. Dunn, J. E., and J. M. Connolly: Proc. 3rd Internat. Conf. on Hyperb. Med., Wash., p. 447, 1966.
3. Fasano, V. A., G. Broggi, R. Urciuoli, T. Denunno, and G. F. Lombardi: Proc. 3rd Internat. Cong. Neurol. Surg., Excerpta Med. Series Chirurgica **110**, 502 (1966).
4. Jacobson, I., A. M. Harper, and D. G. McDowall: The effects of oxygen under pressure on cerebral blood-flow and cerebral venous oxygen tension. Lancet II, 549 (1963).
5. Lambersten, C. J., R. H. Kough, D. Y. Cooper, A. L. Emmal, H. H. Loeschke, and C. F. Schmidt: Oxygen toxicity. Effects in man of oxygen inhalation at 1 and 3.5 atmospheres upon blood gas transport, cerebral circulation and cerebral metabolism. J. appl. Physiol. **5**, 471 (1953).
6. Langfitt, T. W., J. D. Weinstein, and N. F. Kassell: Cerebral vasomotor paralysis produced by intracranial hypertension. Neurology **15**, 622 (1965).
7. Miller, J. D.: Experimental studies in intracranial pressure and hyperbasic oxygen. J. Neurol. Neurosurg. Psychiat. **32**, 66 (1969).
8. —, W. Fitch, and B. Cameron; 1969, in press.
9. Rosomoff, H. L.: Experimental brain injury during hypothermia. J. Neurosurg. **16**, 177 (1959).
10. Sukoff, M. H., S. A. Hollin, D. E. Espinosa, and J. H. Jacobson: The protective effect of hyperbaric oxygenation in experimental cerebral edema. J. Neurosurg. **29**, 236 (1968).
11. Zwetnow, N., A. Kjällquist, and B. K. Siesjö: Cerebral blood flow during intracranial hypertension related to tissue hypoxia and to acidosis in cerebral extracellular fluids. In: Cerebral Circulation, W. Luydendijk (Ed.), p. 87. Amsterdam: Elsevier 1968.

rCBF during Hyperbaric Oxygenation

R. WÜLLENWEBER, U. GÖTT, and K. H. HOLBACH

Department of Neurosurgery, University of Bonn

Since 1966 40 hyperbaric oxygenation treatments (HOT) have been performed in cases with severe head injuries, marked posttraumatic or postoperative cerebral edema, acute cerebrovascular occlusions or grave meningitis. We have used a one man pressure chamber.

Hyperbaric oxygenation was applied in order to diminish or avoid hypoxic or anoxic brain lesions by interrupting the vicious circle: brain edema → circulatory disturbances → hypoxydosis. Most of the cases improved following hyperbaric oxygenation, some showed an amelioration already during the treatment. To elucidate the effect of HOT upon the injured brain we measured blood pH, blood gases, lactate and pyruvate concentrations and registered EEG and EKG.

We also registered rCBF of the affected cerebral region using thermoprobes.

The results of 4 cases (measured twice from the surface and twice from the depth of the brain by needle shaped probes) were analysed.

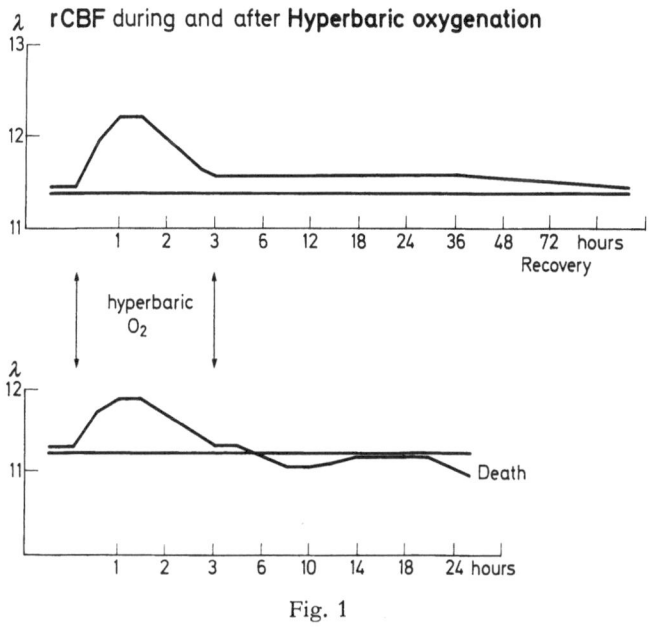

Fig. 1

In contrast to some published results, in these 4 cases rCBF increased during hyperbaric treatment (Fig. 1), but the increase of rCBF continued only until a pressure of 2.5 ata was reached.

At pressure over 2.5 ata no further rise of rCBF occurred. During the different periods of hyperbaric treatment blood pressure did not change. While arterial pO_2 rose to 900 the 1100 mmHg, the venous pO_2 only increased slightly. Arterial pCO_2 and pH remained normal.

The concentrations of lactate and pyruvate dropped corresponding to the rising arterial pO_2. Frequency analysis of the EEG showed an increase in α-waves corresponding to the hyperbaric phases while the amount of δ-waves did not alter.

When the O_2-pressure was lowered the temporarily improved values became worse again. But in the cases showing an improvement of the clinical state, a slightly increased rCBF persisted after decompression (Fig. 1).

In one severely injured patient rCBF rose only a little during the hyperbaric period and then, during decompression, returned to the level recorded before treatment. Furthermore, the decrease in rCBF continued after the treatment and the patient died about 24 h later.

In view of the increase of rCBF seen in these cases, and the positive clinical results, we believe that hyperbaric oxygenation is indicated in cases of severe brain lesions.

References

1. Gött, U. and K. H. Holbach: Hyperbare Sauerstofftherapie bei neurochirurgischen Patienten. Der Anaesthesist, 1969, in press.
2. Holbach, K. H., and U. Gött: Erste Beobachtungen und Erfahrungen mit der hyperbaren Sauerstofftherapie bei Hirngeschädigten. Deutscher Luft- und Raumfahrt-Forschungsbericht, 1969, in press.
3. — — Electroencephalographical and clinical findings in hyperbaric therapy. Joint congress of the Belgian and German Neurosurgical Societies. Knokke, September 1968.

Comments to Chapter VII

Anesthesia and Therapy

This session considered three general topics of both theoretical and clinical importance: 1. factors controlling cerebral circulation during anesthesia; 2. the effects of hypocarbia upon regional circulation in the normal and abnormal brain; and 3. the effects of hyperoxygenation upon regional circulation in the normal and abnormal brain.

The first three papers pointed to several factors which control cerebrovascular resistance. The effects of carbon dioxide on cerebral vessels, however mediated, are probably the most important in both awake and anesthetized man. The impression is created that the influence of carbon dioxide on cerebral vessels cannot be easily modified. A second influence, that of blood pressure, results in the autoregulatory phenomena which are well known in normal awake man. Studies presented here demonstrated that autoregulatory mechanisms are intact under most circumstances during anesthesia. However the controlling effect of oxygen tension on cerebral circulation is changed considerably. As anesthesia deepens, increasing cerebral blood flow and decreasing oxygen consumption result in a considerable increase in venous and presumably tissue oxygenation, which apparently does not result in cerebral vasoconstriction. Therefore, of these three factors controlling cerebral blood flow, oxygen is the one whose influence can be easily disturbed. The fact that cerebral oxygenation increases during deep anesthesia may be a fortunate one in that it provides a degree of reserve against some situations in which cerebral flow may be decreased, and other situations where cerebral metabolic requirements may increase. Just such a situation was discussed in the paper on Ethrane and cerebral seizures. The increase in cerebral oxygenation during moderately deep Ethrane anesthesia was sufficient to provide for the higher metabolic rate occurring during grand mal seizures.

The effect of hypocarbia on regional cerebral circulation in the normal and in the diseased brain has received considerable attention. To date, the hard facts are limited to a few controlled studies in animals. These suggest that under certain circumstances, where a localized cerebral lesion exists, one can make therapeutic use of the inverse cerebral steal. That is, hyperventilation ought to be good treatment for certain cerebrovascular disorders. However clinical data have not yet rigorously confirmed this suspicion. We have a few suggestive pieces of evidence indicating that hyperventilation has helped some patients; we have some evidence that hyperventilation can be harmful to others. What we need are a number of controlled studies. We must find out, in carefully matched series, whether high, normal, or low tensions of carbon dioxide in the arterial blood are the most beneficial therapeutically. We must find out whether it is simply mechanical ventilation, with its improved pulmonary function and improved arterial oxygenation, which is providing the benefit. We must define the types of patients in whom hyperventilation can help and those in whom it can be harmful. The immense amount of patient material available in some institutions should be thoughtfully subdivided into classes and scientifically studied, utilizing standard statistical procedures and principles of experimental design and objective observation.

The field of hyperbaric oxygenation and its effects on local cerebral circulation and intracranial pressure is another promising area. We are in essentially the same state here as in our understanding of high and low carbon dioxide tensions. Animal studies provide us with suggestive data, indicating that perhaps some patients could be helped with hyperbaric oxygenation. A few anedoctes derived from patients tend to add credence to this suspicion. Again what we need are controlled studies in patients. We must know what kinds of cerebral lesions will benefit from high pressures of oxygen; what the potiential harm is in certain cerebral diseases; whether the increase in arterial oxygenation is sufficient to overcome the decrease in cerebral blood flow induced by high oxygen pressures; what the useful doses of high pressure oxygen are; and what the best dosage schedules might be. Again, as in the case of hyperventilation, we will benefit little from further anecdotal reports. Well-designed and controlled scientific studies in patients are vitally needed now.

H. WOLLMAN

Closing Remarks I

Theories and Methods for CBF Measurement

S. S. Kety

Department of Psychiatry, Massachusetts General Hospital, Boston

What has impressed me most about this meeting has been the changes which have occurred in the field of cerebral circulation and the great strides which have been made in the past two decades and especially in the past few years.

I think this field made a major inflection some 6 or 7 years ago when Ingvar and Lassen developed a regional blood flow technique using the clearance of radioactive Krypton and Xenon from the brain. This gave a new dimension to the field because it produced some results which have been of diagnostic, prognostic and therapeutic significance in individual patients. One could previously study a series of 10 patients and find out a great deal about the normal physiology of the cerebral circulation or what a drug did to it, but I have always been somewhat chagrined by how infrequently measurement of the global cerebral circulation was of diagnostic value in a particular patient.

I think that a great deal of the activity in this field and the reason that most of you are here at this meeting, has been because of the possibility of studying regional circulation in the individual patient and the important clinical implications which stem from it.

I would like to comment largely upon the first day's session, which needs perhaps a summary more than the others which are still fresh in our minds. It was there that methodology was largely discussed. A field is no better than the methods which it uses and I want to comment upon a number of very interesting and cogent methodological developments which were presented.

Dr. Hutten started the session with a comparison of the stochastic analytical technique, and the exponential analysis of clearance curves. He presented evidence which, I believe supports a notion that I have had for some time. The stochastic analysis theoretically avoids most of the assumptions which exponential analysis requires. In fact, as Zierler is fond of saying: the stochastic technique requires no information about diffusion or about equilibration between venous blood and tissue. It is a complete black box that requires only information about what goes in and what comes out. That is of course true, and I think that Dr. Hutten showed with his models and with his sophisticated mathematical analysis that the stochastic analysis does not require any assumptions. On the other hand, he showed just as well, with his experiments that, in a natural situation the stochastic method tends to break down a little more readily than does the exponential method. In the stochastic method in actual application, it is difficult to obtain the height of the curve precisely, because it is difficult to get an input function which is sharp enough. Then, of course, at the other end of the curve, in order to get the complete area under the curve, one has to extrapolate the tail out to the very end and one is not quite sure how that extrapolation is made. If diffusion is slow or if there is a large poorly perfused sement, it isn't possible to obtain the area under the curve. I have the feeling that when the stochastic analysis can be used with validity, the exponential method gives equally good results, because under those conditions the assumptions which the exponential method makes are borne out. In addition to which the exponential techniques give one more information because they yield information on the 2 phases, which of course are pure fictions of the computation in curve analysis. Yet, if these fictions can be shown and have some physiological validity, then it is useful to have that information. Dr. Reivich's paper illustrated that these exponential phases do have a real validity in the case of the normal brain which happens unfortunately to be composed of distinct populations of rapid flow and slow flow; thus, it has all the requirements which permit the two-phase model to represent with a remarkable degree of precision, the means of these perfusion rates which actually exist in the brain.

Dr. Reivich also expanded a bit upon his original bimodal Gaussian analysis of these functions. It was interesting that he came up with the findings that in most instances in the normal brain the more precise Gaussian examination of the situation and the simple 2 phase exponential give essentially the same answer, and that this answer is remarkably close to the actual situation.

I was impressed with the accumlating evidence that we have an internal consistency among the various methods which are being used and have been used in the study of cerebral circulation. I was especially impressed with the internal consistency of the 2 approaches based upon exchange of diffusible tracers, but which are based upon quite different assumptions.

I here refer, on the one hand, to those techniques, such as the nitrous oxide technique which are based upon measurements of concentrations in the venous blood, and, on the other hand, to those techniques such as the Xenon clearance which are based upon the tissue concentrations. Each of these techniques makes quite different assumptions about diffusion and equilibration in the brain and it is interesting that the values come out to be as close as they do. For example, the Ingvar and Lassen regional clearance techniques give results which are quite compatible with the nitrous oxide-overall-techniques. The autoradiographic technique, if one averages the individual values approximately according to their weights in the brain, gives a value for the whole brain which is quite compatible with that which is obtained by the inert gas venous blood exchange and by the hydrogen electrode. The particle distribution technique which we have heard about is based upon quite a different principle. It is fair to point out that the particle distribution technique is in fact Sapirstein's original principle, but finally applied in a way which its assumptions require. As you recall, Sapirstein's important contribution to this field was the concept that if a substance was completely extracted from the blood in one circulation then its distribution in the tissues would be a reflection of the partition of the cardiac output among those tissues. Unfortunately, the substances which were originally chosen did not fulfill those requirements, and the data was not always valid. But these microspheres, which do clog up the vessels and stick there, although they may not permit much in the way of clinical use, nevertheless do fulfill the Sapirstein criteria. It is interesting that in most instances but not all, there is considerable agreement between that method and those which are based upon inert gas exchange.

Dr. Nilsson, some years ago presented evidence that there was a monoexponential clearance in the cortex and Dr. Rosendorf at this meeting with a very ingenious device which can be implanted into an animal and used repeatedly over a period of time, also got surprisingly uniform exponential decays which suggest perhaps that in small regions in the brain assumption of the rapid equilibration between brain and blood may actually be fulfilled.

I think Drs. Angoli, Fieschi and their collaborators reported an approach which, although crude at the present time and perhaps open to the objections which Dr. Lassen likes to raise against methods which insist on taking the needle out of the carotid artery, nevertheless, indicate, that it may be possible to obliviate the carotid puncture and still have valid measurements.

By externally monitoring the arterial curve and deconvoluting the brain curves on the basis of the arterial curve, it should be possible to compute blood flow. Dr. VEALL in England and ANGOLI and FIESCHI in Italy have now shown this to be possible. This does not, however, take care of the problem of extracerebral contamination. Here I think we can see in the future some indirect possibilities that may take this into account. If one uses better collimation and more penetrating gamma rays (Krypton 85m described by Dr. GLASS may be such a more suitable isotope) it should be possible to cut the contribution from the skull and scalp. SOKOLOFF some years ago calculated that for Krypton 79m, the extracerebral contamination could be brought down to less than 10%, and then demonstrated this with phantoms. If one uses iodinated antipyrin instead of a gas, one could eliminate the respiratory component of the contamination, which in the case of gases is considerable. With the use of arterial monitoring and deconvolution there is no reason why iodinated antipyrin would not be a suitable substance to use. I think that extracerebral contamination will be gotten rid of eventually by resolusion in the 3rd dimension by one or another technique which will permit one to see not only through the brain, but also to look at various depths within the brain. I mentioned the time-of-flight techniques which are being developed by Dr. BROWNELL, but I learned yesterday on the Rhine-trip that there are many other techniques which are possible, some of them even today. There is a technique which Dr. FEUERHACKE told us about using gamma rays in coincidence which although not very efficient, can permit 3-dimensional counting at the present time. Then there is a technique which I think was developed by Dr. TER-POGOSSIAN in the conversation on that boat trip. We will have to call this technique if it finally becomes developed the technique of "TER-POGOSSIAN am Rhine". This would involve the use of the annihilation gamma rays of a positrone emitter to fix a line across the brain, and then to use the direct gamma emission which comes off at a different angle to fix a particular distance along that line where emanation has taken place. I suspect that the time-of-flight technique has greater efficiency than the other two possibilities, however. In the meantime, I wonder why one cannot get rid of extracerebral contamination by some kind of subtraction method. By injecting iodide or sodium or some other labelled ion which will pass freely across the capilcaries in the extracerebral tissues but which will be kept out of the brain barrier and with the use of gamma rays spectroscopy, it should be possible continously to subtract the extracerebral contamination. Thus, building on Dr. ANGOLI and FIESCHI's suggestions of yesterday, it may eventually be possible to obtain values as conveniently as Dr. OLDENDORF's method for cerebral blood flow without the necessity of putting a needle into the carotid artery.

I should like to close with a comment upon what I think was the most brilliant methodological contribution presented at this conference. That was the paper by TER-POGOSSIAN on the use of oxygen 15. For a long time those of us who have worked with blood flow have recognized that although it would be possible to measure regional blood flow in man and, with O_2 electrode, oxygen consumption of a region in animals, the possibility of combining these techniques to obtain regional O_2 consumption in man seemed practically nil. The use by TER-POGOSSIAN of oxygen 15 to measure not only the oxygen utilization but the blood flow simultaneously was a major intellectual and technical triumph. One sees no major conceptual problem now that lies in the way of the application of this method to measurement of regional blood flow and oxygen consumption in man. All one needs is a cyclotron in the basement and perhaps a few more years of work by a number of people to rule out the possible sources of error and to establish the validity of the technique. That is relatively easy to accomplish once the intellectual leap has been made and it will not be many years hence that at a symposium such as this one, several papers will be given on regional oxygen consumption in the human brain measured safely and accurately by modifications of this technique.

Clinical Implications of Cerebral Blood Flow Studies

N. A. Lassen

Department of Clinical Physiology, Bispebjerg Hospital, Copenhagen

This short review is not confined to summarizing papers presented at the Mainz symposium in 1969. Rather it aims at giving a more general comment on the clinical usefulness of CBF studies and of the various concepts derived from such studies.

Many noxious stimuli such as hypoxia (occlusive cerebrovascular disease), trauma (head injury or intracranial surgery) or compression (tumor, intracerebral hematoma, compression during surgery or due to edema) cause a stereotype alteration of the brain tissue. Some of these alterations are:

1. Tissue acidosis. Metabolic acidosis of the brain tissue [7] has been measured or suggested on indirect evidence in most of the above listed conditions and it probably constitutes a common denominator [5]. A reflection of brain tissue acidosis is conveniently obtained from cerebrospinal fluid samples especially as collected by suboccipital puncture. It appears that lactic acid is in all cases the cause of the acidosis.

The importance of tissue lactacidosis in relation to circulatory abnormalities is best explained by recalling the striking cerebral vasodilatation normally caused by high aPCO$_2$ values. As *isolated* (metabolic) *pH changes of the blood* do not influence the cerebral resistance vessels, the aPCO$_2$ effects are considered to be mediated by tissue pH changes, most probably by pH changes around the brain's arterioles [5, 6, 11]. This tissue pH depends on local pCO$_2$ and on local HCO$_3$ and a loss of bicarbonate (metabolic tissue acidosis) will consequently be expected to dilate just as hypercapnia normally does.

2. Loss of autoregulation. The normal variations in the diameter of the cerebral arterioles counteracting variations in driving pressure are lost. Thus tissue perfusion tends passively to follow variations in driving pressure (see e.g. the studies in brain tumors by Palvölgyi, by Oeconomos et al., by Brock et al. and the experimental study of Reivich et al.).

3. Loss of regulation of intracranial pressure. Normally the intracranial pressure (i.c.p.) varies only a little when CBF (and hence, secondarily, intracranial blood volume) varies. If autoregulation of blood flow is lost in larger tissue areas (by virtue of passive variations in intracranial content of blood and in edema fluid) then i.c.p. varies abnormally with arterial blood pressure. Similarly, variation in aPCO$_2$ also causes marked fluctuation in i.c.p. as do other cerebral vasoactive drugs: Theophyllamine constricts and must be expected to decrease intracranial pressure, papaverine and most volatile anaesthetics (except N$_2$O) dilate cerebral vessels and increase i.c.p. [4].

The loss of the normal regulation of i.c.p. in more extensive acute injury of brain tissue is of the most direct clinical importance. The pressure in the brain's veins is almost the same as the i.c.p. and hence the latter constitutes, together with the arterial pressure, the driving pressure head of the cerebral circulation. In addition, the i.c.p. increase reflect an increase in intracranial – intraspinal total mass resulting in possible tissue distortions.

4. Loss of CO_2 *reactivity.* As already alluded to, hypercapnia will, in presence of marked tissue acidosis, be expected not to be able to dilate the (already dilated) resistance vessels. On the other hand hypocapnia may, especially if the tissue acidosis is not too severe, be expected still to be able to constrict cerebral arteries. This latter effect has been found experimentally [1] and is presumably of importance for the purported beneficial action of hyperventilation in patients with acute brain injuries.

The situation here commented on, viz. the coexistence of moderate and of severe tissue injury is typical. This seems to underly the paradoxical effects of pCO_2 changes, as characteristically seen in patients with intracranial tumors: The *"steal effect"* of cerebral vasoconstrictors (cf. PALVÖLGYI's communication).

As discussed by PALVÖLGYI, by LASSEN and PAULSON and by FIESCHI et al., loss of CO_2 reactivity as measured by the rCBF method may be dissociated from a loss of autoregulation – the latter changes of always being more widespread. It will be of interest to investigate if autoregulation is regained during hypocapnia in areas with such a dissociation in responses. Since the various regulatory mechanisms are not necessarily lost simultaneously it is perhaps a little misleading to use the term "vasomotor paralysis" to describe the syndrome of acute brain injury here called *"bad brain"* or *"tissue acidosis"* or *"luxury perfusion syndrome"*.

5. Loss of metabolic control. Normally the cerebral blood flow is regulated so as to vary in parallel with the oxygen uptake of the tissue. In other words, the cerebral $(A-V)O_2$ is normally relatively constant at about 7 vol% for an $aPCO_2$ of 40 mmHg. The cerebral venous blood has correspondingly a fairly dark colour in normal man.

In the *"bad brain"* this metabolic control is upset and tissue perfusion tends to exceed the tissue demands so that other factors remaining equal, the cerebral venous blood is abnormally red. This bright red venous blood has been seen around tumors or following a temporary occlusion of a cerebral artery. This condition of relative hyperemia, was called "Luxury Perfusion" in order to stress that the tissue in question had a high PO_2 [5].

The relative hyperemia *may or may not* be absolute (higher than normal). Indeed it appears clear now that the *"bad brain"* runs in a complete spectrum: Following a brief arrest of cerebral circulation a marked absolute hyperemia is seen for 1–2 hours, following more prolonged anoxia with irreversible tissue damage a subnormal blood flow may be found – in all probability related to tissue edema increasing the vascular resistance of the microcirculation.

6. Brain edema. Pathological fluid accumulation in the brain tissue is characteristic of the acute brain injury syndrome. Factors of importance are undoubtedly an increase in tissue osmotic and oncotic pressure (e.g. due to lactic acid and protein-rich edema fluid) and an increase in capillary hydrostatic pressure. An increase in permeability of the normal blood: brain: barrier probably also is involved. Finally it can be mentioned that intracellular edema might result from deficient active transport of electrolytes out of the cells, a mechanisms which acidosis must be assumed to impair.

Clinical Comments

The above summary of some aspects of the *bad brain* syndrome is so brief that many finer details have been left out, as has indeed references to most of the original papers. Otherwise a very voluminous review would have been necessary. On the other hand, even if most of the above outline may seem well-known to the reader, it was felt impossible not to have this background against which to make the following comments on clinical implications.

One more general point will be made, i.e. with regard to space occupying lesions in general. Be it a rapidly growing tumor with manifest intracranial hypertension, a brain contusion with edema, a brain infarct with edema, or an intracerebral hematoma, the clinical problems are much the same. In effect, the latter lesions function as "acute brain tumors". For this reason the clinical comments regarding these various diseases have many similarities.

I. Neurosurgical Technique

It is a direct corollary of the extreme vulnerability of brain tissue to trauma that neurosurgical technique must be gentle. Great caution is to be exerted in any type of compression or "pushing aside" of tissue. If it cannot be avoided then very slow (10–15 min?) and gentle steady compression gradually enlarging the field of operation may be preferable to more resolute retraction. Extirpation of tissue is probably in some conditions better than use of compression.

These comments are addressed to the goal of reducing the amount of *bad brain* created by the neurosurgeon (some damage will unavoidably be made). The comments are admittedly inadequate. Consequences as to size of craniotomy (fairly large) and route of access also might be made. But this lies outside my field of competence as do references to other modes of reducing intracranial hypertension. The basic point made here is simply that of stressing the unavoidable trauma of surgery. It is made within the frame of reference of tissue acidosis, tissue hyperemia and tissue edema. Swelling of traumatized tissue may easily compromize the anatomically perfect neurosurgical intervention by creating an uncontrollable intracranial "tumor" by tissue edema. Our neurosurgical colleagues have long been aware of the importance of the avoidance of brain trauma during neurosurgery. All the CBF-concepts can do, is to emphasize this point.

II. The Anesthesiological Care during Neurosurgery

Since acute brain injury either pre-exists and/or is created by the intervention some general principles are valid.

a) *The blood pressure* should not be allowed to rise (risk of hyperemia and edema in "bad brain") and more a pronounced pressure drop is to be avoided (risk of pressure-passive flow decrease in marginally perfused tissue resulting in hypoxia, acidosis etc.). In fact, the spontaneous blood pressure level (often somewhat elevated) should in general not be interfered with.

These comments could start off a very extensive discussion of the often advocated use of controlled hypotension. Perhaps there are surgical procedures which *per se* necessitate marked hypotension. What is said here shall only be understood in the sense that "other factors equal", the use of more marked hypotension cannot be recommended. If used, it should be of brief duration and the simultaneous use of hypothermia should be considered as is indeed the practice in many centers.

b) *The* aPCO$_2$ *should be low – at about 25–30 mmHg.* This is now general practice in most clinics as an adjunct to the other measures taken to reduce brain bulk (hyperosmolar therapy, steroid medication, ventricular drainage, extirpation of edematous tissue or tumor tissue etc.). Again a long discussion could be made regarding the vasoconstriction elicited in normal brain tissue by such treatment: could it be harmful? It shall only be stated that so moderate a degree of hyperventilation* does not have known noxious effect on normal brain, while it tends to induce a more normal pH and circulation in acidotic tissue. Hyperventilation does depress brain function, but this need not be taken to constitute a harmful effect [3].

c) *The anesthetic agent.* The cerebral vasodilator action of most volatile anesthetic agents speaks against their use in neuroanesthesia. Nitrous oxide is the drug of choice since it does not influence CBF. It is satisfactory to note that N_2O is widely recognized as a safe drug in neuroanesthesia (attention will be called to the problem of N_2O swelling the ventricular system if air [after pneumoencephalogram] resides in the ventricles. This may be controlled, if necessary, by ventricular puncture). Especially if neuromuscular blocking agents are used, it is important by proper addition of analgetics to avoid pain-and-arousal induced blood pressure rises during the anesthesia.

A special warning seems here appropriate: if an acute brain injury is suspected (e.g. brain contusion), even a brief general anesthesia for any other clinical reason, as for repositioning a fracture or in conjunction with a diagnostic procedure as angiography may be critical. Precisely the same principles as outlined above must be followed. For example, Halothane anesthesia without controlled respiration resulting in 1. hyperperfusion, 2. hypercapnia, 3. hypotension may fatally aggravate the consequences of a brain injury.

III. The Postoperative Treatment after a Neurosurgical Intervention

Having advocated moderate assisted hyperventilation during surgery it is logical to advice continued hypocapnia in the immediate post-operative period. In many cases, continued active hyperventilation for 24 hours or more seems indicated. This may involve tracheostomy and the appropriately restrained use of neuroleptics and of analgesis.

The continuation of hyperventilation does not necessarily mean a continuation of all the beneficial effects. Especially the vasoconstriction in more healthy brain tissue may disappear due to adaptative phenomena. However, exactly because of such phenomena, then abrupt return to normocapnia (say a rise in $aPCO_2$ from 30 to 40 mmHg) results in a rebound phenomenon in form of hyperemia and marked rise of i.c.p. It is because of this effect, and also in order to avoid more marked hypercapnia and hypoxia, that a more prolonged period of post-operative moderate hypocapnia seems logical in severe cases as proposed by BOZZA-MARRU-BINI and co-workers at this symposium. A definite proof of the value of this treatment is, however, lacking as will also be commented on below.

IV. Conservative Treatment of Patients with Intracranial Tumor

Ventricular drainage, osmolar or steroid therapy will not be commented on. It shall only be stressed that the already mentioned principles can be successfully employed even in severely ill patients: Artificial hyperventilation, analgesics, and mild sedation if necessary. In non-comatose patients surprisingly good temporary remissions sometimes occur even in apparently intractable cases where surgical decompression (external or internal) has already been previously performed and further neurosurgical treatment is considered unadviseable.

---

* In hypothermia the lowered CO_2 production implies that the ventilation must be lower in order to avoid excessive hypocapnia.

V. Intracerebral Hematoma

In conjunction with craniotomy or as a conservative therapeutic measure assisted artificial hyperventilation would seem indicated along with the other measures designed at reducing the intracranial pressure. OLESEN found very severe tissue acidosis in the brain tissue overlying a hematoma, a lesion where also a marked "peritumor" hyperemia exists in some regions [8]. Just as mentioned in relation to the post-operative care it is rational to continue the assisted hyperventilation until the condition is stabilized.

VI. Severest Form of Head Injury (Contusion and Brain Stem Injury)

Present knowledge of rCBF and its derangement in the damaged and acidotic traumatized tissue (see also the important paper of BALDY-MOULINIER and co-workers) gives amble support to the clinical impression of BOZZA-MARRUBINI and co-workers regarding the usefulness of prolonged assisted hyperventilation in such cases [9]. As with all the other clinical comments made here, the final proof must be supplied by the clinician in controlled studies.

Hyperventilation, lasting e.g. one week or more, and necessitating tracheostomy, involves other clinical problems than those concerning cerebral circulation (lung complications e.g.). Hence it is not possible to state that the value of this treatment is already "established". What can be stated, however, is that *other factors remaining equal* it appears likely to be a rational treatment (it was suggested on theoretical grounds independently of clinical experiences [5]).

Two important points: a) Adequate ventilation (with hyperventilation according to my conviction) must start *as early as at all possible*, b) no attempt must be made, using inadequate anesthesiological technique, of repairing other simultaneous lesions: As already stated the anesthesia to be used is that which the brain damage demands (if at all such secondary lesion needs to be treated acutely).

VII. Spontaneous Subarachnoidal Bleeding with or without Angiographically Proven Arterial Aneurysm

Comments have already been made regarding surgical treatment and choice of anesthesia and controlled hypotension was considered together with hypothermia.

In the conservative treatment following acute bleeding prolonged controlled hypotension has been used in some centers. As a preventive measure against renewed bleeding this undoubtedly is rational. But, if signs of acute brain tissue injury exist (focal neurological symptoms, depressed level of consciousness, elevated blood pressure, increased intracranial pressure, so called "spasms" on angiography) then a lowering of blood pressure may be harmful (cf. the discussion of blood pressure in relation to neurosurgical anesthesia). Such complications must be treated along the same lines as discussed in relation to severe head injury.

VIII. Apoplexy

The finding of SOLOWAY and co-workers]12[of a reduction by hyperventilation of the size of experimental infarction to $1/40$ of its normocapnia size is in consonance with what has already been mentioned. BATTISTINI and co-workers reported at the Mainz Symposium some further experimental data. Their series need, however, to be extended. SKINHØJ commented on the clinical application of this form of treatment and advocated the simultaneous study of a control patient hyperventilated with added CO_2 to give normocapnic values. By such measures the therapeutic value of this rather expensive and complex therapy can be evaluated critically. This comment is also relevant to the head trauma cases discussed above. However, perhaps it is most pertinent in relation to apoplexy: these patients are numerous and generally very old (may not tolerate prolonged hyperventilation too well). Attention may also be called, in this connection, to the observation of BROCK and coworkers in a young girl with occlusion of the middle cerebral artery. This suggested that constriction of collateral vessels could result from active hyperventilation, certainly not the type of response desired!

IX. Carotid Surgery

The present symposium contained a very interesting discussion of CBF in relation to carotid surgery. Although the hemodynamic importance of a carotid stenosis can, theoretically, be measured accurately by studying rCBF during maximal relaxation of the arteriolar resistance (LOU and VON WOWERN) the clinical value of such measurements is doubtful. This is so, since there is evidence that unilateral carotid stenoses practically never compromise the adequacy of tissue perfusion: Nevertheless, they seem to be of pathogenetic importance presumably because of distal emboli, forming at the stenotic site.

In agreement with this it is now quite clear that pre- and post-operative studies of CBF during normocapnia and normotension are of no clinical value. In patients without neurological defects global as well as local CBF must be expected to remain normal under such conditions as was also found by BERGENTZ et al. and by O'BRIEN and VEALL.

The use of rCBF measurements during carotid endarterectomy to assess adequacy of hemispheric perfusion during temporary complete block represents a rational approach (BOYSEN et al.). It would seem perhaps to constitute a more specific test than EEG and jugular venous PO_2 measurements.

A comment on the anesthesia to be used for carotid surgery will be made. Of prime importance must be the maintenance of an adequate perfusion pressure (cf. e.g. 2). Several authors have advocated the use of halothane anesthesia combined with hypercapnia. This cannot be considered safe, specially if the blood pressure is not kept up. The use of local anesthesia, allowing continuous check of brain function, perhaps in combination with a moderate dose of a suitable pressor drug seems more rational.

X. Hypoxic Brain Damage and "Brain Death"

A rational treatment of post-hypoxic brain disease would seem to be to follow the pattern already described: prolonged assisted hyperventilation, anti-edema (e.g. osmotic), anti-convulsant and mild sedation if necessary.

That cerebral circulatory and metabolic studies could be of value in deciding the severity of hypoxic brain damage seems *a priori* reasonable. BÈS and colleagues reported on the *red cerebral venous* blood in patients subsequently dying from such irreversible damage, a cerebral $(A-V)O_2$ below 2.0 vol% in all cases being associated with rapid demise. In this situation only a *relative* hyperemia exists as the CBF is much reduced. BROCK and coworkers stressed this point by advocating surgical exposure of the internal carotid artery to assure injection of ^{133}Xenon solely to the brain (for hemispheric CBF measurement). As low $(A-V)O_2$ and low CBF coexist the product of these two factors (the cerebral oxygen uptake, $CMRO_2$) is extremely low as has been pointed out by SHALIT and coworkers [10]. The heterogeneity of brain tissue perfusion in such conditions implies that the overall reduction of CBF and $CMRO_2$ is even more pronounced that the measured values.

Concluding Remarks

This compressed survey of clinical implications runs the risk of over-emphasizing some topics while neglecting others. This risk must be run as must also the risk of extending some theoretical principles too far into clinical practice: only hereby can the possible therapeutic benefit be derived, benefits which certainly are the ultimate (and direct) aim of the endeavours presented at this symposium. If the attempt made here should arouse counterarguments and induce the necessary clinical trials, this is one reason for its having been made. If its various proposals improve the therapeutic results this is a central aim.

Ending the presentation two clinical topics not readily fitted into the above discussion will be touched upon. The first is the *adaptation of CBF* in chronic hypo- or hypercapnia. This adaptation seems to parallel that of the cerebrospinal acid-base balance in consonance with the concept of CSF-pH's central role in CBF control (cf. the study of FIESCHI et al.). These adaptative processes imply that a rapid normalization of chronically abnormal aPCO$_2$ is not advisable (12—24 hours of gradual normalization appearing more reasonable, the precise time course of adaptation not being precisely known yet).

The other topic is that of the direct use of rCBF in diagnostic clinical work, viz. to the therapeutic benefit of the individual subject studied. Assessment of safety of carotid occlusion constitutes one such example. The early diagnosis of intracranial processes, e.g. an intracranial tumor, constitutes another possibility. Regrettably, no series of observations of autoregulation of rCBF in small, hard-to-detect tumors was reported at the Mainz symposium. It is quite possible, however, that even conventional angiographic technique can yield much more in "early" or "borderline" cases if advantage is taken of comparing control films to films taken during induced hypertension and/or hypercapnia. HACKER reported in the discussion a convincing example of the value of this approach in unmasking a tumor on an angiogram.

Summary

The central theme in this review is that of brain tissue acidosis with associated circulatory phenomena in form of abolition of active vasomotor responses (in more severe cases paradoxical flow changes are typical) and associated brain edema. These alterations are found in apoplexy, tumors and trauma and they are to be considered *a non-specific response to acute brain injury*. Such injury runs the whole spectrum from the most mild and completely reversible tissue damage to irreversible tissue necrosis.

Metabolic acidosis of the brain may be considered more important clinically than the metabolic acidosis of the blood so universally studied clinically over many years, e.g. diabetic keto-acidosis, uremic accumulation of fixed acid, and various forms of systemic lactacidoses. If therapeutic consequences as here outlined especially regarding prolonged artificial hyperventilation become more clearly established it is safe to predict a steady rise not only in clinical exploitation of CBF studies but also of the rather unique accessibility of the brains extracellular fluid (the CSF) wherein the brain tissue's gas tensions and acid/base conditions are reflected.

References

1. BALDY-MOULINIER, M., and PH. FRÈREBEAU: Blood flow of the cerebral cortex in intracranial hypertension. Scand. J. clin. Lab. Invest. Suppl. 102 (III International Symposium on Cerebral Blood Flow and Cerebro-Spinal-Fluid, Lund-Copenhagen 1968), V:G (1968).
2. FARHAT, S. M., and R. C. SCHNEIDER: Observations on the effect of systemic blood pressure on intracranial circulation in patients with cerebrovascular insufficiency. J. Neurosurg. 27, 441 (1967).

3. FROMAN, C.: Adverse effects of low carbon dioxide tensions during mechanical over-ventilation of patients with combined head and chest injuries. Brit. J. Anesth. **40**, 383 (1968).

4. JENNETT, W. B., J. BARKER, W. FITCH, and D. G. McDOWALL: Effect of anesthesia on intracranial pressure in patients with intracranial space-occupying lesions. Lancet *I*, 61 (1969).

5. LASSEN, N. A.: The Luxury-perfusion Syndrome and its possible relation to acute metabolic acidosis localized within the brain. Lancet *II*, 1113 (1966).

6. — Brain extracellular pH: The main factor controlling cerebral blood flow. Editorial, Scand. J. clin. Lab. Invest. **22**, 247 (1968).

7. LEUSEN, J., J. WEYNE, and G. DEMEESTER: Acid/base and lactate/pyruvate changes in CSF and brain. Scand. J. clin. Lab. Invest. Suppl. 102 (III International Symposium on Cerebral Blood Flow and Cerebro-Spinal-Fluid, Lund-Copenhagen 1968), I:G (1968).

8. PAULSON, O. B., S. CRONQVIST, J. RISBERG, and F. I. JEPPESEN: Regional cerebral blood flow, a comparison of 8-detector and 16-detector instrumentation. J. Nuclear Med. **10**, 164 (1969).

9. ROSSANDA, M.: Prolonged hyperventilation in treatment of unconscious patients with severe brain injuries. Scand. J. clin. Lab. Invest., Suppl. 102 (III International Symposium on Cerebral Blood Flow and Cerebro-Spinal-Fluid, Lund-Copenhagen 1968), XIII:E (1968).

10. SHALIT, M. N., A. J. BELLER, M. FEINSOD, and S. COTEV: Cerebral oxygen consumption another indicator of brain death. In Der Hirntod (Symposium 14, December 1968 in Bonn). Ed. Georg Thieme Verlag, Stuttgart, p. 56 (1969).

11. SIESJÖ, R. K., Å. KJÄLLQUIST, U. PONTÉN, and N. ZWETNOW: Extracellular pH in brain and cerebral blood flow. In "Progress in Brain Research – vol. 30 – Cerebral Circulation" Ed. W. Leyendijk, Elsevier Publ., Amsterdam, p. 87 (1968).

12. SOLOWAY, M., W. NADEL, M. S. ALBIN, and R. J. WHITE: The effect of hyperventilation on subsequent cerebral infarction. Anesthesiology, **29**, 975 (1968).

Subject Index

Typesetting and printing: Universitätsdruckerei Mainz GmbH